T0133502

Security, Privacy and Reliability in Computer Communications and Networks

RIVER PUBLISHERS SERIES IN COMMUNICATIONS

The "River Publishers Series in Communications" is a series of comprehensive academic and professional books which focus on communication and network systems. The series focuses on topics ranging from the theory and use of systems involving all terminals, computers, and information processors; wired and wireless networks; and network layouts, protocols, architectures, and implementations. Furthermore, developments toward new market demands in systems, products, and technologies such as personal communications services, multimedia systems, enterprise networks, and optical communications systems are also covered.

Books published in the series include research monographs, edited volumes, handbooks and textbooks. The books provide professionals, researchers, educators, and advanced students in the field with an invaluable insight into the latest research and developments.

Topics covered in the series include, but are by no means restricted to the following:

- Wireless Communications
- Networks
- Security
- Antennas & Propagation
- Microwaves
- Software Defined Radio

For a list of other books in this series, visit www.riverpublishers.com

Security, Privacy and Reliability in Computer Communications and Networks

Editors

Kewei Sha

University of Houston
Clear Lake, USA

Aaron Striegel

University of Notre Dame, USA

Min Song

Michigan Tech, USA

Published, sold and distributed by:
River Publishers
Alsbjergvej 10
9260 Gistrup
Denmark

River Publishers
Lange Geer 44
2611 PW Delft
The Netherlands

Tel.: +45369953197
www.riverpublishers.com

ISBN: 978-87-93379-89-3 (Hardback)
978-87-93379-90-9 (Ebook)

Contents

PART II: Vulnerabilities, Detection and Monitoring

PART III: Cryptographic Algorithms

Preface

Future communication networks aim to build an intelligent and efficient living environment by connecting a variety of heterogeneous networks to fulfill complicated tasks. These communication networks bring significant challenges in building secure and reliable communication networks to address the numerous threat and privacy concerns. New research technologies are essential to preserve privacy, prevent attacks, and achieve the requisite reliability.

This book studies and presents recent advances reflecting the state-of-the-art research achievements in novel cryptographic algorithms, intrusion detection mechanisms, privacy preserving techniques and reliable system protocols. It is ideal for personnel in computer communication and networking industries as well as academic staff and collegial, master, Ph.D. students in computer science, computer engineering, cyber security, information insurance and telecommunication systems.

The book is organized into four parts. Part I presents to two novel schemes to preserve privacy in wireless ad hoc networks. In Part II, three creative approaches to detect instructions, service violations, and vulnerabilities in various networked systems. Part III introduces four interesting algorithms to improve security. One chapter evaluates a set of encryption tools in cloud computing. The other chapter designs encryption algorithm for low-energy devices. How to construct secrecy is also discussed in this part. Finally, reliable system design mechanisms are demonstrated in Part IV. Three different routing protocols are presented to improve the reliability and robustness in the communication. A prediction scheme is proposed to ensure Quality of Service in IEEE 802.11.

Acknowledgments

First and foremost, we would like to express our warm appreciation to University of Houston – Clear Lake, University of Notre Dame, and Michigan Technological University. Special thanks go to our authors who contributed excellent book chapters. We would like to express our warm appreciation to the River Publisher staffs who allowed us to publish our work and gave their valuable time to review our book. We would also like to thank the reviewers who provided feedback and suggestions for our book. Finally, we want to thank our families who supported and encouraged us in spite of all the time it took us away from them. Last and not least, we beg forgiveness of all those whose names we have failed to mention. Any suggestions, comments, and feedback for further improvement of the text are welcome.

List of Contributors

Abhishek Parakh, *University of Nebraska at Omaha, Omaha, NE 68182, USA*

Ahsan-Ul-Ambia, *Department of Computer Science & Engineering, Islamic University, Kushtia, Bangladesh*

Ali Hamieh, *American University of Technology, Halat, Lebanon*

Antonio Garrido, *High-Performance Networks and Architectures (RAAP), Albacete Research Institute of Informatics (I3A), University of Castilla-La Mancha, Albacete, Spain*

Bo Wu, *Electrical and Computer Engineering, Clemson University, Clemson, SC 29634, USA*

Chao Zheng, *State Key Laboratory of Information Security, Institute of Information Engineering, Chinese Academy of Sciences, Minzhuang Rd. 89-A, Haidian District, Beijing, 100093, People's Republic of China*

Cong Guo, *National Meteorological Information Center, Beijing, China*

Estefanía Coronado, *High-Performance Networks and Architectures (RAAP), Albacete Research Institute of Informatics (I3A), University of Castilla-La Mancha, Albacete, Spain*

Gabriel Labrador, *Department of Information Sciences and Technology, Pennsylvania State University, Altoona-PA, USA*

Haiying Shen, *Electrical and Computer Engineering, Clemson University, Clemson, SC 29634, USA*

Henrik Holm, *Forest Glen Research, LLC, 2412 Bronson Boulevard Kalamazoo, MI 49008, USA*

Huan Wang, *University of Chinese Academy of Sciences, Yuquan Rd. 19-A, Shijingshan District, Beijing, 100049, People's Republic of China*

Jon R.Ward, *Department of Computer Science and Electrical Engineering, University of Maryland, Baltimore County, USA*

José Villalón, *High-Performance Networks and Architectures (RAAP), Albacete Research Institute of Informatics (I3A), University of Castilla-La Mancha, Albacete, Spain*

Jun Ho Huh, *Honeywell ACS Labs, Golden Valley, MN, 55422, USA*

Kang Chen, *Electrical and Computer Engineering, Southern Illinois University Carbondale, Carbondale, IL 62901, USA*

Kelsey Karpinski, *Department of Information Sciences and Technology, Pennsylvania State University, Altoona-PA, USA*

Leonora Gerlock, *University of Nebraska at Omaha, Omaha, NE 68182, USA*

Liehuang Zhu, *College of Computer Science & Technology, Beijing Institute of Technology, China*

Maciej Korczyński, *1. Rutgers University, New Jersey, USA*
2. Delft University of Technology, 2628 CD Delft, The Netherlands

Mario Gerla, *Department of Computer Science, University of California, Los Angeles, CA, USA*

Matthew Battey, *University of Nebraska at Omaha, Omaha, NE 68182, USA*

Md. Alam Hossain, *Department of Computer Science & Engineering, Jessore University of Science & Technology, Jessore, Bangladesh*

Md. Al-Amin, *Department of Computer Science & Engineering, Jessore University of Science & Technology, Jessore, Bangladesh*

Mohamed Younis, *Department of Computer Science and Electrical Engineering, University of Maryland, Baltimore County, USA*

Nina H. Fefferman, *Rutgers University, New Jersey, USA*

Paulo Montezuma, *1. Instituto de Telecomunicações (IT), Av. Rovisco Pais 1, 1049-001 Lisboa, Portugal*
2. DEE, Faculdade de Ciências e Tecnologia (FCT)-Universidade Nova de Lisboa (UNL), Monte de Caparica, 2829-516 Caparica, Portugal
3. UNINOVA, Campus da FCT/UNL, Monte de Caparica, 2829-516 Caparica, Portugal

Rahamatullah Khondoker, *Fraunhofer Institute for Secure Information Technology, Darmstadt, Germany*

Rui Dinis, *1. Instituto de Telecomunicações (IT), Av. Rovisco Pais 1, 1049-001 Lisboa, Portugal*
2. DEE, Faculdade de Ciências e Tecnologia (FCT)-Universidade Nova de Lisboa (UNL), Monte de Caparica, 2829-516 Caparica, Portugal

Ruolin Fan, *Department of Computer Science, University of California, Los Angeles, CA, USA*

S. Raj Rajagopalan, *Honeywell ACS Labs, Golden Valley, MN, 55422, USA*

Suho Oh, *Department of Mathematics, Texas State University, San Marcos, TX 78666, USA*

Syed Rizvi, *Department of Information Sciences and Technology, Pennsylvania State University, Altoona-PA, USA*

Tuanfa Qin, *School of Computer and Electronic Information, Guangxi University, Nanning, Guangxi, China*

Weichao Wang, *Dept. of Software and Information System, UNC Charlotte, Charlotte, NC 28223, USA*

Whitney Hernandez, *Department of Information Sciences and Technology, Pennsylvania State University, Altoona-PA, USA*

William Mahoney, *University of Nebraska at Omaha, Omaha, NE 68182, USA*

Xiao Chen, *Department of Computer Science, Texas State University, San Marcos, TX 78666, USA*

Xiongwei Xie, *Dept. of Software and Information System, UNC Charlotte, Charlotte, NC 28223, USA*

Yongbin Zhou, *1. State Key Laboratory of Information Security, Institute of Information Engineering, Chinese Academy of Sciences, Minzhuang Rd. 89-A, Haidian District, Beijing, 100093, People's Republic of China*
2. University of Chinese Academy of Sciences, Yuquan Rd. 19-A, Shijingshan District, Beijing, 100049, People's Republic of China

Yuan Xu, *Department of Computer Science, Texas State University, San Marcos, TX 78666, USA*

Yu-Ting Yu, *Qualcomm Research, Bridgewater, NJ, USA*

Zijian Zhang, *College of Computer Science & Technology, Beijing Institute of Technology, China*

List of Figures

List of Tables

List of Algorithms

List of Abbreviations

AAET	Acknowledgement-Aware Evidence Theory
AAR	Active Area-based Routing
Ack	Acknowledgement
ACs	Access Categories
AES	Advanced Encryption System
AI	Artificial Intelligence
AIFS	Arbitration Inter-Frame Spacing
AIFSN	Arbitration Inter-Frame Space Number
AP	Access Point
API	Application Program Interface
ASIC	Application Specific Integrated Circuit
AWGN	Additive White Gaussian Noise
BE	Best Effort
BES	BlackBerry Enterprise Server
BIC	Bayesian Information Criterion
BK	Background
BLID	Benchmark Local Intrusion Detectors
BPSK	Bi Phase Shift Keying
BSS	Basic Service Set
C&C	Command-and-Control
CA	Certificate Authorities
CBR	Constant Bit Rate
CDF	Cumulative Density Function
CDMA	Code Division Multiple Access
CMS	Cryptographic Message Syntax
CPA	Correlation Power Analysis
CSET	Client Side Encryption Tool
CSI	Channel State Information
CSP	Cloud Service Provider
CU	Coordination Unit
CW	Contention Window

DAAR	Distributed Active Area-based Routing
DCF	Distributed Coordination Function
DDoS	Distributed Denial of Service
DDR	Double Data Rate
DHT	Distributed Hash Table
DIAMoND	Distributed Intrusion/Anomaly Monitoring for Nonparametric Detection
DiBAN	Distributed Beamforming protocol for Increased Base Station Anonymity
DK	Derivation Key
DLP	Data Leakage Prevention
DM	Data Mining
DNS	Domain Name System
DOMINO	Distributed Overlay for Monitoring InterNet Outbreaks
DoS	Denial-of-Service
DPA	Differential Power Analysis
DSRC	Dedicated Short-Range Communication
DTN	Delay/Disruption-Tolerant Networks
DTNs	Delay Tolerant Networks
DTQ	Disruption Tolerant Queue
DU	Detection Unit
ECC	Elliptic Curve Cryptography
EDCA	Enhanced Distributed Channel Access
E-Multi-SoSim	Enhanced Social Similarity-based Multicast
ET	Evidence Theory
FIPS	Federal Information Processing Standard
Flooding	Flooding Algorithm
FN	False Negative
FP	False Positive
GB	Gigabyte
GMSK	Gaussian Minimum Shift Keying
GPCR	Greedy Perimeter Coordinator Routing
GPSR	Greedy Perimeter Stateless Routing
GPU	Graphics Processing Unit
GSIS	Group Signature Identity-based (ID-based) Signature
GSM	Global System for Mobiles
HCA	Hybrid Coordination Function
HCCA	HCF Controlled Channel Access
HMAC	Hash Message Authentication Code

HP	Hewlett Packard
HQ	High Quality
IaaS	Infrastructure as a Service
IDS	Intrusion Detection System
IEEE	Institute of Electrical and Electronics Engineers
iOS	iPhone Operating System
IoT	Internet of Things
IP	Internet Protocol
IT	Information Technology
JTAG	Joint Test Action Group
KB	Kilobyte
KS test	Kolmogorov-Smirnov test
KSA	Kolmogorov-Smirnov Analysis
KSM	Kernel Same-page Merging
LRU	Least Recently Used
MAC	Medium Access Control
MAM	Mobile Application Management
MAWI	Measurement and Analysis on the WIDE Internet
MB	Megabyte
MCC	Mobile Cloud Computing
MDM	Mobile Device Management
MI	Mutual Information
MIA	Mutual Information Analysis
MIMO	Multiple Input Multiple Output
MNO	Mobile Network Operator
MPC-KSA	Multiple p-value and cumulative partition method based KSA
MSNs	Mobile Social Networks
MULE	Mobile Ubiquitous LAN Extension
Multi-FwdNew	Variation 1 of the Multi-SoSim Algorithm
Multi-SoSim	Social Similarity-based Multicast
Multi-Unicast	Variation 2 of the Multi-SoSim Algorithm
NAC	Network Access Control
NIDS	Network Intrusion Detection Systems
NIST	National Institute of Standards and Technology
NL	Nonlinear
OBU	OnBoard Unit
OMSNs	Opportunistic Mobile Social Networks
OQPSK	Offset Quadrature Phase Shift Keying

OS	Operating System
P2P	Peer-to-Peer
PaaS	Platform as a Service
PBKDF	Password-Based Key Derivation Function
PCA	Principal Component Analysis
PCF	Point Coordination Function
PDF	Probability Density Function
PKS	Partial Kolmogorov-Smirnov Analysis
PMS	Point Mobile Security
PRF	Pseudorandom Function
PRNG	Pseudo Random Number Generator
QAM	Quadrature Amplitude Modulation
QoS	Quality of Service
QPSK	Quadrature Phase Shift Keying
RAM	Random Access Memory
RBS	Reference Broadcast Synchronization
RF	Radio Frequency
RSU	Utilizes RoadSide Unit
SaaS	Software as a Service
SC	Secrecy Capacity
SCA	Side-Channel Analysis
SDK	Software Development Kit
SDN	Software Defined Network
SHA	Secured Hash Algorithm
SIEM	Security Information and Event Management
SIFS	Short Inter-Frame Space
SIM	Subscriber Identity Module
SLA	Service Level Agreement
SMS	Short Message Service
SPM	Social Profile-based Multicast
SSL	Secure Sockets Layer
SSN	Social Security Number
STIX	Structured Threat Information eXpression
TA	Trust Authority
TAXII	Trusted Automated eXchange of Indicator Information
TCP	Transmission Control Protocol
TDMA	Time Division Multiple Access
TN	True Negative
TP	True Positive

TPSN	Time-Synch Protocol for Sensor Networks
TTL	Time To Live
TXOP	Transmission Opportunity
Ups	User Priorities
USR	User Recovery Key
V2V	Vehicle-to-Vehicle
VANET	Vehicular Ad-hoc Network
VDTN	Vehicle Delay-Tolerant Network
VI	Video
VM	Virtual Machines
VNET	Vehicle Network
VO	Voice
WLANs	Wireless Local Area Networks
WSN	Wireless Sensor Network

PART I

Privacy

1

Distributed Beamforming Relay Selection to Increase Base Station Anonymity in Wireless Ad Hoc Networks

Jon R. Ward and Mohamed Younis

Department of Computer Science and Electrical Engineering,
University of Maryland, Baltimore County, USA

Abstract

Wireless ad hoc networks have become valuable assets to both the commercial and military communities with applications ranging from industrial control on a factory floor to reconnaissance of a hostile border. In most applications, nodes act as data sources and forward information to a central base station (BS) that may also perform network management tasks. The critical role of the BS makes it a target for an adversary's attack. Even if an ad hoc network employs conventional security primitives such as encryption and authentication, an adversary can apply traffic analysis techniques to find the BS. Therefore, the BS should be kept anonymous to protect its identity, role, and location. Previous work has focused on countering the threat of traffic analysis at the network and link layers. In this chapter, we promote a new methodology in which countermeasures are applied at the physical layer. Basically, distributed beamforming is applied to prevent an adversary from associating communicating nodes and make traffic analysis inconclusive. We demonstrate that our distributed beamforming-based technique is effective at boosting the BS anonymity. We further show that the increased anonymity and corresponding energy consumption depend on the quality and quantity of selected helper relays, and we present a novel, distributed approach for determining a set of relays per hop that boosts BS anonymity using evidence theory analysis while minimizing the energy consumption. The identified relay set is further prioritized using local wireless channel statistics. The simulation results demonstrate the effectiveness our approach.

Keywords: Anonymity, Relay selection, Location privacy, Distributed beamforming, Wireless ad hoc networks, Evidence theory, Traffic analysis.

1.1 Introduction

The continual advancement in wireless technologies has enabled networked solutions for many non-conventional civil and military applications. In recent years, ad hoc networks have attracted increased attention from the research and engineering community, motivated by applications such as situational awareness, asset tracking, and border protection [1–4]. Ad hoc networks are particularly well suited for automated monitoring and command and control applications that involve minimal human interaction; consequently, ad hoc networking protocols must be as energy efficient as possible to maximize the functional lifetime of the network. Wireless ad hoc networks have become prevalent in a variety of applications that involve or employ wireless sensor networks (WSN) and Internet of Things (IoT) [1, 5, 6]. A WSN is composed of many sensor nodes, each with a processor, memory, power supply, radio, and some variety of mechanical, thermal, biological, chemical, optical, or magnetic sensor, deployed over an area to measure the surrounding environment [7]. In most topologies, measurements collected by each sensor node are forwarded to a central base station (BS). Similarly, the notion of IoT seeks to equip many conventional data collection and control devices with wireless communication capabilities to enable the networking of these devices to serve new applications [1, 5, 6]. The operation model of many IoT applications involves the dissemination of data collected by the various devices over multihop paths to a BS that processes the data. In both cases, the data carried by the network may be sensitive, and thus, preventing data collection or data delivery may be of interest to a nefarious adversary.

In wireless ad hoc networks, the BS represents a natural focal point, due to its unique role, for an adversary that seeks to disable the network with the least effort. That is, the adversary assumes that achieving a denial of service (DoS) attack against the BS would cripple the larger network since the BS not only serves as the data sink, but also provides other basic control and management features such as protocol synchronization, serves as a gateway to other networks, and passes operator notifications; without these services, the network ceases to function. Such a special role makes the BS a critical node to ensure proper operation of the network. An adversary can nullify the value of a network by simply disrupting or physically damaging the BS. Any disruption of the BS will negatively degrade the network utility with negative effects ranging from loss of unprocessed data to a complete shutdown of the network. Prior to an attack, the adversary must first distinguish the BS from other nodes in the network and discover its location. The most successful protection for the BS against a malicious adversary's attack is to remain anonymous in role, identity, and location and therefore thwart the adversary's attempts to uniquely pinpoint the BS; however, such protection is not provided by conventional ad hoc security mechanisms [8]. The majority of research in the field of anonymity to date has focused on routing algorithms that attempt to hide true routes from source to sink [9]. Although anonymous routing methods may largely mitigate the threat of discovering data paths based on an inspection of packet headers, these approaches do not mitigate the threat of an adversary that may deduce significant information by analyzing link layer, pairwise node relationships from which the location and role of the BS can be inferred [10, 11].

From an adversary's point of view, the problem of finding the BS location is facilitated by the inherent traffic pattern in the network. The primary flow of traffic is from all nodes toward the BS over multihop paths [1]. In the case of a WSN, as sensor nodes monitor the surrounding environment for events such as motion, acoustics, temperature, light, and pressure, they report events to the BS over wireless links using multihop routes to conserve their limited onboard energy supply. The natural ingress of traffic destined for the BS creates a confluence in the area surrounding the BS as the nodes close to the BS transmit comparatively more packets than other nodes. The presence of increased traffic in the vicinity of the BS becomes noticeable to a passive, eavesdropping adversary. We illustrate this behavior in Figure 1.1, where the area surrounding the BS is shown as a red "hotspot" of increased traffic flow. In Figure 1.1, the ad hoc network is a WSN monitoring two events, a fire and an intruder. These events generate alarm messages that are forwarded toward the BS. In Figure 1.2, we illustrate the result of applying an ideal anonymity-boosting technique to the scenario illustrated in Figure 1.1 such that the adversary no longer observes a hotspot of activity, but instead is forced to randomly select the BS from one of the nodes in the network. Anonymity-boosting techniques remove the concentration of traffic around the BS, and consequently, the adversary is unable to distinguish the BS. In this example, the result of applying the proposed techniques is obfuscation of the true path endpoints (i.e., the BS).

In [12, 13], we leveraged distributed beamforming as an anonymity-boosting technique to successfully increase the BS anonymity. Distributed beamforming has recently received attention as a method for improving the communication range, data rate, energy efficiency, reducing interference in

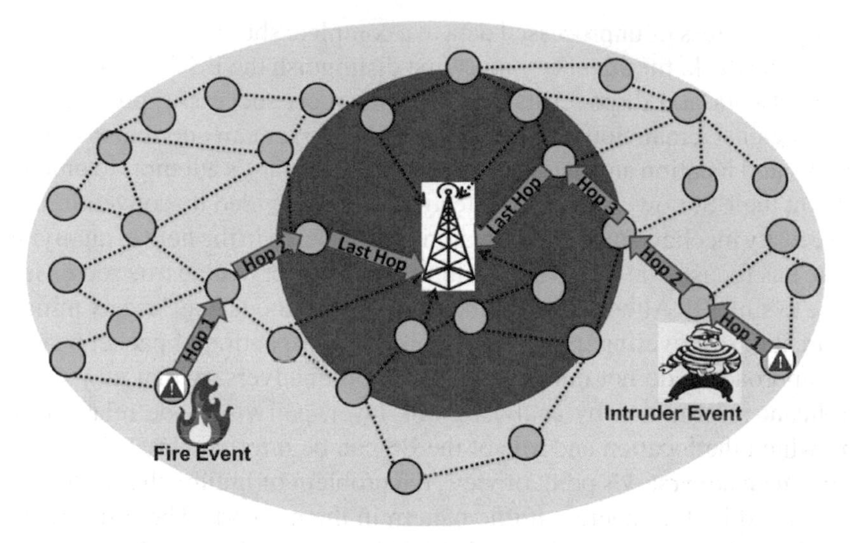

Figure 1.1 Example ad hoc network with low BS anonymity.

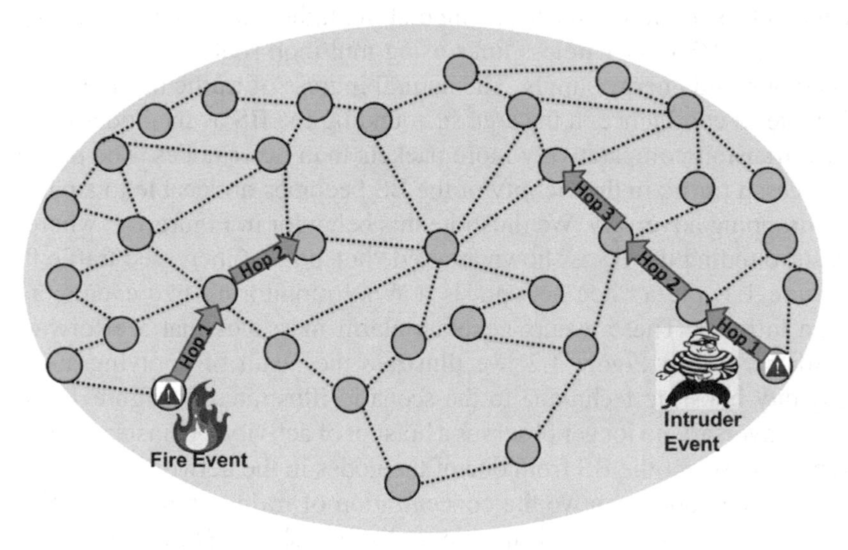

Figure 1.2 Example ad hoc network with high BS anonymity.

distributed wireless networks, and physical layer (PHY) security [14, 15]. In distributed beamforming, multiple network nodes cooperate and share their single antennas to form a virtual, multiantenna system. In wireless networks that employ distributed beamforming, multiple nodes transmit simultaneously, accounting for wireless channel conditions, and precisely control the signal phase, such that all signals constructively combine at the destination. For example, under ideal conditions, N transmitters that send identical messages using equal power while incurring equal path loss when transmitting to a common destination will achieve a factor of N increase in power and signal-to-noise ratio (SNR) at that destination. This property has been demonstrated to increase BS anonymity when applied to a wireless ad hoc network using our distributed beamforming protocol for increased BS anonymity (DiBAN) [1, 12, 13]. We illustrate a distributed beamforming system delivering a single transmission to the BS in Figure 1.3.

A fundamental requirement of distributed beamforming is the ability to recruit a set of helper relay nodes $R_j \in L$ at each hop, where the cardinality of the relay set $|L|$ specifies the number of participating relays; however, there are no suitable relay recruitment approaches for wireless ad hoc networks that consider the trade-off between anonymity and energy consumption. In this chapter, we present a novel relay selection approach that increases BS anonymity by iteratively recruiting relays within a configurable transmission range of each hop on the data route such that the energy advantage of distributed beamforming is sustained. We demonstrate that by maintaining transmission power during relay recruitment below levels that reach the next-hop destination D_i, we boost BS anonymity over a wide range of $|L|$. Our relay selection approach uses the node density in the direction of the next-hop

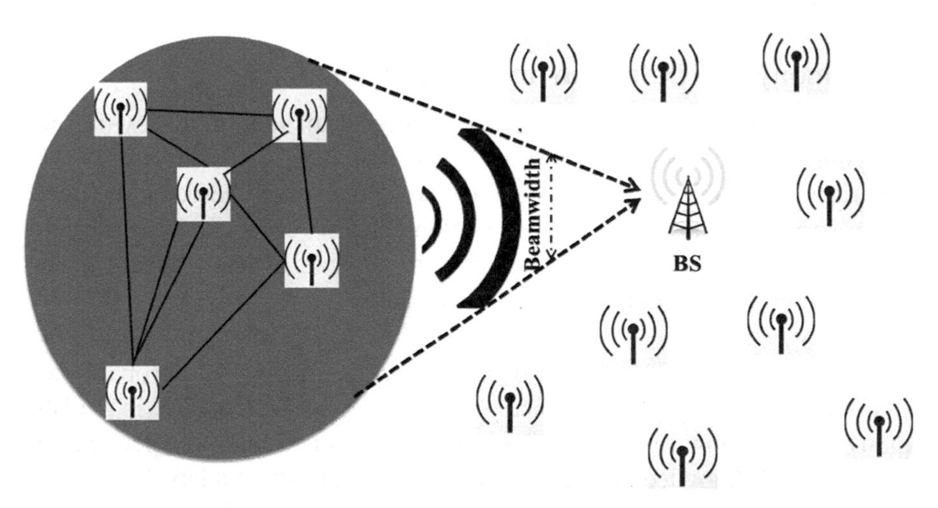

Figure 1.3 Illustration of distributed beamforming nodes directing transmissions toward a BS.

destination D_i to determine a target number of helper relays to be recruited while controlling transmission power levels to prevent the adversary from collecting evidence that implicates the true BS location. Our relay selection approach incorporates channel state information (CSI) measurements to select $R_j \in L$ corresponding to the $|L|$ highest gain paths to most reliably reach D_i. We present simulation results that validate the effectiveness of our relay selection algorithm.

This chapter is organized as follows: Section 1.2 describes anonymity metrics and antitraffic analysis measures. Section 1.3 discusses the system and adversary models including the adversary's evidence theory (ET) analysis framework. Section 1.4 summarizes the DiBAN protocol to increase BS anonymity. Section 1.5 presents an anonymity-aware relay selection algorithm. Section 1.6 reports the validation results. Finally, Section 1.7 concludes the chapter with a summary and a list of topics in the area of anonymity that are worth further investigation.

1.2 Anonymity Definition, Metrics, and Contemporary Measures

This section surveys published work on the topic of anonymity and highlights the aspects of distributed beamforming on which the community has focused. We divide this section into two subsections: BS anonymity definition and assessment and antitraffic analysis techniques.

1.2.1 Anonymity Definition and Assessment

Anonymity can be defined as the state of being not identifiable within an anonymity set $S_U(t)$ at a given time t [11, 16]. The anonymity set $S_U(t)$ comprises all K entities in the system that are still not deciphered or classified as non-suspect by the adversary. All K elements that compose $S_U(t)$ should have similar characteristics such that prior to network analysis the adversary cannot do better than randomly selecting an element $\kappa \in S_U(t)$ as the BS with probability $p_\kappa = \frac{1}{|S_U(t)|}$ [16]. At a time t following the interception of transmitted packets and further analysis, the adversary may update probabilities p_κ to indicate additional confidence gained as to which element κ is the BS. Anonymity is defined from the adversary's perspective and is dependent on the adversary's knowledge at time t and his or her ability to adapt. Therefore, depending on the adversary's level of sophistication, an entity may achieve a different level of anonymity relative to an adversary's capabilities. There are two classes of adversary that are frequently referenced in the literature: the global adversary that has the capability to eavesdrop on the entire network at any time instant and the local adversary that can eavesdrop only on nodes in a small region [16].

When applied to a communication system, anonymity refers to concealing the properties of the nodes that compose an ad hoc network [11]. In this context, anonymity may be applied to the identity, location, or role of a node κ within the network. We define these three anonymity attributes as follows:

1. Identity anonymity: When there is more than one entity in a system, each is referred to by a unique name. In the context of a network, nodes are identified by a static MAC address and a dynamic Internet Protocol (IP) address. Thus, achieving identity anonymity requires preventing an adversary from obtaining the addresses associated with particular network nodes. In many situations, it is advantageous that a node's identity not be revealed to an external or, in some cases, even internal entities, to avert security attacks on the entity or to protect its privacy [11].
2. Location anonymity: The concealment of a node's position at time t is termed location anonymity. An adversary may be interested in locating a particular node $\kappa \in S_U(t)$, regardless of its identity. For example, consider a border protection scenario where no trespassing of any entities is allowed. In such a scenario, the identity of entities is irrelevant. There is a clear distinction between identity and location anonymity; for example, the adversary may know the IP address of a mobile node without being aware of its location. Conversely, in an ad hoc network, an adversary may employ signal detection and traffic analysis techniques to estimate the position of a node without determining its identity [11].
3. Role anonymity: Role anonymity refers to hiding the particular activities of nodes within the network. For example, a node may be a gateway to the larger network, a node along a critical route, or a BS that provides protocol and time synchronization and resource assignments. The importance of some nodes within the network elevates the need of such anonymity. Both location and role anonymity, as applied to the BS node, are the primary types of anonymity considered in this chapter.

A number of quantitative metrics have been proposed to evaluate the anonymity level of the BS and the effectiveness of antitraffic analysis measures. The following are the popular metrics used for quantifying the anonymity of a network node:

Entropy: Serjantov and Danezis [17] and Diaz et al. [18] proposed using entropy as an anonymity measure based on Shannon's work in information theory [19]. An adversary attempts to identify the BS by dividing the network deployment area into N_C cells and assigns to each cell a probability that it contains the BS. Let the probability at time t that a cell i contains the BS be p_i. Then, the entropy $H(x)$ of the network at time t is defined as $H_t(x) = -\sum_{i=0}^{N_C-1}(p_i \log_2 p_i)$. Initially, each cell has an equal probability of containing the BS, i.e., each cell will have probability of $\frac{1}{M}$. Hence, the maximum initial entropy of the system is $H_{\max}(x) = \log_2(N_C)$, which represents the highest achievable level of anonymity because the adversary must randomly select the cell that contains the BS from a set of N_C cells.

The degree of anonymity is defined as the ratio $d_t = \frac{H_t(x)}{H_{\max}(x)}$. By analyzing the network traffic, an adversary learns new information about the cells and the probability assignments correspondingly change at each time step t and updates d_t. When the adversary successfully identifies the cell containing the BS, the probability of that cell becomes 1 and d_t becomes 0. It is important to note that entropy is a measure of anonymity for the entire network that is weighted based on the probabilities p_i derived by a global adversary. Achara and Younis [11] specify a method for the adversary to update p_i at each time t based on the number of transmitted packets observed in each cell, $P_i(t)$. The entropy at time t is therefore given by $H_t(x) = -\sum_{i=0}^{N_C-1}\left(\frac{P_{i(t)}}{T(t)}\log_2\frac{P_i(t)}{T(t)}\right)$, where $T(t) = \sum_{i=0}^{N_C-1} P_i(t)$, the total packets observed by the adversary in all cells at time t, and the anonymity of the BS is measured as $d_t = \frac{H_t(x)}{\log_2 N_C}$ [11].

GSAT Score: Deng et al. [20] proposed the GSAT score as an anonymity measure that relies heavily on the fact that due to the multihop routing, the area in the vicinity of the BS exhibits more frequent packet transmissions than the rest of the network. The GSAT score is used to measure how quickly an adversary detects the BS based on observations from a local eavesdropper who starts at a random position and monitors traffic within the adversary's local reception area. The local adversary attempts to identify the BS by dividing the ad hoc network deployment area into N_C cells. In a greedy way, the adversary moves to the area with the greatest radio transmissions and receptions, hoping to find the BS. If the adversary becomes trapped in local maxima that do not contain the BS, he or she moves in a random direction and resumes the search based on the observed traffic. The algorithm continues until the BS is found, which assumes a secondary method of BS confirmation such as visual inspection. The number of hops until reaching the BS represents the GSAT score and is indicative of the adversary's required effort. A large number of hops indicate that it is difficult for the adversary to locate the BS since the network traffic is more evenly distributed.

To make the GSAT analysis more robust in an ad hoc network where confirmation of the BS through visual inspection is not possible, researchers have proposed GSAT extensions [21]. The modified version of GSAT counts the number of total adversary moves $\aleph$ and the number of times the adversary selects an individual cell i as ν_i to determine the GSAT score $G(i,t) = \frac{\nu_i}{\aleph}$ at time t. The GSAT score for all cells at time t should sum to 1 (i.e., $\sum_{i=0}^{N_C-1} G(i,t) = 1$). GSAT identifies hotspots of transmission activity in the ad hoc network and allows the local adversary to focus on areas that are more likely to be of interest for traffic analysis. Intuitively, an adversary may conclude that a cell i that is visited more frequently than the other cells likely contains the BS [21].

Belief: Huang [10] first proposed evidence theory (ET) as a network traffic analysis model for measuring anonymity in ad hoc networks. In this model, the global adversary considers each captured packet as an evidence of a communication link between a source S_i and a destination D_i. The adversary applies ET by observing the network over a period of time, considering all possible recipients of a transmission, correlating the detected transmissions, and accumulating evidences that implicate a certain node as the BS. Typically, two pieces of prior information are required by the adversary to apply ET: a method to localize all transmitting nodes and an understanding of the RF propagation environment such that the adversary can determine which nodes are within the reception range of all transmissions based on RSS. The Belief metric is a statistical measure derived from the collected evidences and reflects the adversary's confidence that the BS is located in a certain region. ET has received attention from researchers because it provides a global adversary analysis framework that identifies a node or cell with the highest likelihood of containing the BS. ET and the associated Belief metric are the fundamental methods used in this chapter to evaluate BS anonymity. We discuss ET and the Belief metric in detail in Section 1.3.3.

1.2.2 Antitraffic Analysis Measures

A variety of researchers have proposed techniques to improve either source or sink (BS) anonymity of a network. While BS and source anonymity are fundamentally different problems that often have conflicting design goals, there is value in surveying source anonymity techniques to gain insight into how these approaches may be applied to or impact BS anonymity. The primary design objective for source anonymity measures is to reliably and efficiently deliver information from source to sink while preventing the adversary from uncovering the identity and location of packet sources. Published source anonymity techniques have focused on a “panda–hunter” scenario, where an animal poacher plays the role of the adversary and seeks to determine the location of a mobile, tagged panda [22]. In this context, the objective of source anonymity measures is to mitigate the poacher’s threat to a panda. The traffic analysis countermeasures proposed to improve the source anonymity have taken the form of traffic-shaping methods that employ routing loops, the establishment of virtual, fake data packet sources, and random triggering of transmissions to confuse the adversary [22–24]. Source anonymity techniques are not typically applicable to the problem of BS anonymity under the accepted ET attack model. This is because ET analysis makes the basic assumption that the source of a transmission is known, an assumption that is not upheld when source anonymity techniques are applied to the network [10, 11, 24–26]. While source anonymity techniques are not directly applicable to improving sink anonymity, they are referenced here for completeness as they represent the foundation of work in the general area of anonymity.

Foundational work on the topic of BS anonymity has primarily focused on link layer and network layer techniques, especially routing approaches to increase an adversary’s ambiguity. A number of techniques have been proposed to increase the ambiguity of a network’s topology as observed by an eavesdropper performing traffic analysis such that identifying the BS location becomes more difficult [20, 27]. For example, Deng et al. [20] proposed a set of techniques that make the traffic pattern appear more dispersed and introduce random fake paths to confuse the adversary starting from random positions in the network. While such an approach confuses a local adversary, a global adversary applying traffic analysis to the network would still be able to identify the BS [11]. Conner et al. [27] propose a decoy BS that diverts the adversary’s attention away from the real BS. All data packets are first forwarded to the dummy BS and then re-routed to the real BS. Mehta et al. [28] goes further to propose the creation of multiple fake sinks that are spread evenly throughout the network. The concept of a decoy BS implicitly requires that the decoy node has comparable processing and storage capabilities to the true BS. Furthermore, in the case of a single decoy BS, the adversary may still achieve his or her goal of disabling the network by identifying and attacking the decoy [11]. While the adversary's threat is somewhat mitigated by the use of multiple decoy BSs, the adversary may still disable the network by attacking all decoy BSs, which represent a smaller set of targets than all nodes S_U in the network.

Acharya and M. Younis [11] proposed two techniques to boost the BS anonymity level. The first technique requires that the BS retransmits a subset of the packets it receives with different time-to-live in terms of the number of hops these packets are relayed. The goal is to make the BS appear to the adversary as any other node. The second technique assumes a mobile BS that can relocate itself to a more secure location [11]. Techniques proposed in [22, 29, 30] increase anonymity by strategically inserting deceptive packets into the network such that the perceived traffic flow by the eavesdropping adversary is manipulated and attention is directed away from the BS. Meanwhile, the approach of [31] opts to increase the computational complexity of an adversary's ET-based analysis by increasing the transmission power used by all target network nodes. While these techniques have been shown to increase the BS anonymity, they primarily considered an adversary that analyzes data traffic; the impact of inherent traffic patterns associated with control traffic on anonymity has rarely been addressed. Furthermore, because prior antitraffic analysis techniques exist at the link layer and above, novel PHY techniques to boost BS anonymity provide an exciting opportunity for cross-layer antitraffic analysis measures. Our DiBAN approach has distinct advantages over the aforementioned network and link layer antitraffic analysis techniques that we describe in detail in Section 1.4. Figure 1.4 summarizes published anonymity-boosting approaches and opportunities for cross-layer enhancement.

1.3 System Assumptions and Attack Model

In this section, we highlight the underlying capabilities of the ad hoc network and the assumed attack model used to evaluate the BS anonymity throughout this chapter. We also describe the basic assumptions associated with our network model to incorporate distributed beamforming, which are consistent with the related literature. In Section 1.3.3, we provide a detailed description how evidence theory is used by the adversary for traffic analysis and the associated Belief metric that is used to evaluate the BS anonymity [10, 32].

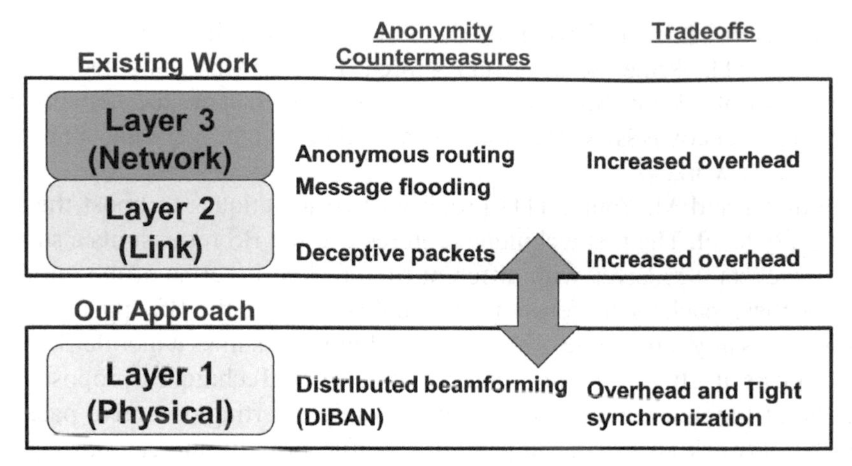

Figure 1.4 Summary of anonymity-boosting approaches and cross-layer enhancement opportunities.

1.3.1 Network Model

We consider a typical homogeneous ad hoc network model where all nodes have similar capabilities including battery life, computation capacity, radio type, and network protocols. The BS serves as a sink of all data traffic originated at the nodes. Only one BS is present in the network. We assume that nodes know their positions relative to the BS and their neighbors [10, 11]. Furthermore, nodes know the transmission power level required to reach all direct (next-hop) neighbors. Least cost, multihop routes are pursued for delivering data frames to the BS, where the required transmission power to reach each destination is used as the link cost. This limits the communication overhead required by any single node since communication is the primary source of energy consumption in a node [33]. The onboard radio is assumed to use matched filter receivers that make received symbol decisions based on the received energy in a symbol period [33–35]. We also assume that all nodes and the BS use omnidirectional antennas and that the RF propagation environment is characterized by the Friis free space path loss equation, $\ell_{\text{FSPL}} = \frac{(4\pi d)^2}{\lambda}$, where λ represents the wavelength and is related to the speed of light $c = 3 \times 10^{-8}$ m/s and carrier frequency f by $c = \frac{\lambda}{f}$ [33]. This assumption establishes a node's method for determining the nodes from which it can receive transmissions and could be augmented with other propagation models with no modification to the PHY and MAC antitraffic analysis measures that we consider in this chapter [33].

Precautionary measures are assumed to be employed in the design and the operation of the BS to avoid exposing it to the adversary. For example, a BS maintains a transmission power level equivalent to other nodes in the network and limits its involvement in control and management traffic such as new route discovery, in order to keep itself undistinguishable via RF analysis from other nodes. The BS sends periodic beacons into the network of frame length equivalent to other message sizes using a random transmission power level selected from a uniform distribution on the interval of power to reach its single- and double-hop neighbors. These periodically transmitted beacons prevent the BS from standing out as a receive-only node within the network. We assume that physical camouflage techniques have been applied to the BS such that an adversary cannot identify it visually, even if he or she is in the vicinity of the BS [28]; hence, the BS may only be located using traffic analysis (i.e., evidence theory).

Unless otherwise stated, we assume that all transmissions contain encrypted headers and payloads. We assume a time division multiple access (TDMA) MAC protocol with sufficient time synchronization to operate all nodes within a tolerable guard time of slot boundaries. Such synchronization precision is well within the specified performance of ad hoc network timing protocols such as reference broadcast synchronization (RBS), timing-synch protocol for sensor networks (TPSN), or equivalent [33, 36, 37]. Furthermore, we assume that sufficient TDMA timeslots exist to allow interference-free coordination between all transmitters wishing to access the medium. No specific PHY protocol is assumed for the target network; however, design principles have been incorporated into the protocol stack to leverage the IEEE 802.15.4 PHY [38].

1.3.2 Adversary Model

The target ad hoc network is assumed to be carrying out critical tasks that a highly motivated, well-equipped adversary is eager to disrupt. After identifying and exploiting the BS, the adversary's objective is to achieve a DoS attack against the BS at all costs, including physically destroying the BS. The adversary is assumed to be a completely passive eavesdropper with global presence throughout the network [23, 28]. Such a global presence could be accomplished practically by deploying antennas in the vicinity of the target network. Under ideal conditions, the adversary has the ability to localize the source and destination of all radio communication in the network deployment area to the resolution of a square cell using well-known signal localization techniques such as angle of arrival (AoA) and the received signal strength (RSS) [33, 39–42]. Although the adversary intercepts data frames, it is assumed that the cryptosystem is sufficiently robust that the adversary cannot apply cryptanalysis to recover the underlying payload or header contents.

The adversary collects the following information about the target network, which we represent as a 3-tuple (η, t, e), where η represents the adversary's location of interception, t represents the time of transmission interception, and e represents the intercepted transmission in its entirety [23]. We acknowledge the possibility that a motivated adversary may be able to determine that the network is using antitraffic analysis measures and adapt the analysis in an attempt to defeat them. We note, however, that our research results [11, 29, 31] and those of other researchers [28, 43–45] have shown that there is not a single countermeasure that provides sufficient protection to the network to prevent a motivated adversary from adapting his or her attack strategy to defeat the countermeasure. The DiBAN countermeasure is a physical layer approach, and we show that it withstands conventional attacks using sophisticated methodologies like ET when the adversary is unaware that the network employs distributed beamforming as an anonymity-boosting technique. An adversary would require additional information, e.g., from an insider, to learn that distributed beamforming is employed and correspondingly extend his or her attack model. Such a possibility is considered well beyond the scope of this chapter.

1.3.3 Evidence Theory and Belief Metric

An adversary may use ET as a traffic analysis model to evaluate the BS anonymity level. To apply ET to a network, the adversary first intercepts all point-to-point transmissions, represented as evidence $E(U)$, where U represents a point-to-point link between two nodes, source S_i and destination D_i. The adversary then derives composite paths by correlating all pairwise evidences. An end-to-end path that contains two or more nodes is represented by V and its associated evidence $E(V)$ is calculated as follows:

$$E(V) = \min_{U \subseteq V}\{E(U)\}, \quad |V| \geq 2, \tag{1.1}$$

where $|V|$ represents the cardinality of end-to-end path V or the number of nodes that compose the inferred path [10]. The normalized evidence $m(V) = E(V)/\sum E(V)$ expresses the proportion to which all evidences collected by the adversary support the claim that a particular element of the set of all nodes

S_U is part of the communication path V [10, 11]. The weighted Belief metric $B(u)$ represents the adversary's confidence in the existence of a path of length n that ends at a specific node u and is represented as follows:

$$B(u) = \sum_{U|u\subseteq V} n\, m(U). \tag{1.2}$$

The Belief metric is the method by which we assess the BS anonymity. A small Belief metric corresponds to decreased adversary confidence or higher anonymity in a BS location, and conversely, a larger Belief metric corresponds to increased adversary confidence or lower anonymity in a BS location. The Belief calculation can be computationally expensive since the number of mathematical operations scales at the rate of the number of nodes S_U factorial in the network. We express this maximum complexity as follows:

$$\max(\text{Complexity}) = O\left(\sum_{k=2}^{S_U}\left(\frac{S_U!}{(S_U-k)!}\right)\right). \tag{1.3}$$

We derive (1.3) from the fact that all paths of length $|V| \geq 2$ must be traversed by the adversary to calculate Belief over the full set of evidences. The deployment of S_U nodes can be viewed as a graph, where the total number of evidences m taken k at a time is represented by $\frac{m!}{(m-k)!}$. Although many evidences $E(V)$ may equal zero and simplify the adversary's calculation, the set of pairwise and derived evidences grows quickly with a large number of S_U nodes. To decrease the computational complexity of (1.3), we further impose a restriction that the adversary partitions the target area into a $M \times M$ grid composed of N_C square cells, which is equivalent to relaxing the adversary's destination localization requirements to only the fidelity of a cell [11, 12, 31, 32]. In this case, the maximum number of computational operations is reduced to scale with N_C independent of S_U, where we replace S_U with N_C in Equation (1.3) and define a square cell to be of size $\Delta \times \Delta$ m^2, where $\Delta = \frac{M}{\sqrt{N_C}}$, hence reducing the complexity of the Belief calculation to $\max(\text{Complexity}) = O\left(\sum_{k=2}^{N_C}\left(\frac{N_C!}{(N_C-k)!}\right)\right)$.

Figure 1.5 shows an illustrative example of ET analysis applied to a seven-node network, where the deployment area is partitioned into $N_C = 9$ cells. Two consecutive frames are transmitted, the first from cells 1 to 4 and the second from cells 4 to 7. The adversary collects pairwise evidences and calculates the derived evidence for composite paths in Table 1.1. Table 1.1 shows the adversary's collected evidence over time, where evidence is collected first

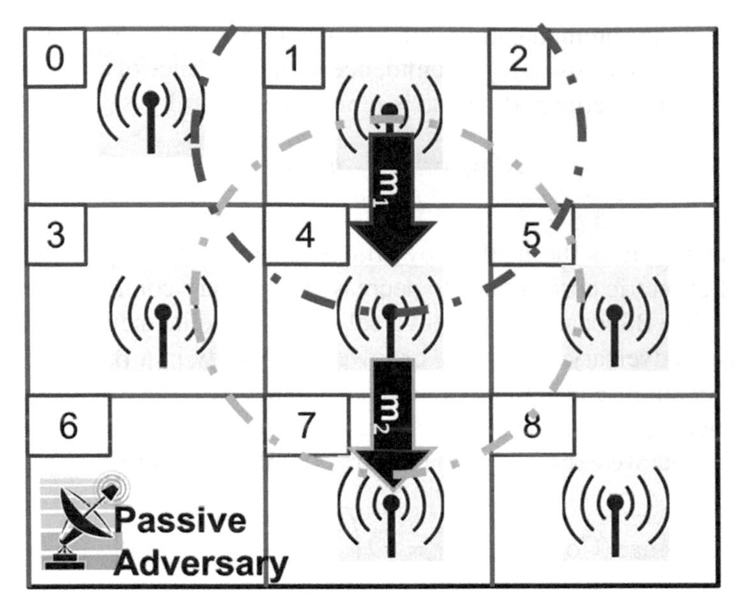

Figure 1.5 Example of an adversary applying ET.

for message m_1 at time t_1 and then for message m_2 at time t_2. The adversary then calculates the derived evidence at time t_3 based on the collection of evidence generated by messages m_1 and m_2. The adversary may build an adjacency matrix directed graph representation of the network filled with source and destination evidences. We force the adjacency matrix diagonal to be zero which imposes the physical constraint that a cell cannot transmit to itself and prevents loops in the ET analysis.

After the adversary observes the network, he or she generates normalized evidence $m(V)$. To generate $m(V)$, the adversary divides each evidence by the total count of evidences at a given time instant. For example, in Table 1.1, the adversary collected four evidences from the transmission of message m_1 at time t_1, each with a count of 1. Since the adversary's collection contains a total

Table 1.1 Collected evidence over time for the illustrative standard ET example

Transmission	Evidence	$E(V)$	$m(V)$
m_1 at time t_1	$E(1,4)$, $E(1,0)$, $E(1,5)$, $E(1,3)$	1	0.25
m_2 at time t_2	$E(4,1)$, $E(4,5)$, $E(4,7)$, $E(4,3)$, $E(4,0)$, $E(4,8)$	1	0.1
Derived at time t_3	$E(1,4,7), E(1,4,5), E(1,4,3), E(1,4,8)$, $E(1,4,0)$	1	0.066

of four evidences, normalization at time t_1 produces normalized evidences of value $\frac{1}{4}$ or 0.25. At time t_2, the adversary's evidence collection contains a total of 10 evidences, each with a count of one. Therefore, normalization at time t_2 results in $\frac{1}{10}$ or 0.1. At time t_3, the adversary then correlates the collected evidence and derives composite evidence for paths of length $|V| \geq 2$. At time t_3, the adversary has collected 15 unique evidences, and therefore, normalization produces unique evidences $m(V)$, each of value $\frac{1}{15}$ or 0.066.

Once the adversary generates a set of normalized evidences $m(V)$, Equation (1.2) is applied to determine the Belief metrics that can be used to assess the BS anonymity. Each Belief metric $B(u)$ is calculated using the simplified cellular-based analysis, where u refers to one of the N_C cells within the adversary's grid instead of a particular node. $B(u)$ represents the adversary's confidence in the existence of a path of length n that ends at a specific cell u. We provide a detailed example of calculating Belief metrics $B(u)$ in this nine-cell scenario in Appendix I. Figure 1.6 illustrates a notional example of the adversary's Belief metric calculation when the network being evaluated employs no anonymity-boosting techniques. We simulated two independent scenarios in this example with $S_U - 100$ uniformly distributed, randomly deployed nodes and an adversary evaluation grid containing $N_C = 9$ cells, where the BS is located in cells 4 and 8. Figure 1.6(i) shows similar Belief metrics for all cells except the one containing the BS, which is cell 4, indicated by the large peak. Similarly, Figure 1.6(ii) shows relatively low-level Belief for all cells except the one containing the BS, which is cell 8, indicated by the large peak. Applying an effective anonymity-boosting technique to the network under evaluation would reduce the BS cell's peak (i.e., decrease the Belief such that the anonymity is increased). The reader may also notice that the Belief associated with cell 4 is higher than the Belief associated with cell 8. This can be explained by the cell's position within the adversary's analysis grid; cell 4 has eight adjacent neighbors, and therefore, the adversary accumulates implicating evidence from all of its surrounding nodes. Conversely, cell 8 is protected in the corner of the adversary's grid and only has three adjacent neighbors to generate implicating evidence. Hence, the Belief associated with cell 8 containing the BS is smaller than the Belief associated with cell 4 containing the BS; however, the important observation from Figure 1.6 is not the absolute difference in the two peak values, but the fact that the ET analysis produces an implicating peak for each scenario.

We remind the reader that a pairwise evidence is defined as existing between two nodes, a source S_i and a destination D_i. The adversary may

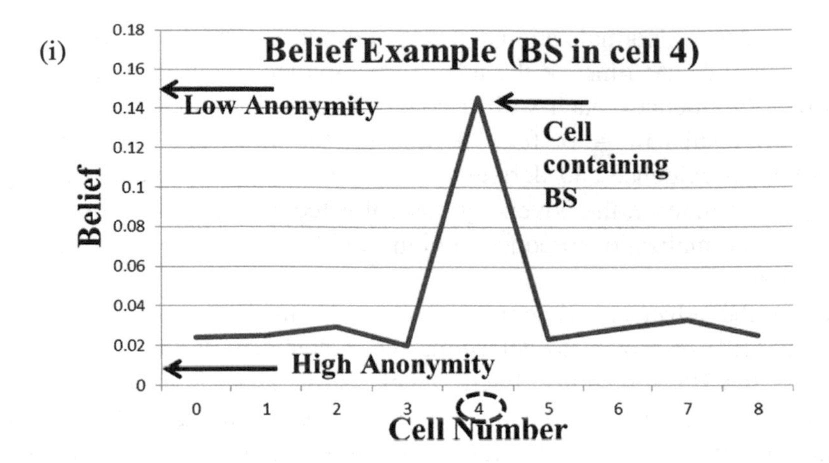

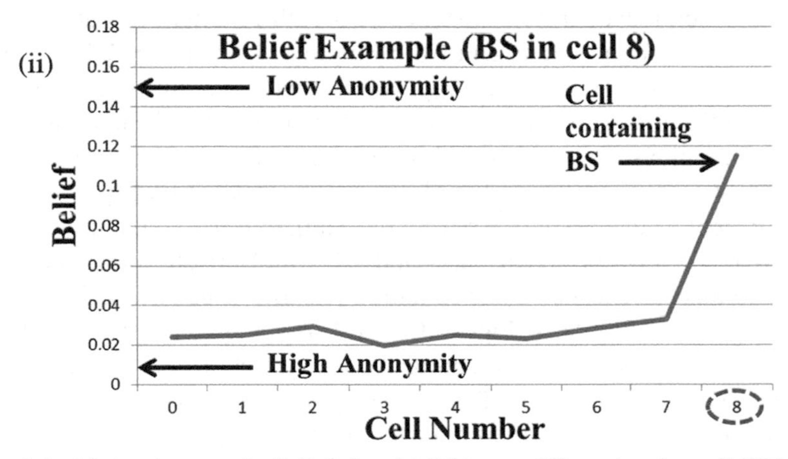

Figure 1.6 Notional anonymity Belief plot with BS in two different locations: (i) BS located in cell 4 and (ii) BS located in cell 8.

only accredit evidences derived from intercepted frames when the source S_i is known and the destination D_i can be determined. Specifically, the imposed cellular-based analysis requires that the adversary determines a destination D_i within a specific cell before evidence $E(S_i,\ D_i)$ is added to the set of collected evidences. That is, if a source or possible recipient of a given transmission is not determined to lie within a particular cell, the adversary associates no evidence with that cell. This becomes an important point when we evaluate the anonymity of a network that employs our DiBAN to boost its anonymity.

1.4 Distributed Beamforming to Increase the BS Anonymity

In this section, we present a PHY approach called DiBAN that leverages distributed beamforming [12, 13]. To the author's knowledge, our DiBAN protocol represents the first time cooperative communication has been applied to a wireless communication system for the purpose of increasing BS anonymity. Researchers have generally considered cooperation in the context of multiple input, multiple output (MIMO) systems and space–time coding algorithms to improve the throughput or link reliability. Such work is not directly applicable to resource-limited, single-antenna nodes, e.g., miniaturized nodes that compose a wireless sensor network. Instead, distributed beamforming is deemed applicable. Mudumbai et al. [46] presented an overview of the current state of the art, challenges, and successful implementations of distributed beamforming systems. In [14], the same authors describe a master–slave carrier frequency synchronization architecture, which is leveraged in our distributed beamforming protocol. Ochiai et al. [47] present an analysis of the achievable directivity with N cooperating nodes. The distributed beamforming beam pattern for N nodes is investigated when phase jitter or estimation errors impair the cooperation process.

Although much of the distributed beamforming work to date has focused on theoretical analysis, Quitin et al. [48] and Rahman et al. [49] present practical implementations and analyses of distributed beamforming systems using GNU radio and universal software radio peripheral (USRP). Oh et al. [50] describes a practical cooperative transmitter experiment with GNU Radio. These SDR implementations provide two features for supporting our work: First, they demonstrate that practical implementations of distributed beamforming systems are achievable using commonplace radio hardware, and second, the proliferation of flexible SDR systems significantly lowers the bar for a motivated global adversary to intercept and localize wireless signals [50].

1.4.1 Overview of the DiBAN Protocol

As discussed in Section 1.2, like other traffic analysis countermeasures, DiBAN requires additional overhead for signaling; however, this additional energy load is shared between the source and helper relays such that each node is able to decrease its transmission power. This decrease in transmission power per node also reduces the overall level of interference in the network since the cumulative power for a given cooperative transmission does not affect nodes outside of the distributed beam pattern. In this section, we demonstrate that beyond its well-known benefits, distributed beamforming also introduces error in the adversary's localization of source nodes and identification of pairwise source and destination relationships. These attributes increase the BS anonymity and are applied to a wireless ad hoc network through our DiBAN. Because the distributed beamforming countermeasure exists at the PHY and link layer, it may be applied to an ad hoc network combined with other higher-layer countermeasures to further boost the BS anonymity.

Distributed beamforming exploits the broadcast nature of wireless transmissions since all frames destined for a particular receiver may be overheard by neighboring nodes. As illustrated in Figure 1.7(i), these neighboring nodes may serve as helper relays by cooperating with a given transmission source S_i such that the transmitted signal arrives at the destination D_i from a diverse set of transmitters. Because each relay R_j transmits the same message as S_i with precise time and carrier synchronization, the signals constructively combine at the destination D_i under ideal timing and carrier synchronization conditions.

There are three primary components of a distributed beamforming system that must be implemented. First, a cross-layer, distributed, relay selection approach is needed that selects a set of relays with the best CSI measurements to maximize BS anonymity while minimizing communication energy consumption. This relay selection approach is described in detail in Section 1.5. Second, timing synchronization is required. All participating helper relays must initialize the transmission process according to a common time reference such that the transmitted waveforms arrive in phase at the intended receiver. The handshaking of DiBAN in addition to the underlying time synchronization algorithm such as RBS and TPSN allows nodes to achieve relative time synchronization by comparing message time stamps [36, 37]. Third, carrier synchronization is required, whereby each relay R_j uses its internal phase-locked loop to synchronize its local oscillator (LO) to an unmodulated carrier transmitted by S_i. Because each node has its own independent LO that is not natively phase-locked to a common reference source, the transmissions of the source S_i and helper relay R_j will not add constructively at the destination D_i and no beamforming gains are achieved without carrier synchronization [14]. Thus, DiBAN allows the relay set $R_j \in L$ to achieve carrier synchronization with the source S_i through the transmission of an unmodulated carrier [12–14, 46].

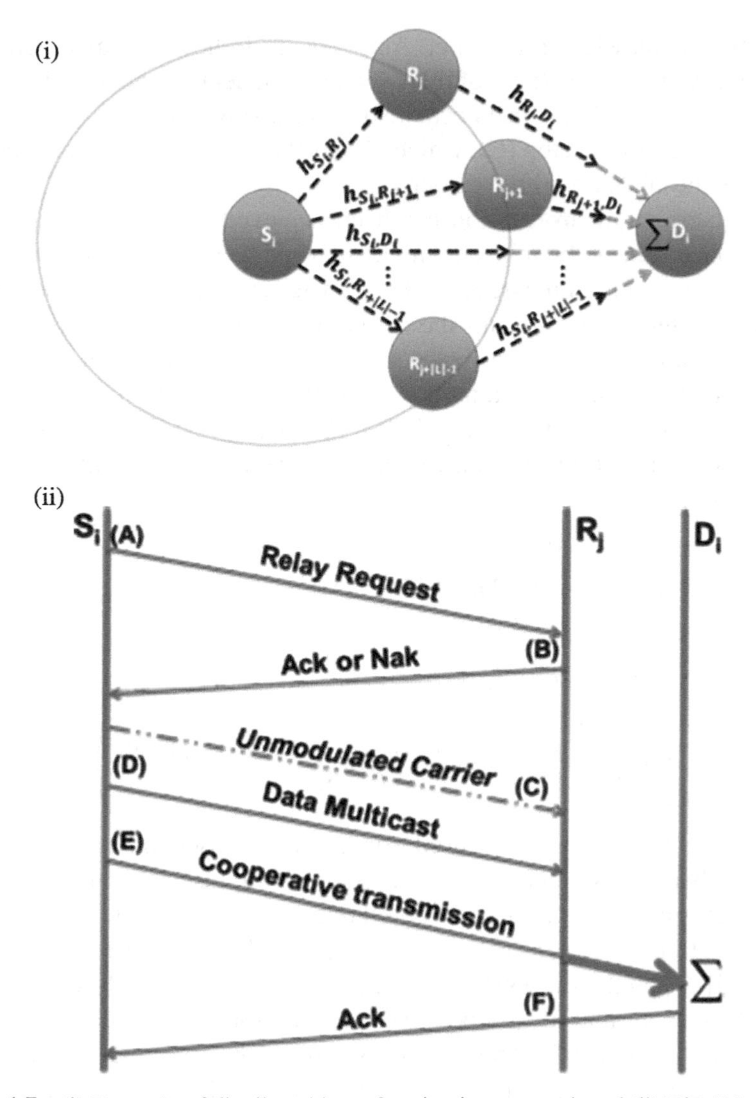

Figure 1.7 (i) Example of distributed beamforming in a network and (ii) DiBAN protocol sequence diagram.

Figure 1.7(ii) illustrates a sequence diagram for the DiBAN protocol. Our relay selection algorithm represents a novel approach to accomplish steps (A) and (B), where the source S_i begins by transmitting a "Relay Request" that is broadcasted with transmission power $P_T^{S_i,R_j}$ to recruit a set of helper relays L within S_i's transmission range. Each R_j responds to S_i with a positive or negative acknowledgement, i.e., Ack/Nak, to indicate its availability to cooperate. We revisit the topic of relay selection in detail in Section 1.5. Once the helper relay set L has been established for a single hop, S_i transmits an unmodulated carrier in step (C), which allows the relay set $R_j \in L$ to achieve carrier synchronization with the source S_i.

Once carrier synchronization has been achieved by $R_j \in L$, S_i multicasts the data frame to be relayed with distributed beamforming to $R_j \in L$ in step (D). This multicast message contains a trigger time stamp for the precise time that S_i intends to transmit the cooperative message to destination D_i. Prior to the simultaneous distributed beamforming transmission, each relay R_j weights its transmitted signal by a channel coefficient w_{R_j,D_i} based on CSI between $R_j \in L$ and D_i [12–14, 46]. Finally, in step (E), S_i and $R_j \in L$ send a cooperative transmission to D_i. The ideal received signal at D_i is represented by:

$$\begin{aligned} r_{D_i}(t) &\triangleq r_{S_i,D_i}(t) + \sum_{j=0}^{|L|-1} r_{R_j,D_i}(t) \\ &= \Re\Big(A_{S_i}(t) w_{S_i,D_i} h_{S_i,D_i} e^{j(2\pi f_c t + \theta(t) + \varphi(t))}\Big) \\ &\quad + \Re\left(\sum_{j=0}^{|L|-1} A_{R_j}(t) w_{R_j,D_i} h_{R_j,D_i} e^{j(2\pi f_c t + \theta(t) + \varphi(t))}\right) + n(t), \end{aligned} \tag{1.4}$$

where $\Re$ is the real operator used for complex baseband notation, $A(t)$ is the baseband pulse shape, h is the channel impulse response, f_c is the carrier frequency, $\theta(t)$ is a phase modulation term, $\varphi(t)$ is an aggregate phase shift term, and $n(t)$ represents the thermal noise present at receiver D_i [34, 35]. Upon the successful receipt of $r_{D_i}(t)$, D_i responds to S_i with an acknowledgement that the cooperative transmission was correctly received, as illustrated in step (F). The Ack is overheard by R_j, and therefore, all cooperating nodes are able to end their dynamic DiBAN relationship at the ith hop. In the case where S_i and $R_j \in L$ do not receive D_i's Ack, one among three retransmission modes may be employed [13]; however, to simplify the presentation, we assume a wireless ad hoc network that implements ideal distributed beamforming, and therefore, retransmission modes are beyond the scope of this chapter.

The significance of (1.4) in the context of BS anonymity is that distributed beamforming allows the received signal component $r_{S_i,D_i}(t)$ from S_i's transmission to be decreased by $\sum_{j=0}^{|L|-1} r_{R_j,D_i}(t)$ and to maintain the same effective non-cooperative received power level at D_i when the phase offset $\varphi(t)$ is held constant. Distributed beamforming therefore prevents the adversary from identifying the evidence $E(S_i, D_i)$. More specifically, if all transmitters S_i and $R_j \in L$ transmit only the required power to reach D_i at a specific SNR, each transmitter may decrease its power by $10 \log(|L| + 1)$ dB.

1.4.2 DiBAN Illustrative Example

Figure 1.8 shows an example of the adversary's ET analysis with and without distributed beamforming applied to a seven-node network, where the target deployment area is partitioned into $N_C = 9$ cells. In Figure 1.8, the baseline transmission is superimposed with the cooperative transmission from DiBAN step (E) in Figure 1.7(ii). Table 1.2 presents the associated collected evidence and Belief calculation at a single hop for the baseline and DiBAN cases, where a node in cell 4 transmits a frame to a node in cell 7.

We compare the baseline results to the DiBAN case, where we show only the cooperative step (E) in Figure 1.7(ii) which occurs after S_i in cell 4 has

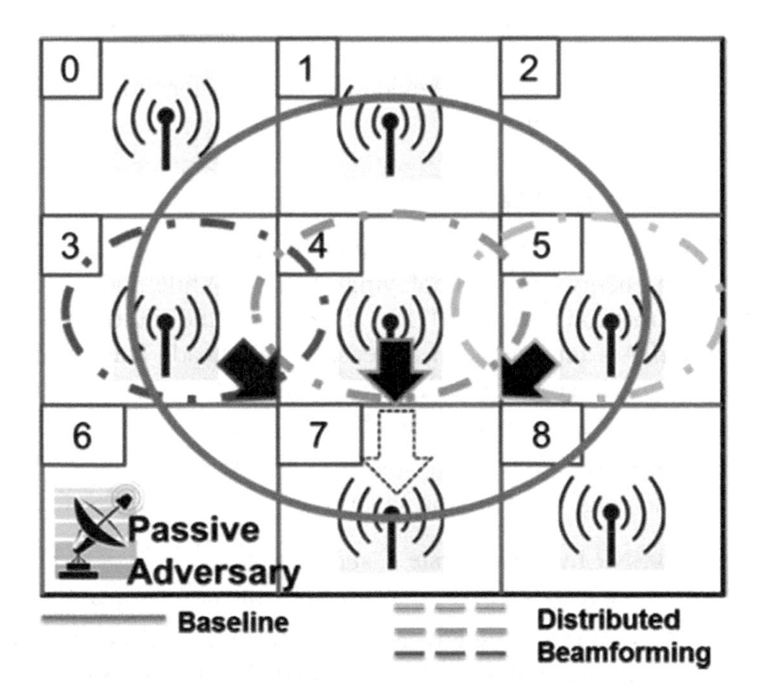

Figure 1.8 Example of an adversary applying ET to a network.

Table 1.2 Collected evidence from baseline and DiBAN example

Transmission	Evidence	$E(V)$	$m(V)$	$B(u)$
Baseline	$E(4,0)$, $(4,1)$, $E(4,5)$, $E(4,8)$, $E(4,7)$, $E(4,3)$	1	0.166	0.166
DiBAN (E)	$E(4,3), E(4,5), E(3,4), E(5,4)$	1	0.25	0.25

already chosen $R_j \in L$, where $L = \{3,5\}$. The three cooperating nodes transmit to D_i in cell 7, and each decreases their transmission power by $10\log(3) = 4.77$ dB, while still reaching D_i in cell 7 with a coherently combined signal that achieves sufficient SNR. In Table 1.2, we observe that not only is the number of evidences decreased for the DiBAN case, but also the generated $B(u)$ applies only to endpoints in cells 3, 4, and 5, not the true endpoint of cell 7. The total number of evidences collected by the adversary is decreased because the range associated with the cooperative transmissions is less than the baseline transmissions. The individual transmissions do not reach D_i, and correspondingly, the pairwise and derived evidences do not implicate D_i as a sink. Distributed beamforming increases the BS anonymity by unlinking S_i and $R_j \in L$ from D_i. We note that the adversary calculates a meaningful $B(u)$ after a collection period much longer than observation of the single round of transmissions shown in Figure 1.8. Moreover, DiBAN steps (A) through (F) also generate evidences that are not shown in Table 1.2 to simplify the presentation.

1.4.3 DiBAN Energy Analysis

The benefit of increased BS anonymity must be balanced with the associated communication energy cost of employing DiBAN. While nodes may enjoy savings in transmission power of $10\log(|L|+1)$ dB, the additional signaling required by DiBAN represents an increase in overhead. This results in an aggregate increase in communication energy usage that scales according to the average data payload size $\overline{\beta}$ and number of participating helper relays $|L|$. The non-cooperative energy per bit ε_b transmitted by each node is calculated by dividing the average transmission power $\overline{P_T^{S_i,D_i}}$ required for S_i to reach D_i at the threshold SNR by the data rate r, such that $\overline{\varepsilon_b^{S_i,D_i}} \triangleq \frac{\overline{P_T^{S_i,D_i}}}{r}$. Therefore, the average communication energy consumed by the non-cooperative system is $\overline{\varepsilon_{\text{base}}} \triangleq \overline{\varepsilon_b^{S_i,D_i}}\overline{\beta}$. Correspondingly, it follows from the DiBAN protocol, as shown in Figure 1.7(ii), that the average expended communication energy at a single cooperative hop is given by:

$$\overline{\varepsilon_{\text{DiBAN}}} \triangleq \left(\overline{\varepsilon_b^{S_i,R_j}} (\Upsilon_{RR} + \Upsilon_{Data} + \overline{\beta} + K) + \frac{\overline{\varepsilon_b^{S_i,D_i}}}{(|L|+1)} \overline{\beta} \right.$$
$$\left. + |L| \overline{\varepsilon_b^{R_j,S_i}} \Upsilon_{Ack} + |L| \frac{\overline{\varepsilon_b^{R_j,D_i}}}{(|L|+1)} + \overline{\varepsilon_b^{D_i,S_i}} \Upsilon_{Ack} \right). \tag{1.5}$$

S_i incurs Υ_{RR} bits of overhead for the "Relay Request" message (A) in Figure 1.7(ii) and $\Upsilon_{Data} + \overline{\beta}$ bits for the data multicast message (D) in Figure 1.7(ii). S_i also incurs the cost of transmitting an unmodulated carrier for t μs, which we represent as the equivalent of K bits (i.e., $K = t \times r$) shown as (C) in Figure 1.7(ii). This overhead cost to S_i results from the point-to-point transmissions between S_i and $R_j \in L$ required for DiBAN, but S_i receives no distributed beamforming power savings for these transmissions at power $\overline{P_T^{S_i,R_j}}$. Consequently, S_i achieves energy savings by recruiting relays positioned as close as possible to S_i to minimize recruiting power $\overline{P_T^{S_i,R_j}}$, which increases proportionally to $d^2_{S_i,R_j}$, where d_{S_i,R_j} is the distance between nodes S_i and R_j. S_i concludes the DiBAN sequence by cooperatively transmitting the data payload of $\overline{\beta}$ bits with power level $\frac{\overline{P_T^{S_i,D_i}}}{(|L|+1)}$, illustrated as message (E) in Figure 1.7(ii). Distributed beamforming allows S_i to potentially decrease its transmission power by a factor of $|L| + 1$ since message (E) is transmitted simultaneously by S_i and $R_j \in L$ [12–14, 46].

Each helper relay $R_j \in L$ incurs the Υ_{Ack} overhead bits of the Ack/Nak message (B) in Figure 1.7(ii) transmitted with power $\overline{P_T^{R_j,S_i}}$. Each $R_j \in L$ transmits the $\overline{\beta}$-bit payload associated with the cooperative transmission to D_i with power $\frac{\overline{P_T^{R_j,D_i}}}{(|L|+1)}$, illustrated as message (E) in Figure 1.7(ii). Finally, the process concludes when D_i receives the cooperative transmission and responds with an Ack to S_i, incurring Υ_{Ack} bits of overhead, illustrated as (F) in Figure 1.7(ii). Two key parameters in (1.5) result directly from a relay selection algorithm, namely $\overline{P_T^{S_i,R_j}}$ and $|L|$. All other terms in (1.5) represent static message parameters or power parameters that result from the particular routing algorithm that selected S_i and D_i. Moreover, $\overline{P_T^{S_i,R_j}}$ and $|L|$ are correlated since increasing $\overline{P_T^{S_i,R_j}}$ provides more opportunity for a node to recruit more relays, hence increasing $|L|$. And conversely, decreasing $P_T^{S_i,R_j}$ results in smaller $|L|$. Because Υ_{Ack} is usually a small message, an energy saving in $\overline{\varepsilon_{\text{DiBAN}}}$ is produced by increasing $|L|$, yet increasing $|L|$ is only energy efficient if $\overline{P_T^{S_i,R_j}}$ is not increased because of the large incurred overhead $\Upsilon_{RR} + \Upsilon_{Data} + \overline{\beta} + K$. To illustrate this point, Table 1.3 evaluates the resulting $\overline{\varepsilon_{\text{DiBAN}}}$ for $|L| = 1,\ 2$, and 5 and provides typical example average values of parameters from (1.5) for node density $\lambda = 0.0001$ nodes per meter-squared.

We observe from Table 1.3 that in this example, we held the energy per bit used for relay recruitment, $\overline{\varepsilon_b^{S_i,R_j}}$, constant across the three $|L|$ values considered to remove the dependency on $|L|$. We account for the cooperative energy savings provided by $|L|$ in the energy per bit quantities. Since the same routing algorithm is used in all cases of $|L|$, the point-to-point Ack sent from D_i and S_i requires the same energy per bit $\overline{\varepsilon_b^{D_i,S_i}}$ to transmit the DiBAN Ack. Table 1.3 clearly illustrates that recruiting more relays (i.e., increasing $|L|$) results in energy savings $\overline{\varepsilon_{\text{DiBAN}}}$ unless we have to increase $\overline{\varepsilon_b^{S_i,R_j}}$ significantly to recruit the desired number of relays.

1.5 Distributed Beamforming Relay Selection Approach

DiBAN requires a recruiting strategy to select a set of relays $R_j \in L$ at each hop that maximizes BS anonymity while minimizing communication energy consumption. Previous work on relay selection has focused on identifying a single best relay from a set of available candidates, where channel quality between both the source and relay and the relay and destination is the primary selection criterion. Bletsas et al. [51] describe an approach to select an optimal

Table 1.3 Example average $\overline{\varepsilon_{\text{DiBAN}}}$ parameters in (1.5)

$\overline{\varepsilon_{\text{DiBAN}}}$ Parameter	$\lvert L\rvert = 1$	$\lvert L\rvert = 2$	$\lvert L\rvert = 5$
$\overline{\varepsilon_b^{S_i,R_j}}$	15.5 nJ/bit	15.5 nJ/bit	15.5 nJ/bit
$\overline{\varepsilon_b^{S_i,D_i}}$	9.8 nJ/bit	6.53 nJ/bit	3.27 nJ/bit
$\overline{\varepsilon_b^{R_j,S_i}}$	0.232 nJ/bit	0.312 nJ/bit	0.658 nJ/bit
$\overline{\varepsilon_b^{R_j,D_i}}$	9.8 nJ/bit	6.53 nJ/bit	3.27 nJ/bit
$\overline{\varepsilon_b^{D_i,S_i}}$	19.6 nJ/bit	19.6 nJ/bit	19.6 nJ/bit
Υ_{RR}	160 bits	160 bits	160 bits
Υ_{Data}	168 bits	168 bits	168 bits
Υ_{Ack}	128 bits	128 bits	128 bits
$\overline{\beta}$	504 bits	504 bits	504 bits
K	800 bits	800 bits	800 bits
$\overline{\varepsilon_{\text{DiBAN}}}$	37.70 µJ/bit	37.61 µJ/bit	36.52 µJ/bit

helper relay using CSI measurements. Our approach leverages the work of Bletsas [51] to determine the helper relays $R_j \in L$ with the best CSI conditions based on the local measurements. Jing et al. [52] formulate CSI-based relay selection as a sequential decision problem. Wang et al. [53] and Ibrahim et al. [54] present relay selection methods that minimize symbol error probability (SEP) in decode-and-forward (DF) cooperative communication systems; however, these approaches only select a single optimal relay and do not directly apply to a distributed beamforming system such as the one we consider in our DiBAN protocol.

Fundamentally, DiBAN requires a distributed relay selection approach for recruiting a set of $R_j \in L$ at each hop that achieves three objectives, namely boosting BS anonymity, sustaining or decreasing the communication energy $\overline{\varepsilon_{\text{DiBAN}}}$ compared to the baseline system $\overline{\varepsilon_{\text{base}}}$ with no distributed beamforming, and having the best instantaneous CSI measurements. Obviously, increasing $|L|$ will diminish the linkability between S_i and D_i and will thus improve the anonymity, but as previously discussed in Section 1.4.3, increasing $|L|$ requires more transmission power $\overline{P_T^{S_i,R_j}}$ during relay recruitment. Our proposed algorithm iteratively attempts to recruit the expected number of potential relays that can be reached by S_i using increasing power levels $\overline{P_T^{S_i,R_j}}$. We maintain BS anonymity by restricting $P_T^{S_i,R_j} < P_T^{S_i,D_i}$ to prevent the adversary from collecting an evidence $E(S_i, D_i)$ that links the source and destination. Our algorithm first analytically determines the number of potential relays $|L_D|$ within a semicircle of radius $d_{S_i,R_j} = \frac{d_{S_i,D_i}}{\delta}$ based on the expected node count within the reception range of

$$P_T^{S_i,R_j} \triangleq SNR + \ell_{S_i,R_j} - 10\log(\kappa TB) - NF, \tag{1.6}$$

where the thermal noise contribution defined by Boltzmann's constant κ, noise temperature T, and signal bandwidth B and noise figure NF remain constant and are specific to the node's receiver [12, 13]. We determine the Friis free space path loss over the distance d_{S_i,R_j} between nodes S_i and R_j as $\ell_{S_i,R_j} \triangleq \frac{\left(4\pi\left(d_{S_i,R_j}\right)\right)^2}{\lambda} \triangleq \frac{\left(4\pi\left(\frac{d_{S_i,D_i}}{\delta}\right)\right)^2}{\lambda}$, but could be augmented with other propagation models [57]. Therefore, $P_T^{S_i,R_j}$ is directly a function of δ since

$$P_T^{S_i,R_j} \triangleq SNR + 20\log\left(\frac{\left(4\pi\left(\frac{d_{S_i,D_i}}{\delta}\right)\right)}{\lambda}\right) - 10\log(\kappa TB) - NF. \tag{1.7}$$

The algorithm iteratively adjusts $P_T^{S_i,R_j}$ using δ, where $\delta > 1$ since $\delta = 1$ is equivalent to $P_T^{S_i,R_j} = P_T^{S_i,D_i}$ and establishes an undesired link between S_i and D_i. The analytical solution for $|L_D|$ is based on nodes deployed according to a random uniform distribution with a mean of λ_U, where the average number of nodes (possible relays) within a semicircle of radius d_{S_i,R_j} will be $\lambda_U \frac{\pi}{8} d^2_{S_i,R_j}$ and the number of desired relays $|L_D|$ can be set such that:

$$|L_D| = \lambda_U \left(\frac{\pi d^2_{S_i,R_j}}{8} \right), \tag{1.8}$$

where λ_U is the node density in the area given by $\lambda_U = \frac{S_U}{M \times M}$.

Our relay selection algorithm operates as follows. First, S_i determines $|L_D|$ and $P_T^{S_i,R_j}$ based on an initial δ and transmits a "Relay Request" message with power $P_T^{S_i,R_j}$. Relays respond with Acks/Naks depending on their willingness to participate in distributed beamforming. If the number of available relays $|L_A|$ that respond positively to the "Relay Request" equals or exceeds the desired number of relays $|L_D|$, then the algorithm terminates. However, if $|L_A| < |L_D|$, S_i recalculates $|L_D|$ using a smaller δ to produce a correspondingly larger $P_T^{S_i,R_j}$ used to retransmit a "Relay Request" to reach more nodes (possible relays). In this case, the algorithm iteratively decreases δ according to a configurable step size δ_{STEP} until the available relays $|L_A| \geq |L_D|$ or $\delta = \min(\delta)$, at which point the algorithm terminates. We represent the smallest value of δ allowed by the network while maintaining unlinkability between S_i and D_i as $\min(\delta)$.

Upon algorithm termination, the set of available relays $|L_A|$ results in one of the three cases. In the first case, the algorithm unsuccessfully recruited a relay set and $|L_A| < |L_D|$. In this case, L_A is the empty set (i.e., $L_A = \emptyset$), and therefore, DiBAN's relay set $L = \emptyset$. The consequence of this outcome is that DiBAN does not have sufficient relays available to cooperate, and therefore, a point-to-point transmission between S_i and D_i is used with no anonymity protection. In the second case, $|L_A| = |L_D|$ and we therefore make $|L_A|$ DiBAN's relay set (i.e., $L = L_A$). In the third case, $|L_A| > |L_D|$ and we need to prioritize DiBAN's relay set L. To achieve relay prioritization, we leverage the relay selection approach in [51] and select $R_j \in L_A$ that corresponds to the $|L_D|$ instantaneous CSI magnitude measurements that achieve the largest α_j, where

$$\alpha_j \triangleq \frac{2}{\frac{1}{\left|h_{S_i,R_j}\right|^2} + \frac{1}{\left|h_{R_j,D_i}\right|^2}} = \frac{2\left|h_{S_i,R_j}\right|^2 \left|h_{R_j,D_i}\right|^2}{\left|h_{S_i,R_j}\right|^2 + \left|h_{R_j,D_i}\right|^2}. \tag{1.9}$$

Each α_j represents the harmonic mean of the channel magnitude-squared $\left|h_{S_i,R_j}\right|^2$ between S_i and R_j and $\left|h_{R_j,D_i}\right|^2$ between R_j and D_i. S_i measures $\left|h_{S_i,R_j}\right|$ directly from the Relay Request Acks/Naks and R_j reports $\left|h_{R_j,D_i}\right|$ to S_i within Relay Request Acks/Naks. Therefore, α_j provides a CSI metric that we use to select the $|L_D|$ highest quality relays $R_j \in L_A$ to compose L. We summarize our relay selection algorithm in the following steps:

1. S_i selects an initial value of δ and analytically determines $|L_D|$ from (1.8). Initially, the DiBAN relay set $L = \emptyset$.
2. S_i transmits a Relay Request message with $P_T^{S_i,R_j}$ calculated based on the current δ to reach $|L_D|$ relays.
3. Nodes that respond to S_i with an Ack become $R_j \in L_A$.
4. One of the three outcomes occurs related to $R_j \in L_A$:
   (a) If $|L_A| < |L_D|$, $\delta = \delta - \delta_{\text{STEP}}$ to increase $P_T^{S_i,R_j}$ and reach more candidate relays on the next iteration. Return to step 2. If $\delta = \min(\delta)$, algorithm terminates with $L = \emptyset$.
   (b) If $|L_A| = |L_D|$, $L = L_A$ and algorithm terminates.
   (c) If $|L_A| \geq |L_D|$, L becomes the $|L_D|$ highest quality relays $R_j \in L_A$ prioritized according to the largest α_j from (1.9).

Our algorithm iteratively recruits the best $|L_D|$ relays, in terms of CSI, while reducing $\overline{\varepsilon_{\text{DiBAN}}}$ to levels equivalent to non-DiBAN systems and boosting BS anonymity. Our approach scales independent of particular network topology since the only user configurable parameter is an initial condition δ that determines the difference in power $P_T^{S_i,R_j}$ for relay recruitment compared to point-to-point transmission power $P_T^{S_i,D_i}$. The lower bound of δ is a direct result of imposing BS anonymity and energy conservation conditions. If $\delta = 1$, $P_T^{S_i,R_j} = P_T^{S_i,D_i}$ and the adversary is able to collect evidence $E(S_i, D_i)$, which negatively impacts BS anonymity and communication energy consumption. The value of $\min(\delta)$ represents the lowest possible value of δ to ensure that $P_T^{S_i,R_j} < P_T^{S_i,D_i}$.

We establish an upper bound of δ by solving (1.8) for δ and setting $|L_D| = 1$, because at least one relay is required for DiBAN. Therefore, $\max(\delta) = \sqrt{\frac{\lambda\pi}{2}} \times d_{\text{max}}$, where $d_{\text{max}} = \sqrt{2} \times M$ is the maximum separation distance of S_i and D_i in a square deployment grid. However, $\max(\delta)$ is an extreme value of δ that will produce a very small $P_T^{S_i,R_j}$. A practical range for δ uses an upper limit based on the expected value of d_{S_i,D_i}, such that $1 < \delta \leq \sqrt{\frac{\lambda\pi}{2}} \times \overline{d_{S_i,D_i}}$. We recommend smaller initial values of δ because the consequence of selecting δ too large is that $|L_D| = 0$, which generates multiple algorithm iterations since the correspondingly small $P_T^{S_i,R_j}$ yields $|L_A| = |L_D| = 0$. That is, the initial value of δ represents a trade-off between anonymity and communication energy. A large δ conserves energy, but may produce too few relays for DiBAN. Conversely, a small δ produces larger $|L_A|$, but at the cost of increased $\overline{\varepsilon_{\text{DiBAN}}}$. We examine this trade-off using simulation in Section 1.6.2.

1.6 Validation Experiments

In this section, we use simulation to evaluate the BS anonymity and energy consumption performance of DiBAN and our relay selection algorithm.

1.6.1 Simulation Environment

We evaluate the effectiveness of our approach using a custom Monte Carlo computer simulation written in C. Results are averaged over 100 independent wireless ad hoc network deployment areas and are subjected to 90 percent confidence interval analysis, resulting in average Belief metrics $B(u)$ and communication energy $\overline{\varepsilon_{\text{DiBAN}}}$ that deviate from the sample mean by a maximum of ± 1.2 percent and ± 1.1 μJ, respectively. S_U nodes are uniformly spread over a grid of size 1000×1000 m^2. The BS location is fixed while randomly triggered nodes send data over multihop paths to the BS. This simulation considers only the link layer relationships of the communicating wireless nodes from the perspective of the adversary. In all cases, the BS is located in cell 35. We apply the AAET enhancements to ET, as described in [32], to properly process the Acks produced by each DiBAN message destination D_i. DiBAN is applied to each hop, and beamforming occurs each time the DiBAN relay count $|L| \geq 1$. All other simulation parameters are specified in Tables 1.3 and 1.4.

1.6.2 Simulation Results

We begin this subsection by presenting results for the Belief metric $B(u)$ or the network's anonymity performance and corresponding average communication energy performance results. We consider two network configurations: a baseline network that employs no anonymity-boosting technique and a network that employs DiBAN at all hops between the initial source and the BS when sufficient helper relays are available. We simulate two static configurations of DiBAN helper relay count, one where $|L| = 1$ and the other where $|L| = 5$.

Table 1.4 Simulation parameters and associated values

Simulation Parameter	Value
Threshold SNR	10 dB [7, 58]
Receiver sensitivity	–100 dBm [7, 58]
Maximum transmission power $\max(P_T^{S_i,D_i})$	30 dBm [7, 58]
Number of cells N_C	36
Number of nodes S_U	50, 100, 150, 200, 250, 300
δ	1.25 to 6
δ_{STEP}	0.5
Carrier Frequency f	2.4 GHz
Data Rate r	250 kbps

In both cases, if at least $|L| = 1$ relay is available at each hop, the message is sent from S_i to D_i using DiBAN. If no relays are available at a given hop (i.e., $|L| = 0$), a point-to-point transmission occurs between S_i and D_i [1, 13].

Figure 1.9 demonstrates the benefit of DiBAN in increasing BS anonymity by comparing the baseline network configuration to DiBAN with static recruitment of $|L| = 1$ and $|L| = 5$ relays. In all cases, the BS is located within cell 35 of the adversary's analysis grid. We first note that DiBAN does not significantly alter the distribution of $B(u)$; the shape of the Belief curve is relatively constant for the baseline and DiBAN configurations which indicates that DiBAN specifically boosts the BS anonymity and not the anonymity of cells that do not contain the BS. Second, we observe that the baseline Belief, i.e., $B(u = 35)$, is approximately 0.1 and the resulting DiBAN Belief is approximately between 0.008 and 0.01, depending on the particular static configuration of $|L|$. These results are typical of DiBAN and represent a significant boost in BS anonymity compared to a baseline network with no anonymity protection. The DiBAN configuration of $|L| = 5$ increases BS anonymity beyond the $|L| = 1$ configuration because the additional helper relays provide additional opportunities to decrease S_i's transmission power and therefore reinforces the unlinkability of paths leading to the BS. We remind the reader that the number of relays $|L|$ in this section corresponds to a static desired quantity at each hop that is only fulfilled if node locations allow relays to be recruited without S_i transmitting sufficient power to establish a link to D_i. In the case where no suitable helper relays exist at a given hop, DiBAN is not employed and the message is delivered using a point-to-point transmission from S_i to D_i [1, 13].

The anonymity performance achieved in the network must be balanced with the amount of energy consumed by the anonymity-boosting technique.

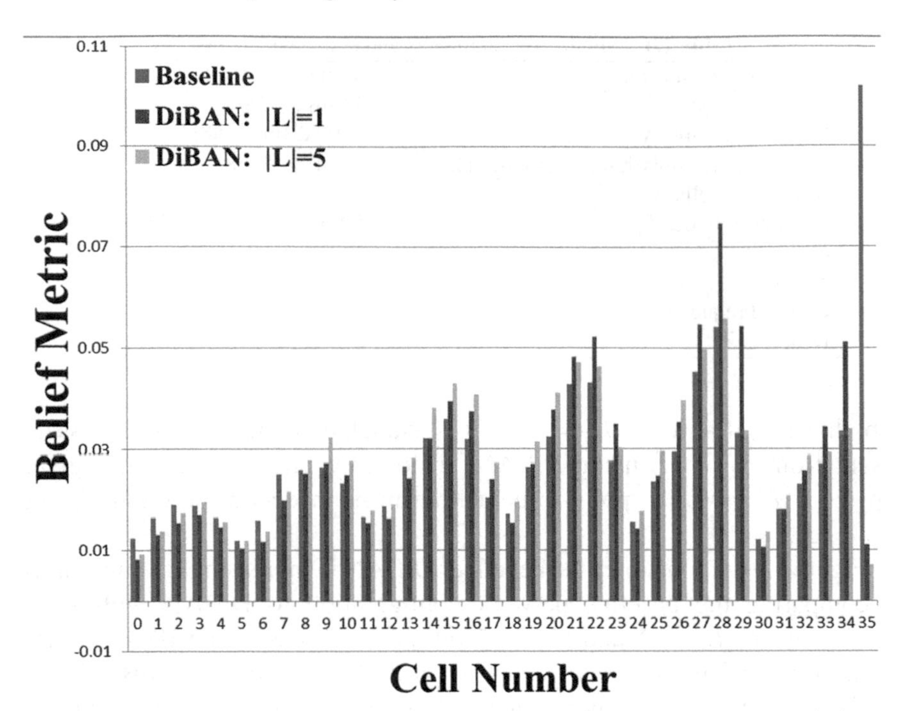

Figure 1.9 Belief per cell $B(u)$ for $S_U = 100$ with no DiBAN and DiBAN $|L| = 1$ and 5 [1, 13].

Table 1.5 summarizes the communication energy consumption associated with the BS anonymity results presented in Figure 1.9. Distributed beamforming allows S_i to decrease its transmission power $P_T^{S_i,D_i}$ by a factor of $|L|$, as shown in Equation (1.5), which is reflected in the decreased average communication energy $\bar{\varepsilon}$ from approximately 30 μJ to 20 μJ. Increasing $|L|$ always results in an average decrease in communication energy consumption when no additional relay recruitment energy is consumed [1, 13]. The increase in DiBAN $\bar{\varepsilon}$ compared to the baseline case is due to the overhead associated with the DiBAN protocol message sequence as described in Section 1.4.3.

Figure 1.10 demonstrates that the benefit of DiBAN in increasing BS anonymity is independent of the number of nodes S_U in the network. Again, we compare the anonymity performance of the baseline network configuration

Table 1.5 Average communication energy consumption $\bar{\varepsilon}$ for $S_U = 100$ [1, 13]

Baseline $\bar{\varepsilon}$ (μJ)	DiBAN $|L| = 1$ $\bar{\varepsilon}$ (μJ)	DiBAN $|L| = 5$ $\bar{\varepsilon}$ (μJ)
15.51	29.24	20.57

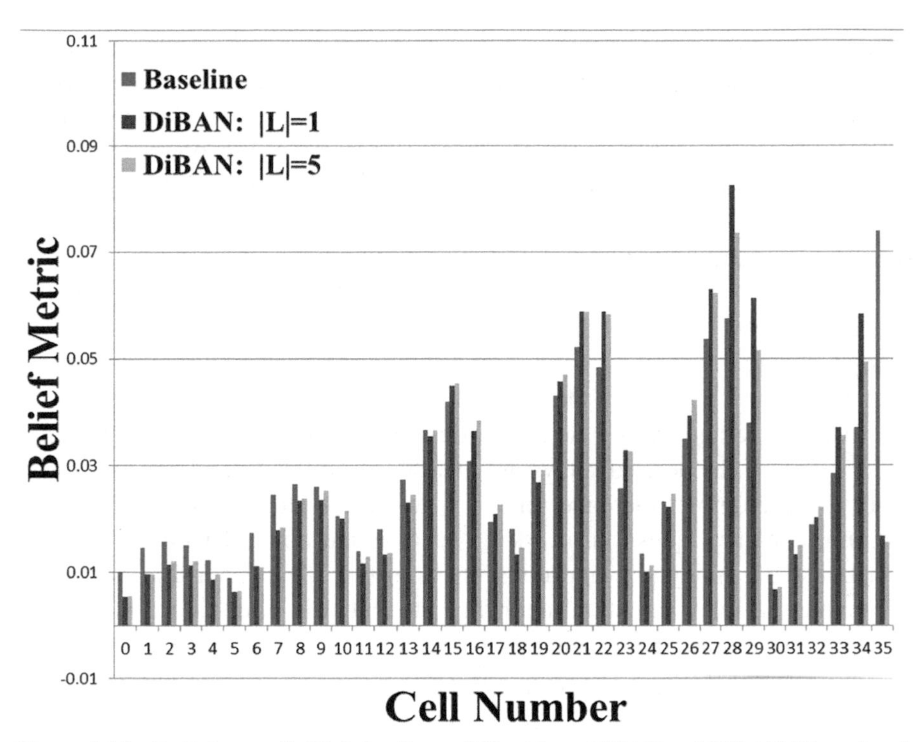

Figure 1.10 Belief per cell $B(u)$ for $S_U = 250$ with no DiBAN and DiBAN $|L| = 1$ and 5 [1, 13].

to DiBAN with static recruitment of $|L| = 1$ and $|L| = 5$ relays, but with $S_U = 250$ instead of $S_U = 100$. In all cases, the BS is located within cell 35 of the adversary's analysis grid. As in Figure 1.9, both configurations of DiBAN significantly decrease the baseline Belief, i.e., $B(u = 35)$ from approximately 0.07 to approximately 0.01. The DiBAN configuration of $|L| = 5$ increases BS anonymity slightly beyond the $|L| = 1$ configuration because the additional helper relays provide additional opportunities to decrease S_i's transmission power and reinforces the unlinkability of paths leading to the BS. Table 1.6 summarizes the communication energy consumption associated with the BS anonymity results presented in Figure 1.10. We observe the same trend as in the $S_U = 100$ configuration, where increasing $|L|$ allows S_i to decrease

Table 1.6 Average communication energy consumption $\bar{\varepsilon}$ for $S_U = 250$ [1, 13]

Baseline $\bar{\varepsilon}$ (μJ)	DiBAN $\lvert L\rvert = 1$ $\bar{\varepsilon}$ (μJ)	DiBAN $\lvert L\rvert = 5\bar{\varepsilon}$ (μJ)
9.36	17.54	12.76

its transmission power $P_T^{S_i,D_i}$, and therefore, the $|L| = 5$ configuration of DiBAN consumes less energy than the $|L| = 1$ configuration. A comparison of Tables 1.5 and 1.6 also demonstrates that as the node density increases in the network (i.e., increasing S_U from 100 to 250), the opportunity for shorter per-hop transmissions also increases. Consequently, lower per-hop transmission power $P_T^{S_i,D_i}$ is required in all cases, resulting in decreased average communication energy $\overline{\varepsilon}$ [1, 13].

The results in Figures 1.9 and 1.10 demonstrate that independent of node density (i.e., S_U = 100 and 200), increasing the static setting $|L|$ boosts BS anonymity and consumes less energy; however, the network's node density must satisfy the number of desired relays. That is, the network is able to decrease its average energy consumption when using DiBAN by increasing $|L|$, but only if the desired relays are available and positioned appropriately relative to the current message sender S_i to participate in distributed beamforming. A natural question to ask is "what is the optimal value for $|L|$ to both boost BS anonymity and conserve energy?" To answer this question, we enable the relay selection algorithm presented in Section 1.5 such that the network is able to dynamically select an optimal value of $|L|$ at each hop. Recall from Equation (1.7) that the relay recruitment power $P_T^{S_i,R_j}$ is a function of the distance parameter δ such that decreasing δ increases $P_T^{S_i,R_j}$, and consequently, more potential relays are reached. Conversely, increasing δ reduces $P_T^{S_i,R_j}$, and therefore, fewer potential relays receive the "Relay Request" message.

Figure 1.11 presents the average available relays $\overline{|L_A|}$ as a function of initial δ, based on Equation (1.7). Clearly, there is a region of δ between 1.25, our choice for $\min(\delta)$, and 4 where $\overline{|L_A|} > 0$ are successfully recruited. We remind the reader that the relay selection algorithm is dynamic and automatically adjusts δ from an initial value until the algorithm terminates with an optimal value for $|L|$. Across the range of simulated node densities, $\delta > 4$ yields no helper relays because $|L_D| = 0$. As pointed out in Section 1.5, there exists a trade-off between $\overline{|L_A|}$ and energy consumption. A small initial δ and correspondingly a large $P_T^{S_i,R_j}$ transmission power used by S_i to send the "Relay Recruitment" message reach more potential relay nodes and therefore increase $\overline{|L_A|}$, yet at the cost of increasing $\overline{\varepsilon}_{\text{DiBAN}}$. This trade-off is illustrated by comparing Figures 1.11 and 1.12. We see that $\overline{|L_A|}$ and $\overline{\varepsilon}_{\text{DiBAN}}$ reach their peak values at an initial δ of 1.25 as a direct result of the increased $P_T^{S_i,R_j}$ required to reach more potential relay nodes.

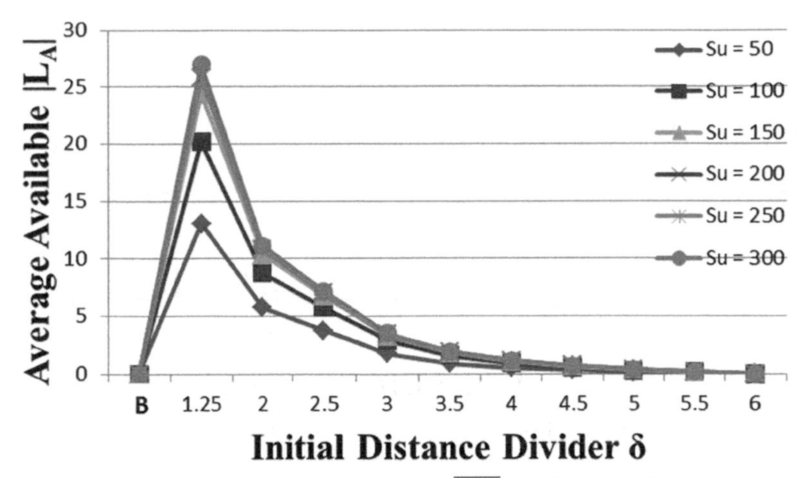

Figure 1.11 Average available relays $\overline{|L_A|}$ as a function of initial δ.

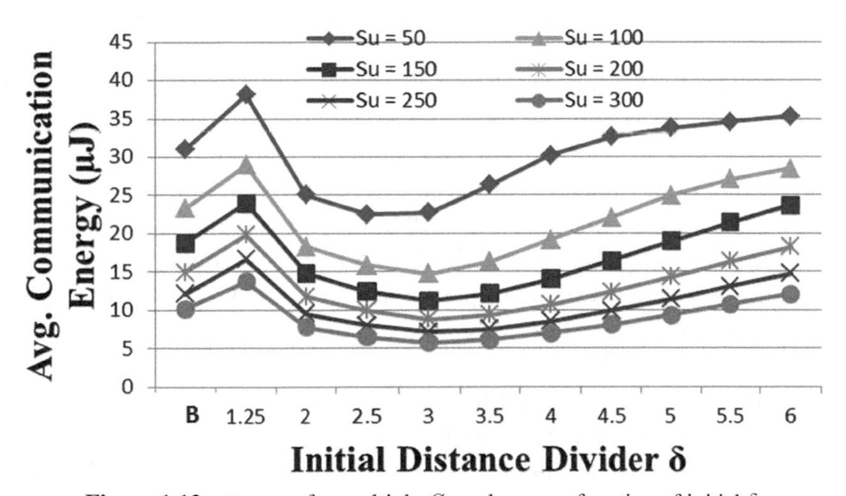

Figure 1.12 $\overline{\varepsilon_{\text{DiBAN}}}$ for multiple S_U values as a function of initial δ.

In Figure 1.12, we observe that the baseline communication energy $\overline{\varepsilon_{\text{base}}}$, denoted as "**B**," decreases as the number of nodes S_U increases. This is because the increased node density increases the size of the anonymity set. For all S_U, the region $1.25 \leq \delta \leq 4$ achieves $\overline{\varepsilon_{\text{DiBAN}}} \leq \overline{\varepsilon_{\text{base}}}$. In the region $4 \leq \delta \leq 6$, we observe $\overline{\varepsilon_{\text{DiBAN}}}$ increasing significantly, to eventually exceed the level of the baseline system. This is because often our algorithm produces no helper relays (i.e., $|L_D| = 0$) because $P_T^{S_i,R_j}$ is too small to reach nodes to recruit.

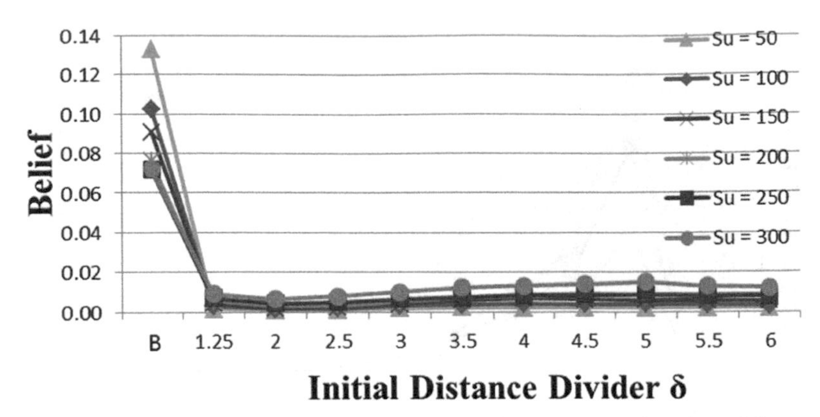

Figure 1.13 Belief of BS cell for multiple S_U values as a function of δ.

In this case, DiBAN cannot be applied and the system must deliver messages using a point-to-point transmission after multiple failed relay recruitment attempts.

In addition to the energy benefits of our relay selection approach, Figures 1.13 and 1.14 demonstrate the BS anonymity benefits using the Belief metric $B(u = 35)$, which represents the Belief of cell 35. In Figure 1.13, we show $B(u = 35)$ as a function of initial δ for a different number of network nodes S_U. Clearly, all values of δ boost BS anonymity, regardless of S_U, when DiBAN is employed at each hop in the network. Even if there are hops where insufficient relays are available to participate in DiBAN, BS anonymity still improves when employing DiBAN compared to a network with no anonymity protection, denoted as **B**.

In Figure 1.14, we compare the results of the baseline network to the case when employing DiBAN under three values of initial δ. We observe that the general shape of the anonymity curve is unaltered by DiBAN, yet the three DiBAN cases significantly decrease $B(u = 35)$ relative to the baseline case. The $\delta = 1.25$ case performs slightly better than the other two DiBAN cases because a larger average number of relays $\overline{|L_A|}$ are recruited. For the $6 \geq \delta \geq 1.25$ range considered in this chapter, our relay selection approach needed an average of 1 to 1.25 iterations to recruit the relay set L.

1.7 Conclusions and Future Work

This chapter has highlighted the threat of traffic analysis in wireless ad hoc networks and summarized anonymity assessment metrics and published anonymity-boosting techniques to prevent an adversary from locating the BS.

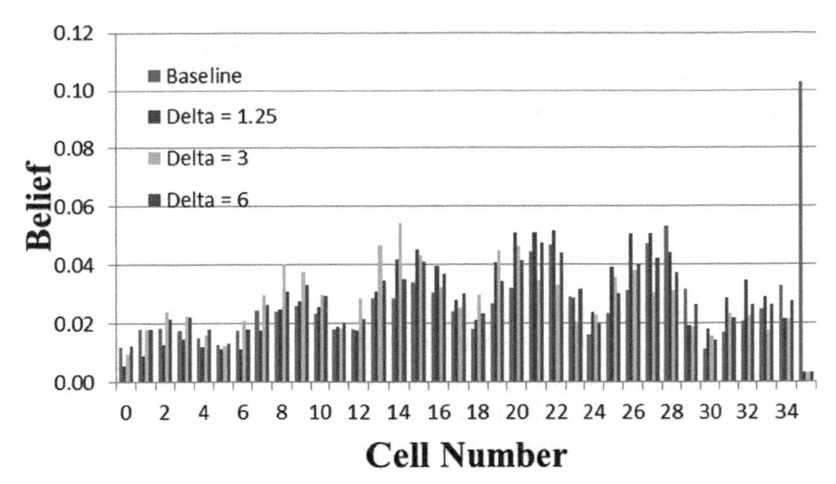

Figure 1.14 Adversary's calculated Belief per cell.

A novel approach, namely DiBAN, has been presented to counter traffic analysis attacks. Unlike contemporary countermeasures, DiBAN exploits the physical layer of the communication protocol stack and enables cross-layer optimization. Basically, distributed beamforming is applied to deprive the adversary of evidences that associate communicating nodes and thus degrades the effectiveness of the traffic analysis. We have also presented a distributed beamforming helper relay selection algorithm for the DiBAN protocol that both increases BS anonymity and decreases the energy overhead. Our anonymity and energy trade-off analysis demonstrated that the relay selection approach allows our DiBAN protocol to boost BS anonymity while decreasing energy consumption to levels less than a baseline, non-DiBAN system. Simulation results have been provided to demonstrate the advantage of DiBAN. Future work shall consider the performance of DiBAN in a fading wireless channel that requires retransmissions, the development of cross-layer techniques that leverage the link and network layers to strengthen and facilitate the applicability of DiBAN, and validation of DiBAN using a hardware-in-the-loop test bed.

Appendix I: Numerical Evidence Theory Belief Calculation Example

The evidence theory analysis presented in [10] and calculation of the Belief metric from Equation (1.2) is more easily understood with a numerical example. In Table 1.7, we provide a detailed, numerical example of how an adversary generates the Belief metrics $B(u)$ corresponding to the evidence collected by the adversary in Table 1.7 at time t_3. For the reader's convenience, we include Equation (1.2), Figure 1.5, and Table 1.1 as Equation (1.10), Figure 1.15, and Table 1.7, respectively.

$$B(u) = \sum_{U|u \subseteq V} nm(U) \tag{1.10}$$

An examination of the evidences in Table 1.7 shows that only cells 0, 1, 3, 4, 5, 7, and 8 are endpoints associated with evidence (i.e., in our notation $E(S_i, \ldots D_i)$, only those cells possibly contain the BS). The BS is the sink for all transmissions within the network. In a network that does not employ anonymity-boosting techniques, an adversary's calculation of $B(u)$ over a

Table 1.7 Collected evidence over time for the illustrative standard ET example

Transmission	Evidence	$E(V)$	$m(V)$
m_1 at time t_1	$E(1,4), E(1,0), E(1,5), E(1,3)$	1	0.25
m_2 at time t_2	$E(4,1), E(4,5), E(4,7), E(4,3), E(4,0), E(4,8)$	1	0.1
Derived at time t_3	$E(1,4,7), E(1,4,5), E(1,4,3), E(1,4,8), E(1,4,0)$	1	0.066

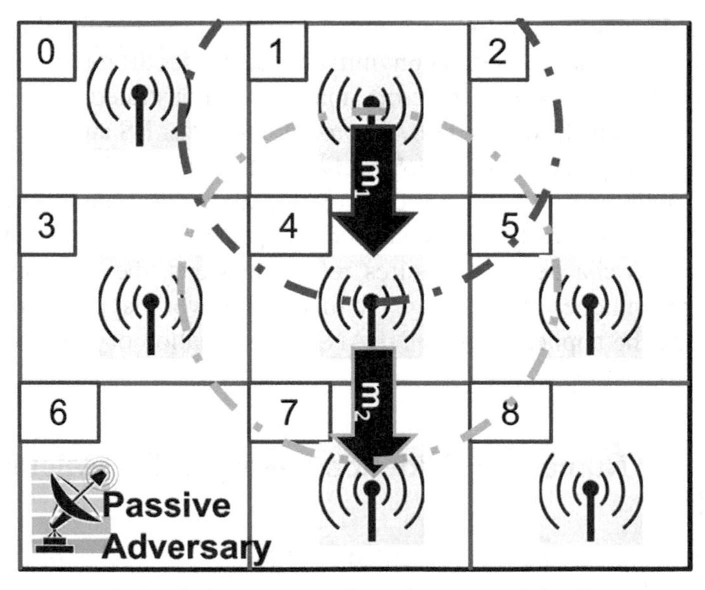

Figure 1.15 Example of an adversary applying ET.

Table 1.8 Detailed belief calculation numerical example for the illustrative standard ET example

Path Endpoint Cell u	Related Evidence	Symbolic $B(u)$ Calculation	Numerical $B(u)$ Calculation	$B(u)$
0	E(4,0), E(1,0), E(1,4,0), E(1,4)	2*m(4,0) + 2*m(1,0) + 3*(m(1,4,0) + m(1,4) + m(4,0))	2*0.066 + 2*0.066 + 3*3*0.066)	0.866
1	E(4,1)	m(4,1)	2*0.066	0.133
3	E(4,3), E(1,3), E(1,4,3),	2*m(4,3) + 2*m(1,3) + 3*(m(1,4,3) + m(1,4) + m(4,3))	2*0.066 + 2*0.066 + 3*(3*0.066)	0.866
4	E(1,4)	m(1,4)	2(0.066)	0.133
5	E(4,5), E(1,5), E(1,4,5)	2*m(4,5) + 2*m(1,5) + 3*(m(1,4,5) + m(1,4) + m(4,5))	2*0.066 + 2*0.066 + 3*(3*0.066)	0.866
7	E(4,7), E(1,4), E(1,4,7),	2*m(4,7) + 3*(m(1,4,7) + m(1,4) + m(4,7))	2*0.066 + 3*(3*0.066)	0.733
8	E(4,8), E(1,4), E(1,4,8),	2*m(4,8) + 3*(m(1,4,8) + m(1,4) + m(4,8))	2*0.066 + 3*(3*0.066)	0.733

large set of evidences collected during an adversary's observation period shall implicate the BS as the largest $B(u)$. The simple example from Figure 1.15 and Table 1.7 is not a sufficiently large set of evidences to produce meaningful $B(u)$, but illustrates the process of adversary Belief metric calculation [1].

To calculate $B(u)$, we first group all of the evidences that are related to a given path endpoint cell u. For example, the claim that cell 0 is a path endpoint is only supported by evidences: E(4, 0), E(1, 0), E(1, 4, 0), and E(1, 4). E(4, 0) and E(1, 0) are pairwise evidences that support the claim that a point-to-point path ends at cell 0. When the adversary correlates all collected evidences, a composite evidence of E(1, 4, 0) is produced from the pairwise E(1, 4) and E(4, 0). Intuitively, if a transmission occurs from cell 1 to cell 4 and a transmission occurs from cells 4 to 0, a multihop path may exist between cell 1 and cell 0. From (1.10), the adversary generates $B(0)$ by summing each normalized evidence $m(V)$ multiplied by the length of the path it represents. That is, $B(0) = 2{*}m(4, 0) + 2{*}m(1, 0) + 3{*}(m(1, 4, 0) + m(1, 4) + m(4, 0)) = 2{*}0.066 + 2{*}0.066 + 3{*}(3{*}0.066) = 0.866$. Calculation of the other Belief metrics $B(u)$ follows the same process and is shown in Table 1.8.

References

[1] Ward, J. R. (2015). *Physical- and MAC-Layer Mechanisms for Increasing Base-Station Anonymity in Wireless Ad-hoc Networks*, University of Maryland, Baltimore County Ph.D. Dissertation, May.
[2] Perkins, C. E. (2001). *Ad Hoc Networking*. New York: Addison-Wesley.
[3] Chong, C.-Y., and Kumar, S. P. (2003). Sensor networks: evolution, opportunities, and challenges. *Proc. IEEE*, 91 (8), 1247–1256.
[4] Akyildiz, I. F., Su, W., Sankarasubramaniam, Y., and Cayirci, E. (2002). Wireless sensor networks: a survey. *Comput. Netw.*, 38 (4), 393–422.
[5] Gubbi, J., Buyya, R., Marusic, S., and Palaniswami, M. (2013). Internet of Things (IoT): a vision, architectural elements, and future directions. *Future Gen. Comput. Syst.*, 29 (7), 1645–1660.
[6] Atzori, L., Iera, A., and Morabito, G. (2010). The internet of things: a survey. *Comput. Netw.*, 54 (15), 2787–2805.
[7] Yick, J., Mukherjee, B., and Ghosal, D. (2008). Wireless sensor network survey. *J. Comput. Netw.*, 52, 2292–2330.
[8] Perrig, A., Szewczyk, R., Wen, V., Culler, D., and Tygar, D. (2001). SPINS: Security Protocols for Sensor Networks. In *Proceedings of the*

7th ACM International Conference on Mobile Computing and Networks (MOBICOM'01), Rome, Italy, July.

[9] Kong, J., Xiaoyan, H., and Gerla, M. (2007). An identity-free and on-demand routing scheme against anonymity threats in mobile ad hoc networks. *IEEE Trans. Mobile Comput.*, 6, 888, 902.

[10] Huang, D. (2006). On Measuring Anonymity for Wireless Mobile Ad-hoc Networks, *Proceedings of the 31st IEEE Conference on Local Computer Networks (LCN'06)*, Tampa, FL, November.

[11] Acharya, U., and Younis, M. (2010). Increasing base-station anonymity in wireless sensor networks, *Ad Hoc Netw.*, 8 (8), 791–809.

[12] Ward, J. R., and Younis, M. (2013). On the Use of Distributed Beamforming to Increase Base Station Anonymity in Wireless Sensor Networks. In *Proceedings of the International Conference on Computer Communications and Networks (ICCCN 2013)*, Nassau, Bahamas, July.

[13] Ward, J. R., and Younis, M. (2015). Increasing base station anonymity using distributed beamforming. *Ad Hoc Netw.*, 32, 53–80.

[14] Mudumbai, R., Barriac, G., and Madhow, U. (2007). On the feasibility of distributed beamforming in wireless networks. *IEEE Trans. Wireless Commun.*, 6 (5), 1754–1763.

[15] Bassily, R., Ekrem, R., He, X., Tekin, E., Xie, J., Bloch, M., Ulukus, S., and Yener, A. (2013). Cooperative security at the physical layer: a summary of recent advances. *IEEE Signal Process. Mag.*, 30 (5), 16–28.

[16] Pfitzmann, A., and Hansen, M. (2005). Anonymity, unlinkability, unobservability, pseudonymity, and identity management – a consolidated proposal for terminology. Working draft. Available at http://dud.inf.tudresden.de/literatur/AnonTerminologyv0.26.doc, September.

[17] Serjantov, A., and Danezis, G. (2002). Towards an Information Theoretic Metric for Anonymity. In Dingledine, R. and Syverson, P. (eds.) *Proceedings of the Privacy Enhancing Technologies Workshop (PET 2002). LNCS 2482*, April. Berlin: Springer.

[18] Diaz, C., Seys, S., Claessens, J., and Preneel, B. (2002). Towards Measuring Anonymity. In Dingledine, R., and Syverson, P. (eds.) *Proceedings of the Privacy Enhancing Technologies Workshop (PET 2002), LNCS 2482*, April. Berlin: Springer.

[19] Shannon, C. E. (1948). A mathematical theory of communication. *Bell Sys. Tech. J.*, 27 (379–423), 623–656.

[20] Deng, J., Han, R., and Mishra, S. (2005). Countermeasures Against Traffic Analysis Attacks in Wireless Sensor Networks. In *Proceedings of*

the 1st International Conference on Security and Privacy For Emerging Areas in Communications Networks, September.

[21] Ren, Z., and Younis, M. (2011). Effect of Mobility and Count of Base-stations on the Anonymity of Wireless Sensor Networks. In *Proceedings of the 7th International Wireless Communications and Mobile Computing Conference (IWCMC 2011)*, Istanbul, Turkey, July.

[22] Mehta, K., Liu, D., and Wright, M. (2007). Location Privacy in Sensor Networks Against a Global Eavesdropper. In *Proceedings of the IEEE International Conference on Network Protocols (ICNP)*, Beijing, China, October.

[23] Ozturk, C., Zhang, Y., and Trappe, W. (2004). Source-Location Privacy in Energy Constrained Sensor Network Routing. In *Proceedings of the Workshop on Security of Ad Hoc and Sensor Networks (SASN '04)*, Washington, DC, October.

[24] Kamat, P., Zhang, Y., Trappe, W., and Ozturk, C. (2005). Enhancing Source-Location Privacy in Sensor Network Routing. In *Proceedings of the 25th IEEE International Conference on Distributed Computing Systems (ICDCS'05)*, Columbus, OH, June.

[25] Raj, M., Li, N., Liu, D., Wright, M., and Das, S. K. (2014). Using data mules to preserve source location privacy in Wireless Sensor Networks. *Perv. Mobile Comput.*, 11, 244–260.

[26] Das, S. K., Kant, K., Zhang, N. (2012). *Handbook on Securing Cyber-Physical Critical Infrastructure*. New York: Elsevier Inc.

[27] Conner, W., Abdelzaher, T., and Nahrstedt, K. (2006). Using Data Aggregation to Prevent Traffic Analysis in Wireless Sensor Networks. In *Proceedings of the International Conference on Distributed Computing in Sensor Systems (DCOSS'06) LNCS 4026*, June. Berlin: Springer.

[28] Mehta, K., Liu, D., Wright, M. (2012). Protecting location privacy in sensor networks against a Global Eavesdropper, *IEEE Trans. Mobile Comput.,* 11 (2), 320–336.

[29] Ebrahimi, Y., and Younis, M. (2011). Using Deceptive Packets to Increase Base-station Anonymity in Wireless Sensor Networks. In *Proceedings of the Wireless Communications and Mobile Computing Conference (IWCMC 2011),* Istanbul, Turkey, July.

[30] Baroutis, N., and Younis, M. (2015). Using Fake Sinks and Deceptive Relays to Boost Base-station Anonymity in Wireless Sensor Network. In *Proceedings of the 40th Annual IEEE Conference on Local Computer Networks (LCN 2015)*, Clearwater Beach, FL, October.

[31] Ebrahimi, Y., and Younis, M. (2011). Increasing Transmission Power for Higher Base-station Anonymity in Wireless Sensor Network. In *Proceedings of the IEEE International Conference on Communication (ICC 2011)*, Kyoto, Japan, June.

[32] Ward, J. R., and Younis, M. (2014). A metric for evaluating base station anonymity in acknowledgement-based wireless sensor networks. In the *Proceedings of the IEEE Military Communications Conference (MILCOM 2014)*, Baltimore, MD, November.

[33] Karl, H., and Willig, A. (2005). *Protocols and architectures for wireless sensor networks*. New York: John Wiley & Sons Ltd.

[34] Couch, L. W. (2001). *Digital and Analog Communications Systems*, 6th ed. Prentice Hall, New Jersey, USA.

[35] Proakis, J. G. (2001). *Digital Communications*, 4th ed. McGraw Hill.

[36] Elson, J., Girod, L., and Estrin, D. (2002). Fine-Grained Network Time Synchronization using Reference Broadcasts. In *Proceedings of the 5th Symposium on Operating Systems Design and Implementation (OSDI 2002)*, Boston, MA, December.

[37] Ganeriwal, S., Kumar, R., and Srivastava, M. B. (2003). Timing-Sync Protocol for Sensor Networks. In *Proceedings of the 1st ACM International Conference on Embedded Networked Sensor Systems (SenSys'03)*, Los Angeles, CA, November.

[38] IEEE 802.15.4-2011, IEEE Standard for local and metropolitan area networks – Part 15.4: Low-Rate Wireless personal Area Networks (LR-WPANs), September 2011.

[39] Kamat, P., Zhang, Y., Trappe, W., and Ozturk, C. (2005). Enhancing source-location privacy in sensor network routing. In *Proceedings of the 25th IEEE International Conference on Distributed Computing Systems (ICDCS'05)*, Columbus, OH, June.

[40] Savvides, A., Han, C., and Srivastava, M. (2001). Dynamic Fine-grained Localization in Ad-hoc Networks of Sensors. In *Proceedings of the 7th ACM Conference on Mobile Computer and Networking (MobiCom'01)*, Rome, Italy, July.

[41] Bensky, A. (2008). *Wireless Positioning Technologies and Applications*, Artech House.

[42] Chandrasekaran, G., Ergin, M., Yang, J., Liu, S., Chen, Y., Guteser, M., and Martin, R. (2009). Empirical Evaluation of the Limits on Localization Using Signal Strength. In *Proceedings of the IEEE Conference on*

Sensor and Ad Hoc Communications and Networks (SECON 2009), New Orleans, LA, June.

[43] Li, N., Zhang, N., Das, S. K., and Thuraisingham, B. (2009). Privacy preservation in wireless sensor networks: a state-of-the-art survey. *Ad Hoc Netw.*, 7 (8), 1501–1514.

[44] Shao, M., et al. (2009). Cross-layer enhanced source location privacy in sensor networks. In *Proceedings of the 6th Annual IEEE communications society conference on Sensor, Mesh and Ad Hoc Communications and Networks (SECON'09)*, New Orleans, LA, June.

[45] Deng, J., Han, R., and Mishra, S. (2006). Decorrelating wireless sensor network traffic to inhibit traffic analysis attacks. *Perv. Mobile Comput. J.*, Special Issue on *Security in Wireless Mobile Computing Systems*, 2, 159–186, April.

[46] Mudumbai, R., Brown, D. R., Madhow, U., and Poor, H. V. (2009). Distributed transmit beamforming: challenges and recent progress, *IEEE Commun. Mag.*, 47 (2), 102–110.

[47] Ochiai, H., Mitran, P., Poor, H. V., and Tarokh, V. (2005). Collaborative beamforming for distributed wireless ad hoc sensor networks. *IEEE Trans. Signal Process.*, 53 (11), 4110–4124.

[48] Quintin, F., Rahman, M. M. U., Mudumbai, R., and Madhow, U. (2013). A scalable architecture for distributed transmit beamforming with commodity radios: design and proof of concept, *IEEE Trans. Wireless Commun.*, 12 (3), 1418–1428.

[49] Rahman, M. M., Baidoo-Williams, H. E., Mudumbai, R., and Dasgupta, S. (2012). Fully Wireless Implementation of Distributed Beamforming on a Software-Defined Radio Platform. In *Proceedings of the 11th ACM/IEEE Conference on Information Processing in Sensor Networks (IPSN '12)*, Beijing, China, April.

[50] Oh, S., Vu, T., Gruteser, M., and Banerjee, S. (2012). Phantom: Physical Layer Cooperation for Location Privacy Protection. In *Proceedings of the IEEE International Conference on Computer Communications (INFOCOM 2012)*, Orlando, FL, March.

[51] Bletsas, A., Khisti, A., Reed, D. P., and Lippman, A. (2006). A simple cooperative diversity method based on network path selection. *IEEE J. Selec. Areas Comm.*, 24 (3), 659–672.

[52] Jing, T., Zhu, S., Li, H., Cheng, X., and Huo, Y. (2013). Cooperative Relay Selection in Cognitive Radio Networks. In *Proceedings of IEEE International Conference on Computer Communications (INFOCOM 2013)*, Turin, Italy, April.

[53] Wang, C. L., and Syue, S. J. (2009). An efficient relay selection protocol for cooperative wireless sensor networks. In *Proceedings of IEEE Wireless Communications and Networking Conference (WCNC 2009)*, Budapest, Hungary, April.
[54] Ibrahim, A. S., et al. (2008). Cooperative communications with relay selection: when to cooperate and whom to cooperate with? *IEEE Trans. Wireless Commun.*, 7 (7), 2814–2827.
[55] IRIS Mote Data Sheet, MEMSIC Inc., www.memsic.com.
[56] Mahmoud, M., and Shen, X. (2012). Cloud-based scheme for protecting source location privacy against hotspot-locating attack in wireless sensor networks. *IEEE Trans. Parallel Distrib. Syst.*, 23 (10), 1805–1818.
[57] Burbank, J. L., Ward, J. R., and Kasch, W. T. (2011). *An Introduction to Network Modeling and Simulation for the Practicing Engineer*, IEEE Press.
[58] Chen, H., and Lou, W. (2015). On protecting end-to-end location privacy against local eavesdropper in wireless sensor networks. *Perv. Mobile Comput.*, 16 (Part A), 36–50.

2

A Privacy-Preserving and Efficient Information Sharing Scheme for VANET Secure Communication

Cong Guo[1], Liehuang Zhu[2] and Zijian Zhang[2]

[1]National Meteorological Information Center, Beijing, China
[2]College of Computer Science & Technology,
Beijing Institute of Technology, China

Abstract

Traffic information sharing is the basic and kernel functionality for vehicular ad hoc network (VANET), but the issue of privacy preservation and efficiency still remains a great challenge during traffic information sharing. In this chapter, we propose a privacy-preserving and efficient traffic jam information sharing scheme named PETS, which utilizes roadside unit (RSU) as an aggregator to securely collect traffic information sent from vehicles with pseudo-identities and semantically aggregate messages to generate a traffic jam message. Then, the traffic jam message is signed by trust authority (TA) and propagated to vehicles to notify drivers, and RSU plays as a proxy to assist message verification when vehicles are busy to prevent denial-of-service (DoS) attack. The proposal achieves both privacy preservation and efficiency at the same time and significantly decreases bandwidth consumption by 10.00%∼69.66% for message collection. Extensive simulation reveals that the novel scheme is feasible and has a better performance than previously suggested counterparts in terms of message loss ratio and delay.

Keywords: VANET, Key agreement, Traffic information Sharing, Privacy-preserving.

2.1 Introduction

Traffic jam is a worldwide problem which leads to air pollution, frequent traffic accidents, problems in personal health, loss productive labor time, and economic losses. Air pollution from traffic congestion in 83 of the largest urban areas in America contributes to more than 2,200 premature deaths annually. A report of 2011 urban mobility in America from Texas Transport Institute showed that 1) waste associated with traffic jam summed to 101 billion dollars; 2) 1.9 billion gallons of fuel was wasted because of traffic jam; and 3) traffic jam caused aggregate delays of 4.8 billion hours. Beijing Transportation Research Center estimated that traffic jam costs 70 billion dollars (7.5% of GDP) in Beijing, China, in 2014. There are 1.2 million people are killed in road accidents every year, over speed resulting from time lost in traffic jams is a major cause for road accidents, and at least 120,000 lives can be saved, every year, if traffic congestions are reduced. VANET is a promising direction that could help to ease traffic jam.

In VANET, each vehicle is equipped with an onboard unit (OBU), together with roadside units (RSUs), and a large-scale self-organized network can be constructed by utilizing dedicated short-range communication (DSRC) [1]. VANET is aiming at improving transportation safety and efficiency through traffic information sharing, e.g., broadcasting vehicle emergency braking information to remind drivers' attention, or propagating traffic jam information to guide surrounding vehicles to avoid jam area.

The emergence of VANET is an important significance to transportation efficiency and road safety. Traffic information sharing in VANET must meet two basic requirements. 1) Security: On the one hand, the wireless communication in VANET is broadcasted in essence, which makes the data easily monitored, altered, and forged. On the other hand, vehicles are located in an open physical space and privacy (e.g., driver's identity, license plate, position, and travel route) leaked in VANET not only exposes privacy information, but also brings threat to the lives and properties of drivers and passengers. For example, adversaries can use legitimate vehicle's identity information to trace the vehicle's travel route and analyze driver's habit, and then, theft or other crimes may be implemented by abusing VANET communications. So security must be guaranteed for traffic information sharing. 2) Efficiency: VANET is large scale, and network recourses are limited. According to DSRC, each vehicle broadcasts a message every 100 ms$\sim$300 ms to its 300 m communication range. A typical OBU with a 400-MHz processor requires about 20 ms to process one message [2]. It is not an issue when the vehicle density keeps low.

But when the vehicle density is high, e.g., 1000 vehicles are in the communication range, each vehicle needs to process 3333∼10,000 messages/s, which causes heavy communication burden for VANET and significant computation overhead for vehicles. The feature of efficiency is important for traffic information sharing. It does not only save network resources, but also avoids possible (DoS) attack.

In [3–8], several schemes based on privacy-preserving authentication have been proposed to tackle the traffic information sharing in VANET. But all of them exchange traffic information directly among vehicles, which leads to duplicate message verification and is vulnerable to DoS attack. In [9], data aggregation is introduced to improve the efficiency of traffic information sharing in VANET. Due to the lack of security consideration, privacy leakage remains a problem.

Traffic information sharing is the basic and kernel functionality for VANET. But the lack of secure and efficient information sharing scheme brings a great threat to VANET, which not only fails to improve the transportation safety, but introduces new security problems. In this chapter, we proposed a privacy-preserving and efficient traffic jam information sharing scheme which, to the best of our knowledge, is the first scheme that achieves both privacy preservation and DoS resistant to traffic information sharing based on secure data aggregation. The main contributions of this chapter are as follows.

We propose a secure data aggregation scheme to improve the information sharing efficiency. The scheme employs RSU as aggregator to collect and aggregate traffic information within its communication range, and then, the semantic aggregated message (traffic jam information) is propagated to the local and nearby districts. Upon reception, vehicles can change their directions to avoid traffic jam. As aggregated message is broadcasted to the vehicles instead of direct vehicle-to-vehicle (V2V) information sharing, duplicated messages are significantly reduced and the experimental result shows that bandwidth consumption is decreased by 10.00%∼69.66%.

We design a vehicle–RSU key agreement protocol that establishes secret key and pseudo-identities between vehicle and RSU without leaking vehicle's identity privacy. The protocol is based on identity-based signature/encryption and Diffie–Hellman protocol and employs trust authority (TA) as a verifier to aid secure and efficient key agreement.

The proposal provides privacy preservation to vehicles and conditional traceability. Even if all the network messages are monitored by adversary, it still cannot take any advantage of linking several messages to a specific vehicle; thus, vehicle's privacy is protected. When a message is in dispute, the trust authority can disclose vehicles' real identities and trace message's original sender. In this way, the feature of conditional traceability is provided.

We evaluate the proposal and the previously presented schemes through extensive simulations on several performance metrics, such as bandwidth consumption, computation overhead, message loss ratio, and network delay. The experiment results reveal that the novel scheme is feasible and has an outstanding performance.

The remainder of this chapter is organized as follows: Some related works about our theme are briefly summarized in Section 2.2. Section 2.3 mainly focuses on the system model and preliminaries. Section 2.4 is the most important part of our study, in which a novel traffic jam information sharing scheme-*PETS* is proposed. Section 2.5 is the security analysis of our scheme. In Section 2.6, we evaluate the performance of our PETS scheme. Some conclusions are summarized in Section 2.7.

2.2 Related Works

Recently, quite a few schemes have been suggested to tackle security and privacy challenges of traffic information sharing in VANET. The classification of these schemes can be divided into two types: 1) privacy-preserving authentication-based schemes and 2) secure data aggregation-based schemes.

Privacy-preserving authentication-based schemes. In [3], Raya et al. proposed a scheme named BP by predistributing a number of anonymous private keys and the corresponding certificates (e.g., 43,800 certificates in [3]) for each vehicle. Message is signed with a randomly chosen private key and be verified by the corresponding anonymous certificate. As the anonymous certificate is generated according to a pseudo-identity, real identity of a sender is not revealed and privacy is protected. The list of anonymous certificates that relate to the real identity of the drivers is kept by the authority to provide conditional traceability. In [4], Raya et al. introduced TPD to the scheme to take care of storing cryptographic material and operations to enhance the security. However, this kind of scheme has some obvious shortcomings: 1) The CRL increases quickly which will take a large storage to store CRL and a long time to do CRL checking before message verification. 2) When the authority needs to revoke a vehicle, it has to revoke all the anonymous certificates that are hold by the vehicle. It causes a lot of bandwidth consumption and increases authority's certificate management overhead.

In [6], Lin et al. suggested a privacy-preserving authentication scheme based on group signature [10, 11] and identity-based (ID-based) signature [12] (GSIS). Group signature is used to anonymously sign messages with private key by senders and verified with the group public key by receivers, while identities of the senders can only be recovered by authorities. ID-based signature is applied by RSUs to digitally sign each message launched by RSUs to ensure its authority, where the signature overhead can be greatly reduced. CRL size of the group signature is linear with the number of revoked vehicles, but the checking operation involves two paring calculations, which would take about 104 times computation cost than a string comparison [8]. In [8], researchers proposed a hybrid scheme by combining pseudonym scheme with group signature. Each vehicle V is equipped with a group signing key gsk_v and a group public key pgk_{CA}. A vehicle can issue a "self-certify" certificate for itself by gsk_v and then signing its message using private key corresponding to the "self-certify" certificate. In such a way, the average overhead of message authentication can be reduced, but the expensive group signature CRL checking still remains a problem.

All the above schemes utilize individual vehicle to generate and authenticate traffic information itself and are directly or indirectly based on the digital signature technology for message signing and verification, which leads to the following drawbacks: 1) These schemes lack adequate efficient message verification, and meanwhile, they are inappropriate for real large-scale VANET deploying and are vulnerable to DoS attack; 2) it leads to duplicated traffic information authentication and high network communication overhead.

Secure data aggregation-based schemes. Secure data aggregation-based schemes can be divided into two classes: 1) Syntactic ones: Compress or encode the data from multiple vehicles in order to fit the data in a unique record or frame and 2) semantic ones: Traffic information from individual vehicle is summarized. Raya et al. proposed an efficient secure aggregation scheme based on message aggregation and group communication in [3]. The scheme utilized three classes of syntactic aggregation technique to perform message aggregation: combined signatures, onion signatures, and hybrid signatures. But Raya's scheme did not provide security feature of vehicle privacy preservation and conditional traceability. In [13], the author proposed a scheme that syntactically/semantically aggregated application data and randomly picked one of the application data for authentication. In this way, both computation and communication overhead are reduced. But probabilistic validation strategy may leak false data verification and take in fake data, which is a great thread to VANET security. In [14], Zhu et al. suggested a scheme that utilized both syntactic and cryptographic aggregations to decrease both communication and computation overhead. Vehicles in VANET dynamically formed a group, and the group cluster played as an aggregator. Syntactic aggregation deleted redundant information and cryptographic (with batch verification technique) aggregates signatures and certificates to reduce packet size and bandwidth consumption. The scheme did not take vehicle privacy into consideration, and certificate revocation/updating caused extra system overhead. In [15], RAISE was proposed and RSU was utilized as aggregator to verify and aggregate messages from vehicles. As the message verification overhead was moved to powerful RSU, computation overheads of vehicles were significantly reduced. Upon receiving a message, a vehicle needs store and wait for aggregation message from RSU. This makes the scheme introduce extra delay and be vulnerable to possible DoS attack.

2.3 System Model and Preliminaries

In this section, system model (network model and attack model), and security requirements are presented.

2.3.1 Network Model

We consider the VANET consisting of three kinds of entities: vehicles, RSUs (partial trusted), and TA.

- Vehicle: In VANET, each vehicle is equipped with an OBU. OBU is employed to send traffic information, store cryptographic materials, and process cryptographic operations. A vehicle generates traffic information every 100$\sim$300 ms and sends it to the neighboring vehicles or RSU through OBU.
- RSU: RSU is a kind of infrastructure deployed on the roadside and can directly communicate with TA and vehicles in its communication range. It is partially trusted and has a powerful communication capability with a transmission range of 3 km. All RSUs constitute a network that covers the whole area. RSU is also powered with sufficient computations and storage resources, and it is responsible for traffic information collection and traffic jam information propagation.
- TA: TA is a fully trusted authority. It is powered with sufficient computations and storage resources, and it is in charge of 1) RSUs and vehicles' registration, 2) RSUs and vehicles' verification, and 3) traffic jam information propagation.

2.3.2 Attack Model

We assume the adversary can control the whole communication channel, and it can monitor all the network communications and can also tamper the messages, drop some packets, and even replace the original messages. Furthermore, the adversary can also capture or corrupt small part of vehicles. All the data transmitted to/through compromised vehicles can be obtained and analyzed by the adversary. The purpose of the adversary is to induce the legitimate vehicles to accept false or harmful messages without being detected and abuse the VANET to maximize its gains (e.g., cheating neighboring vehicles to make a clear path to greedy driver's destination regardless of the cost to the system and snooping legitimate users' privacy).

2.3.3 Security Requirements

The proposed scheme has the following security design goals:

- Authentication: All messages should be authenticated to ensure that these messages are indeed sent unaltered by legitimate entities in VANET. All messages have not been altered in the transmission process. If the message is altered, the receiver can detect this modification.
- DoS resistant: The proposal should be efficient (low storage cost and computation overhead) and balanced (the more resources an entity has, the more workload it is allocated) to avoid possible performance bottleneck and DoS attack.
- Identity privacy preserving: Real identity of a vehicle should be protected, so that adversaries have no knowledge of vehicle's real identity. V2V communication is anonymous, and only TA and RSU (verified by RSU) have the knowledge of vehicle's real identity.
- Unlinkability: The outside observer cannot link multiple messages to a specific vehicle through traffic analysis, which disables the adversary to trace a particular vehicle and incurs location privacy violation [16] problem.
- Conditional traceability: For vehicles, V2V communications are anonymous and unlinkable. But the TA has the ability to verify origin and non-repudiation of a message to ensure that no vehicles can deny the message generated by itself and retrieve a vehicle's real identity when the message is in dispute.

Our goal is to design a traffic jam information sharing scheme with security feature of verification to guarantee message authenticity, DoS resistant, and vehicle identity preservation. In addition, it should keep both unlinkability to outside observers and conditional traceability to TA at the same time.

2.4 The Proposed PETS Scheme

2.4.1 Scheme Overview

At a high level, the proposed scheme includes the following four phases: 1) system initiation; 2) vehicle–RSU key agreement; 3) traffic information collection and aggregation; and 4) traffic jam message propagation. Overviews of these phases are as follows:

1. **System initiation**: TA generates a master key and system parameters. Both RSUs and vehicles register themselves to TA with their real identities to obtain the corresponding secret identity keys.
2. **Vehicle–RSU key agreement**: When a vehicle enters the communication range of a RSU, the vehicle performs key agreement with the RSU to get secret key and n pseudo-identities. The secret key is utilized for secure communication between vehicle and RSU, and pseudo-identities are employed to cover vehicle's real identity.
3. **Traffic information collection and aggregation**: After vehicle–RSU key agreement, vehicles encrypt self-generated traffic information with their secret keys and send these messages to the RSU. The RSU collects thousands of messages from vehicles within its communication range and performs traffic jam detection function to detect the traffic jam. If the traffic jam is detected, it aggregates received messages and produces a traffic jam message.
4. **Traffic jam information propagation**: Once the traffic jam message is produced, the RSU submits it to TA to obtain the corresponding signature. Then, the message and signature are sent to the RSU and its adjacent RSUs. RSU broadcasts the traffic jam message to the vehicles within its communication range instead of sharing the original messages directly, to notice vehicles that a traffic jam occurs.

For convenience, the notations used in the proposed scheme are listed in Table 2.1.

2.4.2 System Initiation

System initiation consists of TA initiation and vehicle/RSU registration.

Table 2.1 Notations

Notations	Descriptions
TA	Trust authority
RSU_i	The i-th RSU
V_i	The i-th vehicle
G	A cyclic additive group
V	A cyclic multiplicative group
m	A message
aP	$P \in G$ is a generator of cyclic additive group G, and aP denotes P^a.
ID_{V_i}	The real identity of V_i
ID_{R_i}	The real identity of RSU_i
ID_{TA}	The identity of TA
PID	Pseudo identity
s_{v_i}	Identity secret key of V_i
s_{v_i,R_x}	Secret key between V_i and R_x
$IBEnc_{ID}(m)$	Identity-based encryption function: Encrypt message m with identity ID [17]
$IBDec_s(c)$	Identity-based decryption function: Decrypt cipher textc with secret identity key s [17]
$Sign_s(m)$	Identity-based digital signature function: Sign message with secret identity key s [18, 19]
$Verify_{ID}(m, \sigma)$	Identity-based digital signature verification function: Verify the signature σ of message m with identity ID [18, 19]
$Enc_k(m)$	Symmetric encryption function: Encrypt message m with symmetric key k
$Dec_k(c)$	Symmetric decryption function: Decrypt cipher text c with symmetric key k
$info_{V_i}$	Vehicle information of V_i
$MAC_k(\cdot)$	Message authentication code computation function using k as a key, such as HMAC [20]
$H(\cdot)$	Hash functionh: $\{0,1\}^* \rightarrow \{0,1\}^n$
$\|\|$	Message concatenationoperation

- TA initiation: Let G be a cyclic additive group of prime order q, $P \in G$ a generator of G and let $e : G \times G \rightarrow V$ be a bilinear map which satisfies following conditions [12]:

1. Bilinear: $e(x_1+x_2, y) = e(x_1, y) \cdot e(x_2, y)$ and $e(x, y_1+y_2) = e(x, y_1) \cdot e(x, y_2)$.
2. Non-degenerate: There exists $x \in G$ and $y \in G$ such that $e(x, y) \neq 1$.

Then, TA generates master key and system parameters as follows:

(1) TA randomly picks integer $\alpha \in \mathbb{Z}_q^*$ as system master key and computes $\beta = \alpha P$ as system public key.

(2) TA computes $s_{\mathrm{TA}} = \alpha H(ID_{\mathrm{TA}})$ as its identity secret key.
(3) $\{\beta, ID_{\mathrm{TA}}\}$ are published to public, and $\{\alpha, s_{ID_{\mathrm{TA}}}\}$ are kept secret.

Vehicle and RSU register themselves to TA as follows:

(1) For a vehicle, represented as V_i, it submits its real identity ID_{V_i} and vehicle information $info_{V_i}$ (e.g., engine serial number, date of manufacture, and vehicle owner) to the TA. Identity secret key of V_i is then given by $s_{V_i} = \alpha H(ID_{V_i})$ which is computed by TA and given to V_i.
(2) For a RSU, represented as RSU_i, it submits its real identity ID_{R_i} and RSU information $info_{R_i}$ (e.g., RSU serial number and location) to the TA. Identity secret key of RSU_i is then given by $s_{R_i} = \alpha H(ID_{R_i})$ which is computed by TA and given to V_i.
(3) TA saves vehicle/RSU-submitted information and their secret keys.

2.4.3 Vehicle–RSU Key Agreement

In our proposed scheme, each RSU is responsible for traffic information collection and traffic jam detection of its communication range. In order to preserve the privacy of vehicles, when a vehicle enters the communication range of a RSU, it should firstly perform key agreement with the RSU to get secret key and pseudo-identities. The key agreement procedure is shown in Figure 2.1.

Vehicle V_i randomly picks $r_1 \in \mathbb{Z}_q^*$ and encrypts $r_1 \,||\, ID_{V_i}$ with identity of TA and generates the signature of r_1 with identity secret key of V_i (steps 1–3 in Figure 2.1). Upon receiving message $\{c_1, \sigma_1\}$, R_x randomly picks $r_2 \in \mathbb{Z}_q^*$ and generates signature of r_2, and then, $\{r_2, ID_{R_x}, \sigma_1, c_1, \sigma_2\}$ is sent to TA

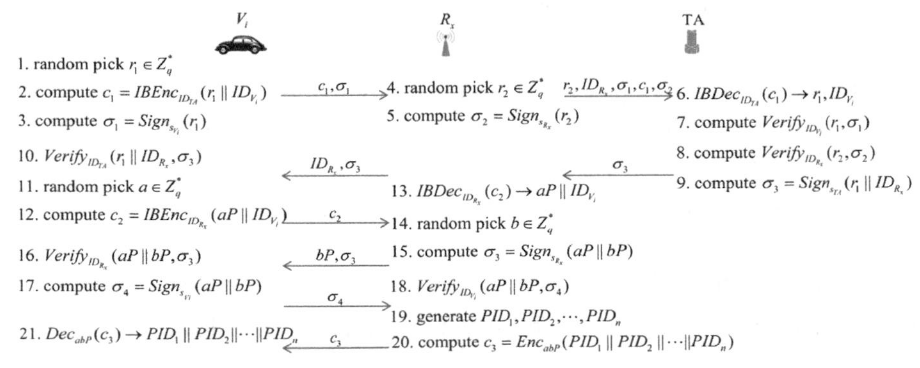

Figure 2.1 Vehicle–RSU key agreement.

(steps 4, 5). TA decrypts c_1 to recover r_1 and ID_{V_i}, and then, TA verifies σ_1 and σ_2, respectively. If both signatures are valid, it means that the request is from a legal vehicle and RSU, and then, TA computes the signature of $r_1 \,||\, ID_{R_x} \,||\, ID_{V_i}$ and sends the signature σ_3 to R_x (steps 6–9). Upon reception, R_x forwards σ_3 with identities ID_{R_x} and ID_{V_i}. V_i verifies σ_3 to make sure that R_x is trusted and valid (step 10). Up to now, entity authentication is finished.

V_i randomly picks $a \in \mathbb{Z}_q^*$ and encrypts $aP \,||\, ID_{V_i}$ with identity of R_x (steps 11, 12). R_x decrypts c_2 to recover $\{aP, ID_{V_i}\}$, randomly picks $b \in \mathbb{Z}_q^*$, and computes the signature of $aP \,||\, bP$. Then, bP, σ_3 is sent to V_i (steps 13–15). V_i verifies the validity of σ_3 and computes the signatures of $aP \,||\, bP$ and sends σ_4 back to R_x (steps 16–17). R_x verifies the validity of σ_4. If σ_4 is valid, the secret key between V_i and R_x is $s_{V_i,R_x} = abP$. Then, R_x generates *n* pseudo-identities and encrypts them with abP. R_x stores $\{ID_{V_i}, PID_1 \,||\, PID_2 \,||\, \ldots \,||, PID_n, abP\}$ and sends c_3 to V_i (steps 18–20). As secret key abP is only known by V_i and R_x, V_i can decrypt c_3 to recover pseudo-identities (step 21). Pseudo-identities are utilized for anonymous traffic information reporting by vehicle. Up to now, the whole key agreement procedure is finished.

2.4.4 Traffic Information Collection and Aggregation

After vehicle–RSU key agreement, V_i begins to send its traffic information to R_x with the secret key between V_i and R_x. It should be noted that if a vehicle is located in the overlapping region of several RSUs, it should report traffic information to RSUs, respectively, so that each RSU could collect enough traffic information for traffic jam detection. R_x collects traffic information from all vehicles located in its communication range for traffic information aggregation and traffic jam detection. The traffic information collection is performed as follows:

1. Vehicle V_i collects and generates its traffic information as {*timestamp, position, speed, direction*}.
2. V_i picks a random integer $r \in \{1, 2, \ldots, n\}$. Let $m = PID_r \,||\, timestamp \,||\, position \,||\, speed \,||\, direction$ and compute $c = Enc_{abP}(m)$, $\rho = \mathrm{MAC}_{abP}(m)$. Each pseudo-identity should only be used for limited times, and V_i should perform another vehicle–RSU key agreement when all pseudo-identities are used up.
3. V_i sends $\{PID_r, c, \rho\}$ to R_x, and then, R_x verifies the message as follows:

```
{PID'_r, timestamp, position, speed, direction} = Dec_abP(c)
If PID_r != PID'_r:
    drop message {PID_r, c, ρ}
else if (current_time - timestamp) > thresholdtime:
    drop message {PID_r, c, ρ}
else:
if (MAC_abP(PID'_r ||timestamp||position||speed||direction)
    == ρ):
    store message {ID_Vi, timestamp, position, speed,
    direction}
else:
    drop message {PID_r, c, ρ}
```

Threshold time is a preset value to prevent the replay attack. V_i reports its traffic information to R_x periodically, and R_x aggregates collected traffic information to detect the traffic jam. Traffic jam is a condition on road networks that occurs as user increases and is characterized by slower speeds, longer trip times, and increased vehicular queuing. So the traffic jam detection strategy is based on the two essential factors: vehicle density and average vehicle speed. Area covered by RSUs is divided into square-shaped districts with sides of several hundred meter length. Districts covered by R_x are shown in Figure 2.2. R_x detects the traffic jam as follows:

1. Select the most recent traffic message from distinct vehicles, and mark the corresponding vehicles on the map.
2. Select and mark districts where vehicle density is bigger than the preset $vehicle_{density}$ and $threshold_{density}$:

```
for each district_i:
if (count(vehicle in district_i)/area(district_i)>
threshold_density):
    mark district_i
end if
end for
```

   If a district is not fully covered by R_x, e.g., $district_{21}$ in Figure 2.2, function area ($district_i$) only calculates the area that is covered by R_x.
3. Compute vehicle average speed of the marked districts. If average speed is lower than a preset speed threshold $threshold_{speed}$, traffic jam in the district is detected and a traffic jam information of the district is generated as $\{timestamp, district_i\}$:

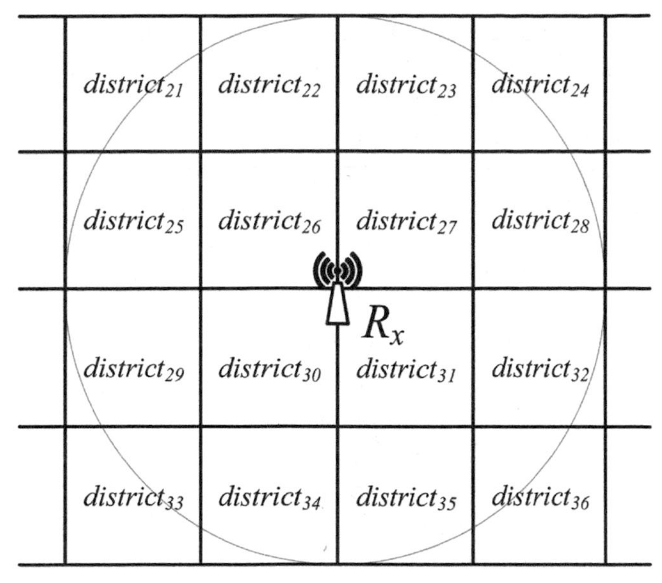

Figure 2.2 Division into districts.

```
for each marked district_i:
if (sum(vehicle speed)/count(vehicle in district_i)>
threshold_speed):
  generate traffic jam message m={timestamp, district_i}
end if
end for
```

2.4.5 Traffic Jam Message Propagation

Upon the detection of traffic jam detected by R_x, the traffic jam message m is propagated to the communication range of R_x and its adjacent RSUs. The message propagation procedure is shown in Figure 2.3.

Firstly, R_x generates the signature of current time and traffic jam message and sends $\{timestamp, m, ID_{R_x}, \sigma_1\}$ to TA (steps 1, 2). TA verifies the validity of σ_1 to confirm that the traffic jam message is from a valid RSU. If the verification passes, TA generates the signature of $timestamp \,||\, m$ and sends the signature to R_x and its adjacent RSUs, R_i, in Figure 2.3 (steps 3, 4). When RSUs receive traffic jam message from TA, they verify the signature and broadcast valid traffic jam message to vehicles in their communication range to notify vehicles. Upon the message reception, vehicles verify the traffic jam message. We take V_i as an instance to illustrate the verification process:

$timestamp_{last}$: the last traffic jam message received time
$threshold_{interval}$: threshold interval of two traffic jam messages
V_i gets current time: $timestamp_{current}$
if $(timestamp_{current} - timestamp_{last}) >= threshold_{interval}$:
if $Verify_{ID_{TA}}(timestamp \,||\, m, \sigma_2)$ is valid:
  notify driver that traffic jam occurs in $district_i$ at time timestamp
else:
  random pick a pseudo-identity PID_r
  compute $c_1 = Enc_{s_{V_i,R_x}}(timestamp \,||\, m \,||\, ID_{R_x} \,||\, \sigma_2)$ send $\{PID_r, c_1\}$ to R_x
  R_x decrypts c_1: $\{timestamp, m, ID_{R_x}, \sigma_2\} = Dec_{s_{V_i,R_x}}(c_1)$
if ($\{timestamp, m, ID_{R_x}, \sigma_2\}$ is verified):
ret = select result from ret_table where message $= timestamp \,||\, m \,||\, \sigma_2$
else:
  R_x computes ret=$Verify_{ID_{TA}}(timestamp \,||\, m, \sigma_2)$
  store ret in ret_TABLE as {result=ret,message=$timestamp \,||\, m, \sigma_2$}
  compute $c_2 = Enc_{s_{V_i,R_x}}(timestamp \,||\, m \,||\, ID_{R_x} \,||\, \sigma_2 \,||\, ret) \,||\, ret$
  R_x sends c_2 to V_i
  V_i decrypts c_2 to get verification result ret
if ret ==1: notify driver that traffic jam occurs in $district_i$ at time timestamp

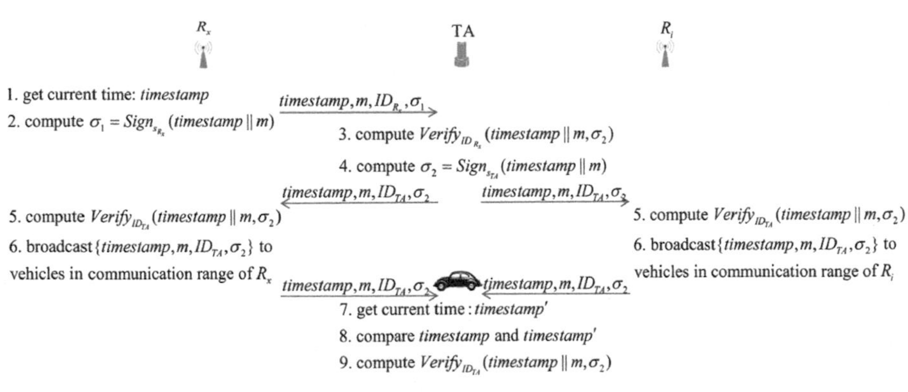

Figure 2.3 Traffic jam message propagation.

It should be noticed that RSU verifies at most a traffic jam message once RSU stores the verification result of a message, when another vehicle sends the same request; it returns the result to vehicle directly. This saves RSU computation recourses and avoids computational DoS attack. In this way, the driver could recount his driving route to avoid traffic jam to get a faster and unblocked route to destination, and traffic jam is eased.

2.5 Security Analysis

According to the security requirements in Section 2.3.3, the security analysis is discussed in the following paragraph. In discussion, we also analyze two representative traffic information sharing schemes (VAST [21], RAISE [15]) to compare with PETS. The comparison result of different security properties that schemes fulfill is demonstrated in Table 2.2.

- **Authentication**: Message authentication code is utilized to traffic information collection. As message authentication code is based on one-way function, it is hard to forge a valid message authentication code without knowing secret key (In Section 2.3.3, the key is abP.). Identity-based digital signature is employed for traffic jam message propagation. Due to computational Diffie–Hellman assumption, it is hard to generate a valid signature without knowledge of identity secret key. Security feature of authentication makes sure that vehicles only accept message without modification during transmission from legal entities.
- **DoS resistant**: After vehicle–RSU key agreement, RSU is in charge of traffic information collection, and vehicles no longer collect traffic information individually. In this way, computation and storage load are moved from vehicles to powerful RSU. Furthermore, traffic information verification only involves symmetric cryptographic operation and message authentication code computation, which makes a great contribution to the efficiency of message collection. Identity-based digital signature is computationally expensive, especially for vehicles that resources

Table 2.2 Comparison of different security properties schemes fulfill

	Property		
Scheme	VAST	RAISE	PETS
Authentication	✓	✓	✓
DoS resistant	✓	×	✓
Identity privacy preserving	×	✓	✓
Unlinkability	×	✓	✓
Conditional Traceability	×	✓	✓

are limited. In traffic jam message propagation phase, a vehicle verifies at most one message in a preset interval, or it requests the powerful RSU to verify the message. If small parts of vehicles are compromised, the adversary still cannot make an effective DoS attack by utilizing compromised vehicles. In RAISE, vehicles need to verify digital signature themselves. The adversary could launch a successful attack by employing compromised vehicles to send invalid signatures at the same time.

- **Identity privacy preserving**: When a vehicle enters the communication range of a RSU, the vehicle performs key agreement with the RSU. During the key agreement, real identity of the vehicle is only known by TA and legal RSU which is authorized by TA. So vehicles have no knowledge of other vehicles' identities. For traffic information collection, a vehicle reports its traffic message encrypted by secret key generated in key agreement along with a pseudonymous identity to a RSU. Then, the identity privacy is preserved even if the adversary could monitor the whole network. In VAST, certificates are used for authentication directly, which leads VAST does not provide security feature of identity privacy preserving, unlinkability and conditional traceability.
- **Unlinkability**: A vehicle randomly picks a pseudo-identity for each traffic information report, and every pseudo-identity is used for limited time, which avoids the adversary link multiple messages to one vehicle. In addition, the traffic information reported by each vehicle is encrypted by secret key between vehicle and RSU, in the sense that the adversary cannot take any advantage of traffic information to help construct the linkability between messages and vehicles.
- **Conditional traceability**: In our scheme, only TA and authorized legal RSUs have the knowledge of real identity of a vehicle. TA can query RSUs to link the message with the corresponding vehicle and disclose the original sender of the message. The feature of conditional traceability is important, for it not only help to trace the message origin, but also be useful to detect greedy and malicious vehicles.

2.6 Performance Evaluation

In this section, we evaluate the performance of the proposed PETS with VAST and RAISE schemes. As traffic information sending/collection and traffic information propagation/verification are the most common and frequent operations in VANET, we analyze the computation and communication overhead of these operations at first, and then, we simulate schemes to give a further evaluation.

Tate pairing [22] is adopted in our evaluation, where G is represented by 161 bits, and the prime order q is represented by 160 bits. Moreover, we utilize AES-128 as $Enc_k(\cdot)/Dec_k(\cdot)$, HMAC as $MAC_k(\cdot)$, and SHA-1 as $H(\cdot)$. Let T_{mul} denotes the time to compute one point multiplication, T_{par} denotes the time to perform one pairing operation, T_{hash} denotes the time of one hash function operation, T_{mac} denotes the time of one message authentication code operation, T_{enc} denotes the time of one encryption operation, and T_{dec} denotes the time of one decryption operation. T_{mul}, T_{par}, T_{hash}, T_{mac}, T_{enc}, and T_{dec} dominate the computation performance of schemes, for simplicity; we only consider these operations for traffic information sending/collection evaluation and traffic jam message propagation/verification evaluation. We run 100 times point multiplication, tate pairing, SHA-1 hash function, AES-128 encryption/decryption, and HMAC operation on a machine equipped with an Intel Core (TM) 2 Duo CPU@2.4GHz, respectively, and the average operation times are 5.4 ms, 40.7 ms, 6 μs, 16.7 μs, 15.8 μs, and 40.7 μs, respectively. The following simulation adopts the measured processing time based on these data.

2.6.1 Traffic Information Sending/Collection Overhead

In VAST, traffic information sending requires one point multiplication and a message authentication code operation, so the computational cost is $T_{\text{mul}} + T_{\text{mac}}$. According to our former experiment, VAST can send:

$$\frac{1}{T_{\text{mul}} + T_{\text{mac}}} = \frac{1}{5.4 * 10^{-3} + 4.07 * 10^{-5}} \approx 183.8$$

messages per second. Traffic information sending/collection overhead computation overhead is shown in Table 2.3. Figure 2.4 illustrates the number of messages that the scheme can send/collect per second. It can be seen that all of the above schemes have a rather fast speed for message sending and collection. But VAST cannot achieve both efficiency and security feature of non-repudiation for message collection, and RAISE does not encrypt the message, which may lead to message abusing by adversaries.

Table 2.3 Computation overhead of traffic information sending/collection

	VAST	RAISE	PETS
Sending	$T_{\text{mul}} + T_{\text{mac}}$	T_{mac}	$T_{\text{enc}} + T_{\text{mac}}$
Collection	$2T_{\text{mac}} + 2T_{\text{mul}}$ (non-repudiation required) $2T_{\text{mac}}$ (no non-repudiation)	$T_{\text{mac}} + T_{\text{hash}}$	$T_{\text{dec}} + T_{\text{mac}}$

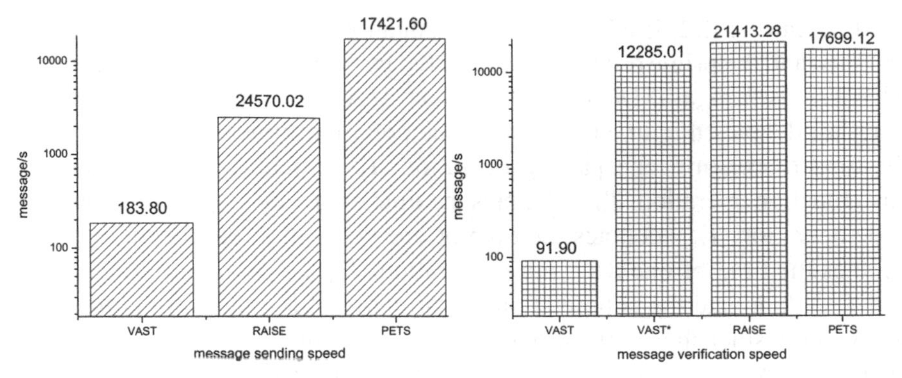

Figure 2.4 Message sending/verification speed.

Note: VAST* is VAST scheme when non-repudiation is not required.

In VAST, the communication overhead consists of 63 bytes certificate, 20 bytes message authentication code, 42 bytes signature, 16 bytes symmetric key, and 4 bytes index ID. Table 2.4 shows communication overhead of sending one traffic information message. It can be seen that PETS has the lowest bandwidth consumption, and it decreases communication overhead by 10.00%∼69.66% compared with VAST and RAISE. When the number of vehicles is big, low bandwidth consumption of PETS contributes to low message loss ratio as we will further simulate it in Section 2.6.3.

2.6.2 Traffic Information Propagation/Verification Overhead

When traffic information is collected, it should be aggregated or forwarded to the other vehicles. As RAISE and PETS utilize powerful RSU to aggregate traffic information, we mainly consider computation and communication overhead at the vehicle side. In PETS, vehicle has two ways to verify traffic information: perform identity-based digital signature verification itself or request RSU to verify traffic information instead. Computation and communication overhead are shown in Table 2.5, and every second the number of messages that the scheme can verify is illustrated in Figure 2.5. It reveals that 1) VAST has a rather high performance when non-repudiation is not required, and RAISE is vulnerable to DoS attack when a vehicle needs

Table 2.4 Communication overhead of traffic information sending/collection

	VAST	RAISE	PETS
Communication overhead (byte)	145	44	40

Table 2.5 Computation overhead of traffic information propagation/verification

	VAST	RAISE	PETS
Computation overhead	$2T_{\mathrm{mac}} + 2T_{\mathrm{mul}}$ (non-repudiation required) $2T_{\mathrm{mac}}$ (no non-repudiation)	$2T_{\mathrm{mul}} + T_{\mathrm{hash}}$	$2T_{\mathrm{par}} + 2T_{\mathrm{hash}}$ (no RSU assist) $T_{\mathrm{enc}} + T_{\mathrm{dec}}$ (with RSU assist)
Communication overhead (byte)	145	$20n + 42$	42 + 20 + 4 (no RSU assist) 20 + 35 (with RSU assist)

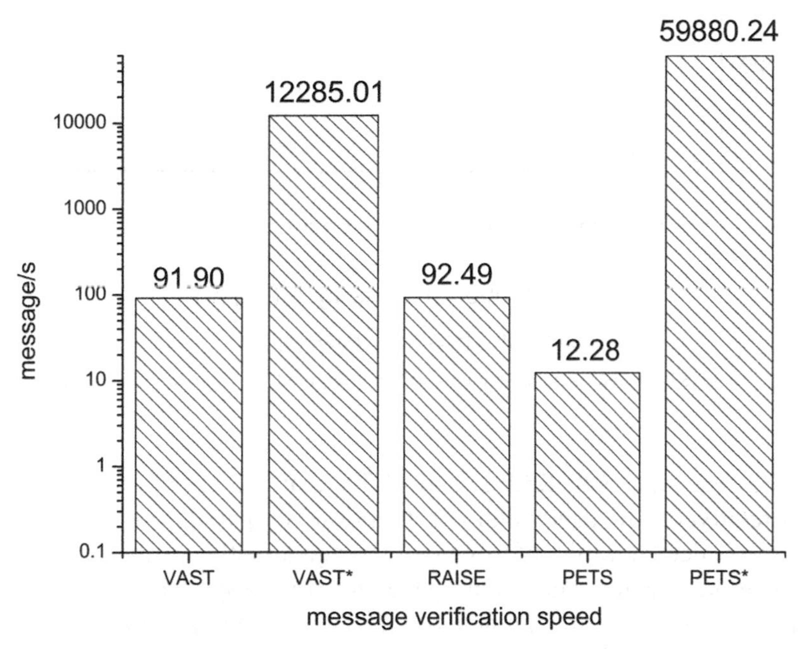

Figure 2.5 Message verification speed per second.

Note: VAST* is VAST scheme when non-repudiation is not required; PETS* is PETS scheme with RSU assist verification.

to verify more than x messages (e.g., adversary utilized several promised vehicles to send a number of invalid messages for verification). Only PETS achieves both efficient message verification and DoS-resistant security feature. 2) We assume that RSU propagates aggregates message every 100 messages in RAISE, and every 10,000 messages in PETS. Figure 2.6 illustrates the relation curves that communication cost varies with the number of messages n.

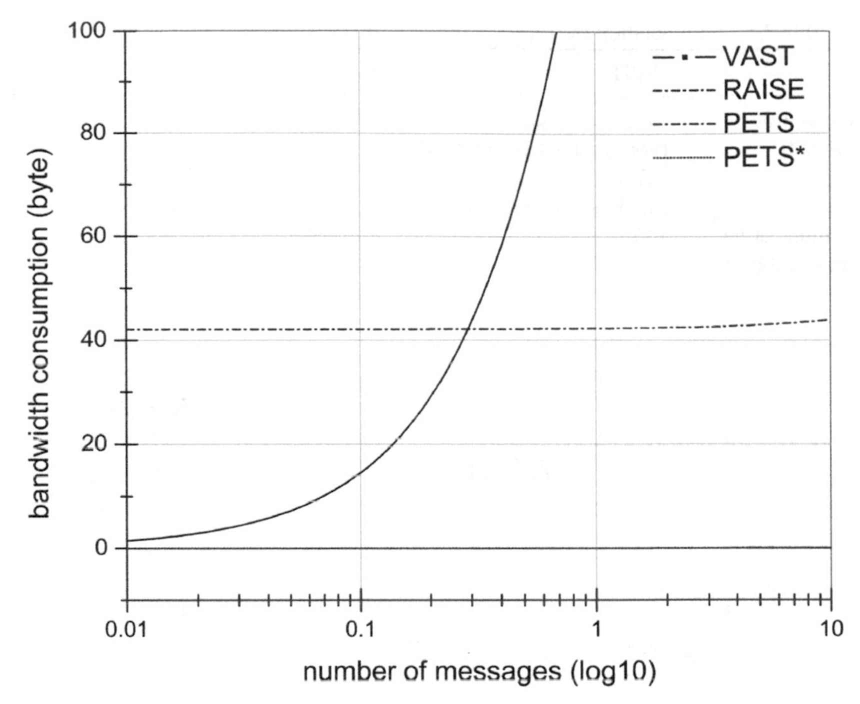

Figure 2.6 Bandwidth consumption.

Note: VAST* is VAST scheme when non-repudiation is not required; PETS* is PETS scheme with RSU assist verification.

It can be seen that both RAISE and PETS that are based on aggregation have a relatively low bandwidth consumption compared with VAST which shares the collected traffic information message directly. Furthermore, PETS is application-oriented, and it only generates a message when traffic jam is detected. So the proposal significantly decreases network communication at the message propagation phase.

2.6.3 Scheme Simulation

In this subsection, we simulate VAST (VAST*), RAISE, and PETS with opportunistic networking environment (ONE [10]). Aiming at estimating real-world road system properly, we select a part from real map of Beijing (northeast corner of area surrounded by the No. 2 Ring Road of Beijing) using OpenJump and import it into ONE as a city street scenario. The adopted map and the user interface of ONE in this chapter are presented as in Figures 2.7 and 2.8.

Figure 2.7 City street scenario corresponding to a roughly.

RSUs are distributed randomly on the roads of the map at the beginning of each simulation at a speed of 0 km/h. The number of RSUs would be sufficient to cover all the vehicles in the map. All vehicles are distributed deliberately on the roads of the map at the beginning of each simulation. Each of them would choose one casual point separately on roads and moves toward it following a specific movement model, at a random speed generated from a range of 10 km/h centered at a velocity value configured in advance. ONE provides several advanced practical movement models to imitate different actual scenarios in life. Hereby, we cautiously equip every vehicle with shortest path map-based movement in which Dijkstra's algorithm is used to find the shortest path along connected road between two random map nodes. Having arrived at the destination, the vehicle waits for a short time, and then, it would pick next random target on some road of the grids and repeat the

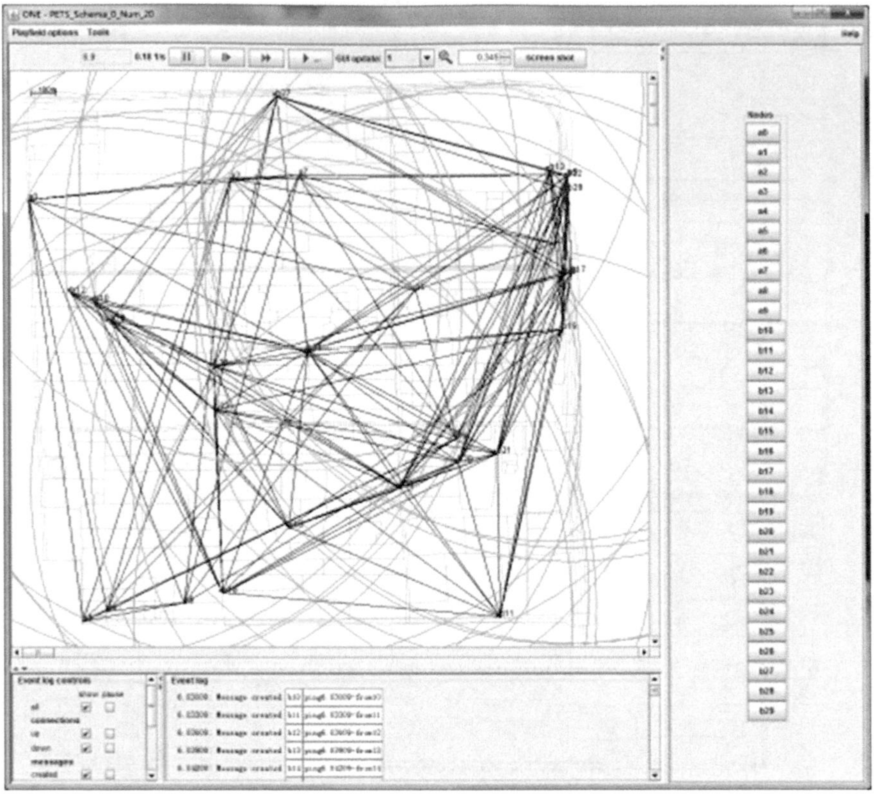

Figure 2.8 ONE user interface square are of size $2250 \times 2250\ m^2$.

aforementioned moving process till the end of this round of simulation. Other essential parameters are listed in Table 2.6.

Metrics for performance evaluation in this chapter are the average message delay, average message loss ratio, and percentage of signature verified, which are represented as $avgD_{msg}$, $avgLR$, and $avgPer\ SV$, correspondingly, and are stated as follows:

$$avgD_{msg} = \frac{1}{N_D \cdot M_{sent\_n} \cdot K_n} \sum_{n \epsilon D} \sum_{m=1}^{M_{sent\_n}} \sum_{k=1}^{K_n} \left(T_{sign}^{n\_m} + T_{transmission}^{n\_m\_k} + T_{verify}^{n\_m\_k}\right) \cdot (L_{n\_m\_k} + 1),$$

where D is the simulation district, N_D is the total number of RSUs and vehicles in D, M_{sent_n} is the number of messages sent by RSU or vehicle n, K_n is the

Table 2.6 Simulation configuration

Simulation Scenario		City Streets
Total of RSUs		10
Communication range	RSU	300 m
	Vehicle	1500 m
Simulation time		200 s
Channel bandwidth (RSU & vehicle)		6 Mbps
Wait time		0~5 s
Buffer size	RSU	100 M bytes
	Vehicle	1 M bytes
Vehicle broadcast interval		0.3 s
Speed		[20 km/h, 100 km/h]

number of RSUs and vehicles within the one-hop communication range of RSU or vehicle n, $T_{sign}^{n\_m}$ represents the time consumed for signing message m by RSU or vehicle n, n_m_k is one message sent by RSU or vehicle n and received by RSU or vehicle k, and $L_{n\_m\_k}$ is the length of the buffer queue equipped in RSU or vehicle k when n_m_k is received by RSU or vehicle k.

$$avgLR = \frac{1}{N_D}\sum_{n=1}^{N_D}\frac{M_{dropped}^{n}}{\sum_{k=1}^{K_n} M_{arrived}^{n}},$$

where $M_{dropped}^{n}$ means the total of dropped messages by RSU or vehicle n in the application layer and $M_{arrived}^{n}$ the number of received messages in the network layer by RSU or vehicle n. Here, consideration of message loss caused by wireless transmission is excluded, as leaving only message loss by security protocol due to full buffer space.

$$avg\ Per\ SV = \frac{1}{N_D}\sum_{n=1}^{N_D}\frac{M_{consumed}^{n}}{\sum_{k=1}^{K_n} M_{arrived}^{n}},$$

where $M_{consumed}^{n}$ means the total of consumed messages by RSU or vehicle n in the application layer. In the following, we conduct a set of experiments to analyze the impacts of different traffic loads.

Simulation results are shown in Figures 2.9–2.11.

It could be seen that with the growth of traffic load, the average message delay is increasing for VAST and RAISE, while it remains stable for VAST* and PETS. Besides, the value for VAST is above one second even with only 10 vehicles in the communication range, which is much larger than the other three and could not be applicable for high traffic density, while for

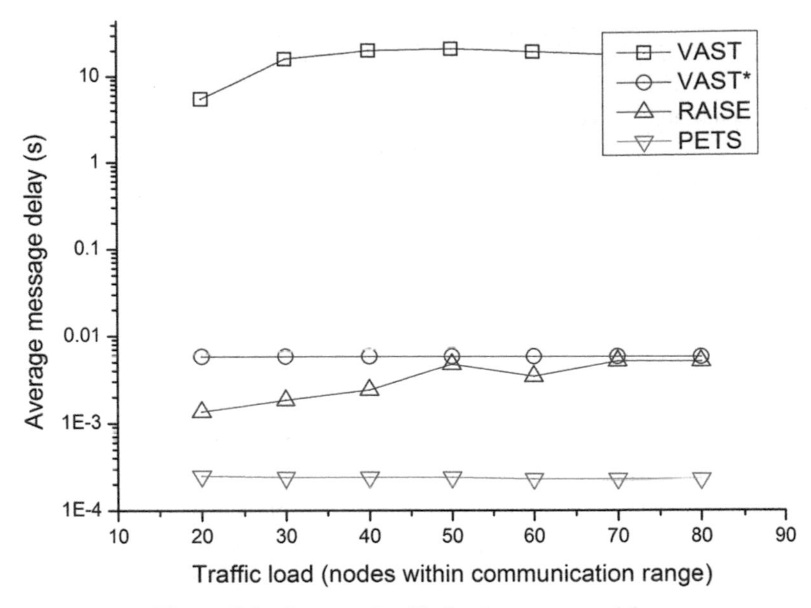

Figure 2.9 Impact of traffic load on message delay.

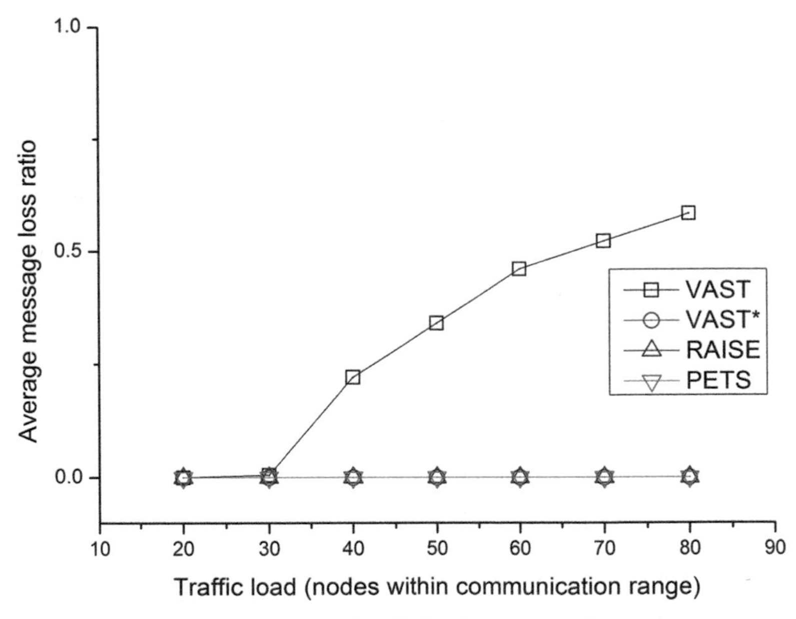

Figure 2.10 Impact of traffic load on message loss ratio.

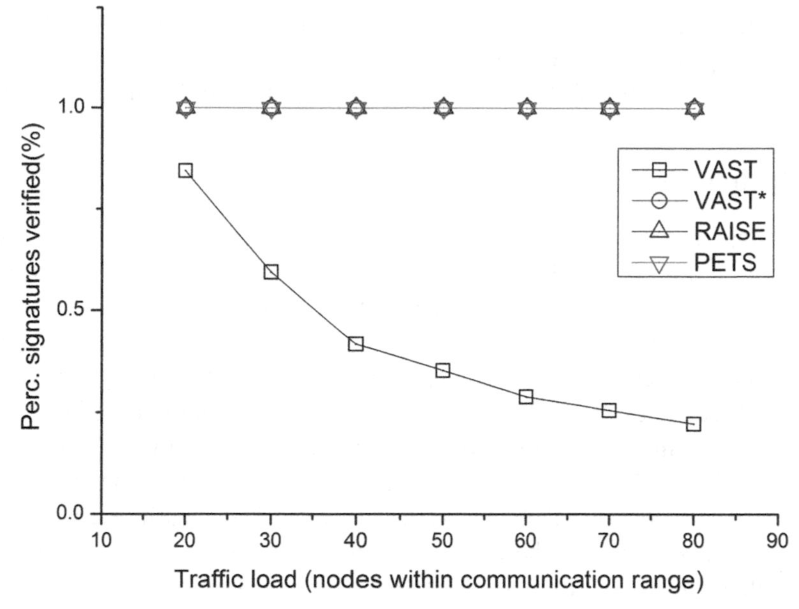

Figure 2.11 Impact of traffic load on percentage of signatures verified.

PETS, the average message delay keeps the lowest of the four schemes in all kinds of traffic scenarios, which is effective to handle scenarios like warning broadcasting with high traffic density.

The average message loss ratio for VAST keeps growing when the traffic load is above 30 and reaches beyond 50% when the traffic load is larger than 70 which is normal in the scenario of traffic jam, while for the other three including PETS, it keeps zero no matter what the traffic load it is now. Actually, considering this metric, PETS is one of the best.

The percentage of signature verified decreases for VAST with the growth of traffic load and drops to less than 50% when the traffic load is larger than 35. This is not tolerable in most of the VANET applications due to that without enough information, the decisions made would be pale and useless for improvement of traffic efficiency, while for the other three schemes including PETS, the metric value keeps nearly 100% all the time. Actually, it is 100% all the time for PETS in original experiment result when the traffic load is lower than 80, while for the other three schemes, signature not verified do exist even in a low traffic density.

To conclude from the above analysis of simulations, PETS turns out to have the lowest average message delay, and its message loss ratio and signature verified percentage are also in the best category of all schemes. Although VAST* also produces a good performance, it does not provide some essential security, such as unlinkability, conditional traceability, and non-repudiation.

2.7 Conclusion

In this chapter, a novel traffic jam information sharing scheme is proposed. The proposal employs RSU as an aggregator to collect and aggregate traffic information sent from vehicles to detect traffic jam. Vehicles use pseudo-identities for information reporting, which avoids identity privacy leaking and adversary tracing. As traffic information is not collected by individual vehicles directly, cost of communication and computation is significantly reduced. When the traffic jam is detected by RSU, it generates a traffic jam message to TA, and TA signs the message and propagates it to the vehicles to notify drivers. To prevent DoS attack, the powerful RSU is utilized as a proxy to assist vehicle to verify the traffic jam message. The proposal is the first scheme that achieves both privacy preservation and DoS resistant for traffic information sharing-based secure data aggregation. In our future work, we will extend the proposal to support more traffic information sharing applications including traffic jam detection. Furthermore, RSU may be not pervasive at the early stage of VANET. So a scheme that utilizes a vehicle as an aggregator will be investigated.

References

[1] Armstrong, L., Dedicated Short Range Communications (DSRC) Home. [Online]. http://www.leearmstrong.com/DSRC/DSRCHomeset.htm

[2] Hsiao, H., et al. (2011). Flooding-resilient broadcast authentication for VANETs. In *Proceedings of the 17th Annual International Conference on Mobile Computing and Networking*, pp. 193–20.

[3] Raya, M., et al. (2005). The security of vehicular ad hoc networks. In *Proceedings of the 3rd ACM Workshop on Security of Ad hoc and Sensor Networks*, pp. 11–21.

[4] Raya, M., et al. (2006). Securing vehicular communications. *IEEE Wirel. Commun.*, 13, 8–15.

[5] Sun, Y., et al. (2010). An efficient pseudonymous authentication scheme with strong privacy preservation for vehicular communications, *IEEE Trans. Veh. Technol.*, 59, 3589–3603.

[6] Lin, X., et al. (2007). GSIS: a secure and privacy preserving protocol for vehicular communications, *IEEE Trans. Veh. Technol.*, 56, 3442–3456.

[7] Wang, F., et al. (2015). 2FLIP: a two-factor lightweight privacy preserving authentication scheme for VANET. *IEEE Trans. Veh. Technol.*

[8] Alexiou, N., et al. (2013). VeSPA: vehicular security and privacy-preserving architecture. In *Proceedings of the 2nd ACM Workshop on Hot Topics on Wireless Network Security and Privacy*. ACM, pp. 19–24.

[9] Shibata, N., et al. (2006). A method for sharing traffic jam information using Inter-vehicle Communication. In *Proceedings of Mobile and Ubiquitous Systems-Workshops. 3rd Annual International Conference*, pp. 1–7.

[10] Chaum, D., et al. (1991). Group signatures. In *Proceedings of Advances in Cryptology—EUROCRYPT*, pp. 257–265.

[11] Boneh, D., et al. (2004). Short group signatures. In *Proceedings of CRYPTO*, pp. 227–242.

[12] Hess, F. (2003). Efficient identity based signature schemes based on pairings, *Selec. Areas Crypt.*, 310–324.

[13] Picconi, F., et al. (2006). Probabilistic validation of aggregated data in vehicular ad-hoc networks. In *Proceedings of the 3rd International Workshop on Vehicular Ad hoc Networks. ACM*, pp. 76–85.

[14] Zhu, H., et al. (2008). AEMA: An aggregated emergency message authentication scheme for enhancing the security of vehicular ad hoc networks communications. In *Proceedings of IEEE International Conference on. IEEE*, pp. 1436–1440.

[15] Zhang, C., et al. (2008). An efficient message authentication scheme for vehicular communications. *Veh. Technol., IEEE Trans.*, 57(6), 3357–3368.

[16] Sampigethava, K., et al. (2006). CARAVAN: providing location privacy for VANET. In *Proceedings of International Workshop on Vehicular Ad-hoc Networks.*

[17] Boneh, D., et al. (2001). Identity-based encryption from the weil pairing. In *Proceedings of Advances in Cryptology—CRYPTO 2001*, pp. 213–229.

[18] Shamir, A. (1985). Identity-based cryptosystems and signature schemes. In *Proceedings of Advances in Cryptology*, pp. 47–53.

[19] Boneh, D., et al. (2001). Short signatures from the weil pairing. In *Proceedings of ASIACRYPT*, pp. 514–532.

[20] Bellare, M., et al. (1996). Message Authentication Using Hash Functions—The HMAC Construction, *RSA Laboratories Crypto Bytes*, 2(1), Spring.

[21] Studer, A., et al. (2008). Flexible, Extensible, and Efficient VANET Authentication. In *Proceedings of the 6th Annual Conference on Embedded Security in Cars*.

[22] Scott, M., et al. (2007). Efficient Implementation of Cryptographic Pairings. (Online). http://www.pairing-conference.org/2007/invited/Scott_slide.Pdf.

[23] Calandriello, G., et al. (2007). Efficient and Robust Pseudonymous Authentication in VANET. In *Proceedings of the Fourth ACM International Workshop on Vehicular Ad hoc Networks*, pp. 19–28.

PART II

Vulnerabilities, Detection and Monitoring

3

DIAMoND: Distributed Intrusion/Anomaly Monitoring for Nonparametric Detection

Maciej Korczyński[1,5], Ali Hamieh[2], Jun Ho Huh[3], Henrik Holm[4], S. Raj Rajagopalan[3] and Nina H. Fefferman[1]

[1]Rutgers University, New Jersey, USA
[2]American University of Technology, Halat, Lebanon
[3]Honeywell ACS Labs, Golden Valley, MN, 55422, USA
[4]Forest Glen Research, LLC, 2412 Bronson Boulevard Kalamazoo, MI 49008, USA
[5]Delft University of Technology, 2628 CD Delft, The Netherlands

Abstract

In this chapter, we describe a fully nonparametric, scalable, distributed detection algorithm for intrusion/anomaly detection in networks. We discuss how this approach addresses a growing trend in distributed attacks while also providing solutions to problems commonly associated with distributed detection systems. We explore the impacts to detection performance from network topology, from the defined range of distributed communication for each node, and from involving only a small percent of total nodes in the network in the distributed detection communication. We evaluate our algorithm using a software-based testing implementation and demonstrate up to 20% improvement in detection capability over parallel, isolated anomaly detectors for both stealthy port scans and DDoS attacks.

Keywords: Intrusion Detection, Anomaly Detection, Distributed Systems, Nonparametric, Collaborative Defense, Network Security.

3.1 Introduction

Cyber attacks are among the top threats facing today's world. Common examples include compromising individual hosts to steal confidential data and commit fraudulent transactions and volume attacks such as scanning large network spaces, spreading worms, or launching distributed denial-of-service attacks (DDoS) to bring down legitimate services. For instance, a DDoS-based DNS amplification attack typically involves an attacker first compromising a large number of hosts to create a coordinated botnet that can be further used to generate huge volumes of spoofed DNS queries (see Figure 3.1). A spoofed IP address is used to forge DNS queries that appear to have come from a target victim host. DNS responses are sent to that spoofed IP address, all hitting the target host. Normal DNS queries and responses are around 60–80 bytes and 100–500 bytes in size, respectively. But when the attacker uses the EDNS0 extension of the DNS protocol along with a special DNS query type, TXT or ANY, much larger DNS responses of size around 4,000 bytes can be generated, amplifying normal responses by a factor of about 70. DDoS-based DNS amplification attacks exploit that amplification feature, relying on open recursive DNS resolvers to amplify the responses and generate huge volumes of attack traffic against victim hosts.

To evade the detection of traditional intrusion detection systems (IDSs), cyber attacks are migrating from centralized into more sophisticated peer-to-peer (P2P) architectures with malicious activities stealthily spread over large number of nodes [1]. For example, in December 2013, Microsoft attempted to shut down the ZeroAccess botnet which had taken control over two million machines worldwide, resulting in financial losses of over $2.7 million each month to search engine advertisers on Google, Bing, and Yahoo [2]. Identifying and shutting down 18 C&C servers disrupted the botnet, but did not take it down completely mostly because the decentralized, distributed

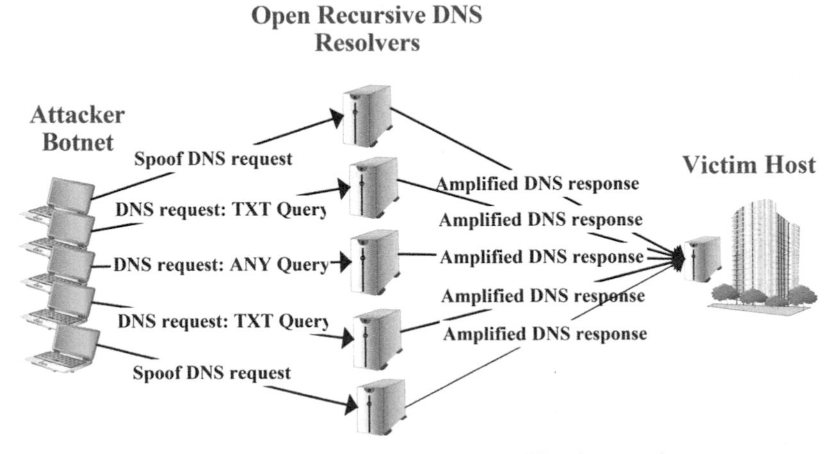

Figure 3.1 DDoS-based DNS amplification attack.

nature of the P2P botnet infrastructure made it very hard for a centralized defense mechanism to effectively thwart it. These sophisticated distributed cyber attacks demand a distributed anomaly detection approach, possible only through information sharing. Machines coordinating in this way can effectively learn and adapt according to any changing environment.

A new frontier in cyber defense that has emerged in recent times is based on the idea of sharing cyber attack information across organizational boundaries so that multiple organizations may collaborate in the rapid detection and thwarting of cyber attacks, especially attacks for which prior knowledge is scant or nonexistent [stix–honied]. Indeed, an entire new infrastructure is being created with new sharing protocols such as STIX [3] and TAXII [4], cyber threat "exchanges" [5], and government backing. Automated cyber data sharing is already being touted as the new defensive strategy against smart and highly distributed adversaries [6]. However, as Serrano et al. pointed out there are at least four fundamental technical challenges that need to be met before, this paradigm can become reality [7]. Our work directly addresses two of the four identified challenges. First, they point out that there are policy issues that prevent sensitive data from being shared between organizations. These policies pertain to privacy issues and/or other sensitivities of information sharing beyond given jurisdictions relating to legal and competitive issues (see, e.g., [8]). The second issue relates to the semantics of the data being exchanged, since the STIX and TAXII protocols are designed to mainly address the syntax of data sharing. IT environments show a tremendous diversity, and cyber attacks are evolving rapidly; as a result, the data captured in one particular environment may be unique and incomparable to similar data from another, vitiating any gains from the data sharing. Any form of detection that relies on comparison of semantically rich data is thus in jeopardy if the data come from sensors in different domains.

Our approach addresses both these challenges by providing a mechanism for cooperation between sensors in an arbitrary virtual topology and does not rely on sharing the specific details of the underlying event, but only the pattern of "excitation" seen in the sensors. By its nature, these data do not contain any personal information, or even any information about the specific attack. We expect that sharing of such data would be far easier to overcome organizational hurdles. For the same reason, our scheme also easily addresses the second challenge; because the data shared are very simple (even the individual threshold values are not shared), there is no question of creating semantic equivalence. Finally, the scheme enables sensors to self-tune their threshold values using the feedback mechanism. When new attack patterns appear, the sensors learn by cooperation to sense them adaptively—it takes some time, but there is no need for prior modeling to be applied to the sensors, which makes our scheme especially appealing in detecting novel attacks.

In this chapter, we propose DIAMoND (Distributed Intrusion/Anomaly Monitoring for Nonparametric Detection): a nonparametric fully distributed coordination framework that decouples local intrusion detection functions from network-wide coordination. DIAMoND first builds coordination overlay networks on top of physical networks. DIAMoND then dynamically combines *direct observations* of traditional localized/centralized NIDS with knowledge exchanged with other coordinating nodes called *neighbors* to dynamically detect the anomalies of underlying physical systems. Specifically, coordinating nodes in DIAMoND exchange generic nonparametric *levels of concern* between neighbors that reflect the observed probability of network attacks without elaborating any further details about the observations themselves. As a result, the coordination layer of the DIAMoND framework can be readily coupled with any local detection schemes without the need for increasing the detection feature sets. The coordination network layer is also decoupled from the underlying physical network layer to facilitate the flexible coordination strategies based on, for example, previously observed correlated behaviors, instead of being artificially limited to direct connectivity or geographic proximity. Interactions inside DIAMoND are limited to local neighborhood (e.g., 1 or 2 hop neighbors) in the overlay network, thus ensuring system scalability linear to the coordination network density instead of network size. The overall architecture of DIAMoND thus allows the preservation of private information of individual participating parties, which eases the deployment of DIAMoND across political and administrative boundaries.

DIAMoND is evaluated by emulation using an OS-level virtualization test bed. Additionally, our prototype shows how a distributed anomaly detector can be implemented in the POX SDN (software-defined network) controller. DIAMoND is released to the community for further works [9].

Our research contributions are summarized as follows:

- We propose a simple, scalable, nonparametric, fully distributed coordination framework that can be coupled with versatile local/centralized detection schemes for enhanced system performance.
- We propose fully automated strategies for creating *mutual, dynamic coordination neighborhoods* to facilitate the detection and fast reaction to network threats as well as to minimize the number of interactions within our system. These coordination topologies are fully decoupled from physical topologies thus not limited by the boundaries of administrative

control or geography and can be based on previously observed, repeated network-wide dispersed attacks.

- We evaluate the system accuracy to maximize sensitivity without compromising specificity by proposing and comparing four nonparametric, naïve *excitation algorithms* that combine the direct observation of a node with a summarized level of concern from local neighbors. In principle, each administrator can independently control the trade-off between sensitivity and specificity by, for example, choosing a more conservative algorithm. Moreover, we calculate the overall system accuracy and we quantify the additional information that is gained by deploying our system on top of local intrusion detectors.
- We demonstrate that even a minimal system deployment rapidly increases the overall detection accuracy and information gain.

Section 3.2 discusses the related work. Section 3.3 presents the architecture and the communication protocol of our solution. Section 3.4 introduces the test bed environments. Performance evaluation is described in Section 3.5, and we conclude our work in Section 3.6.

3.2 Literature Review

There have been several proposals for distributed intrusion detection such as distributed collection and centralized coordination [10, 11], hierarchical [12, 13], and fully distributed systems [14–23].

Grochocki et al. [24] discuss trade-offs between different types of distributed IDS deployment architectures, demonstrating the coverage and detection rate benefits of fully distributed (embedded sensing) systems over semidistributed systems and centralized systems.

In centralized coordination approaches, a correlation unit becomes a single point of vulnerability or failure. Shutting down a central server results in deactivating the detection system. Moreover, with an increasing volume of data to forward from remote collection units, the scalability of such an architecture is reduced significantly because of the required processing capabilities to perform alert correlation and the large volumes of bandwidth. This may result in unacceptable delays in responding to network intrusions or data losses.

Hierarchical detection architecture could scale better, but again, it would still suffer from the same problems; for example, a failure of higher level nodes shuts down the whole subtree of the intrusion detection system.

To overcome some of the limitations of distributed collection and centralized or hierarchical coordination approaches, fully distributed intrusion detection systems have been proposed. Chhabra et al. [15] demonstrate that a fully distributed approach to spatial volume anomaly detection can be just as effective as centralized approaches.

Locasto et al. proposed a fully distributed P2P IDS [16] in which every node is equipped with two principal components: *Worminator* and *Whirlpool*. The first one is responsible for alert correlation based on suspicious IP addresses encoded with Bloom filters, whereas the second one provides a mechanism for effectively conferencing alert-related information with other peers within the so-called federations.

Another P2P approach for collaborative intrusion detection is proposed by Zhou et al. [17]. It implements a distributed hash table (DHT) system to share the detection information. Each peer submits its blacklist to a fully distributed P2P overlay. The participating nodes are notified if other peers are attacked by the same source.

However, both methods use a single traffic feature, which might be too restrictive for detecting some important characteristics of large-scale intrusions. In our work, we advance these methods by proposing DIAMoND in which nonparametric *levels of concern* are exchanged between neighbors without elaborating any further details of the attacks themselves.

DefCOM [25], which is a distributed system for DDoS mitigation, consists of three types of nodes: core, classifier, and alert generator nodes. It implements an overlay communication protocol between source, victim, and core networks to detect and block the attack at the source.

In a distributed IDS proposed by Dash et al. [19], local detectors use a binary classifier to analyze incoming/outgoing host traffic and raise an alarm if a threshold value is crossed. Through their information sharing system (ISS), those alarms are sent to a random set of global detectors that generate a global view of security status of the system being monitored. One of the main drawbacks of their system, however, is the separation of global detectors and local detectors and the need for the ISS to coordinate messages between them. Such complexity would introduce communication and detection delays that are not imposed in our system.

Zhang et al. [26] proposed a distributed detection system for the multilayer network architecture of smart grids. Their IDS, however, is very specific to the network architecture of smart grids and does not really facilitate distributed decision making; rather, each IDS running in each different network layer monitors their own network traffic and, in most cases, makes independent decisions.

Yegneswaran et al. proposed the DOMINO system [18] that consists of three types of participants: *axis overlay nodes* that create a fully distributed P2P architecture, *satellite communities* that create smaller networks of communicating nodes and provide a wide diversity of alert data to axis nodes, and, finally, *terrestrial contributors* that deliver to the system any form of external intrusion data from firewalls, NIDS, etc. Axis nodes participate in a periodic alert data exchange, whereas alerts are generated in majority within small networks or satellites and can be propagated to the axis level. Each axis node maintains local and global views of the current intrusion. The local summaries are based on the direct observation of monitored subnetworks and its satellites, while the global summary is created by the alert correlation from other participating nodes. Open questions in DOMINO remain regarding its ability to preserve data privacy and the efficiency of the distributed architecture. In our work, we propose to exchange nonparametric *levels of concern* between the nearest neighbors which (i) reduce the communication overhead, (ii) preserve data privacy, and (iii) reduce the computation cost at each node due to a simple alert correlation algorithm.

Other security systems propose to detect various network attacks, in particular DDoS by type [27–29]. For instance, a DDoS DNS amplification attack relies on the ability to spoof IP addresses, so source address validation solutions can be effective. These solutions include authentication-based methods [27], traceback-based methods [28], and filtering-based methods [29]. Deployment of such solutions is practically limited because they require fundamental changes in Internet infrastructure. Ingress filtering allows routers to drop packets with source IPs that do not match any entry in the FIB forwarding table. The table lookup time adds an important delay on a router's forwarding time, which is unacceptable on high-speed links. Moreover, this type of filtering is not compatible with "special" services such as mobile IP and may create problems when the end-host is multihomed. Note that amplified answers from DNS resolvers will not be dropped by ingress filtering as they have a legitimate DNS source IP address; thus, this filtering requires a high fraction of deployment.

These solutions require deployment of a new detection system for each type of DDoS attacks. In this chapter, we describe a generic distributed, coordination system to mitigate all types of DDoS attacks that can ideally incorporate any existing detection algorithm.

3.3 System Design

3.3.1 Architecture Overview

DIAMoND is deployed over multiple *nodes* (switches, middle boxes) in a fully distributed architecture. We define a node's *neighborhood* as a subset of all nodes with which it directly exchanges nonparametric alert-related information. Neighborhoods are dynamic and can change over time based on, e.g., previously observed correlated behaviors or topology. Two collaborating nodes enjoy a symbiotic, mutual relationship, which means that both of them need to authenticate each other and agree on joining each other's neighborhoods. Furthermore, each node is equipped with two functional units: a *detection unit (DU)* and a *coordination unit (CU)*. The former is responsible for the data-driven assessment of the so-called *threat level*—the level of likelihood that an intrusion is occurring based on the *direct observation* reported by local NIDS and/or firewall implementation. The latter calculates the *concern level* which is a function of the *threat level* and the *concern levels* of its neighbors (cf. Figure 3.2).

Let us imagine a simple network composed of three nodes in a linear topology (cf. Figure 3.3). Each node is equipped with our distributed system which also includes a rate-limiting algorithm implemented in DU. If the traffic exceeds a predefined *sensitivity threshold* for a given IP address, it is dropped or delayed; otherwise, it is allowed to pass the filter. If we further assume that neighborhoods are based on physical connections, then we obtain the following three dynamic neighborhoods: $A = \{B\}, B = \{A, C\}$, and $C = \{C\}$. Each node shares its level of concern instead of threat level or

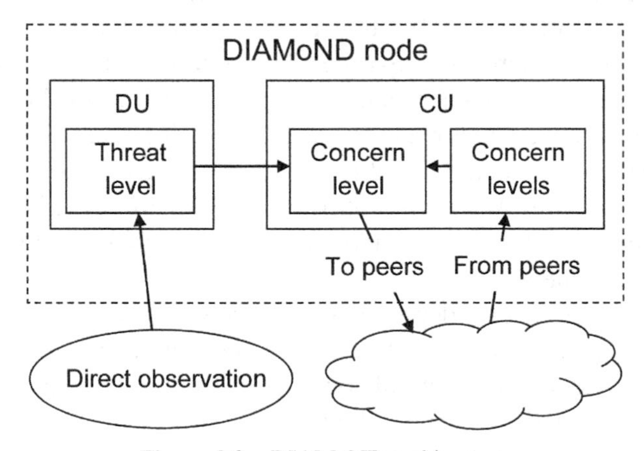

Figure 3.2 DIAMoND architecture.

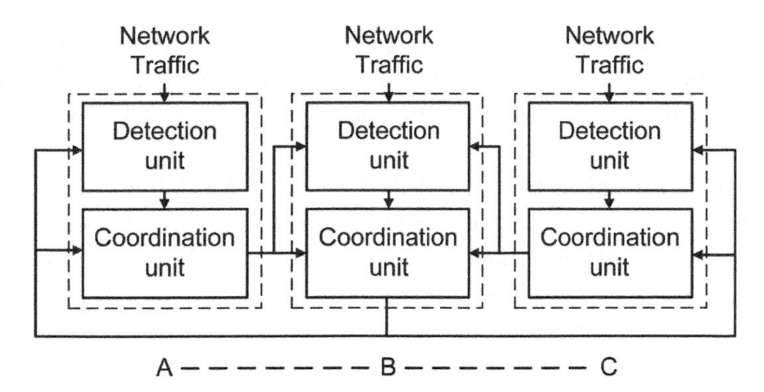

Figure 3.3 An example of node organization.

some specific information derived from the direct observation. Note that a *local concern consensus* of nodes A and C can (i) influence the level of concern of node B and (ii) tune the dynamic sensitivity thresholds of the rate-limiting detection algorithm in a detection unit of node B to react to outbreaks or to long-term network changes.

3.3.2 Detection Unit

NIDS, firewalls, and any other security intelligence can be implemented in DU so long as there is an appropriate plug-in to CU to translate the output of DU to the nonparametric threat level. Additionally, there must be an incorporated appropriate response by the DU to different levels of concerns of its neighbors (e.g., tuning of sensitivity thresholds). To foster interoperability, we do not require the extraction and provision of any potentially sensitive and/or incomparable attack details. In fact, a node may choose from among a set of different local anomaly detection methods, thereby making it difficult for an attacker to manipulate the local anomaly detection's influence on the CU network by making it harder to predict what types of traffic may trigger a local intrusion warning. These features greatly increase the potential of such a system to be able to detect diverse characteristics of large-scale zero-day attacks, depending on a variety of local detection algorithms adapted to DIAMoND.

3.3.3 Coordination Unit

In this section, we provide a description of the key component of CU that locally calculates the levels of concerns based on local information (levels of concerns of its neighbors) and direct observation (outbreaks detected by internal NIDS and/or firewalls). We also discuss the plug-in allowing the dynamic adjustment for detection thresholds of DU according to the local information.

Let us first formalize the DIAMoND system. We define $B = \{b_i, i = 1 \ldots n\}$ as the entire set of all participating nodes. Each node, b_i, has access to some subset of information about the traffic it handles, which is defined as the *observation set*, $o(b_i)$. In particular, a node may have access to all the packet headers in its traffic, but it may also have access to some part of the payloads. For each node, there is also a time *window*, $W_i^{t-w,t}(o(b_i))$, of length w over which the observation set may be analyzed. In our experiments, w is set to 5s for all nodes. Each of the participating nodes has an internal set of "native" detection algorithms, $a_i(W_i^{t-w,t}(o(b_i)), st_{i,t})$, that are a function of the information in the observation set of that node and of the set of detection *sensitivity thresholds* for that node at time *t*, $st_{i,t}$, each associated with a detection algorithm available to that node (i.e., each element of st_i is associated with an element of a_i). These sensitivity thresholds are dynamically updated over time, and there is no *a priori* assumption of uniformity in sensitivity thresholds across nodes. In fact, since each node may employ its own local anomaly detector, these thresholds are completely independent from each other.

At each time *t*, each node computes a function of the observed *threat level*, $T_{i,t}(a_i)$, which is the data-driven assessed level of likelihood that an anomaly is occurring. (For brevity, the (a_i) will henceforth be omitted.)

As discussed earlier, each node, b_i, has an associated set of nodes, called the *neighborhood*, $n(b_i) \supseteq B$, such that there is a path of links in the network from each node in $n(b_i)$ to b_i that influence its sensitivity threshold. Note this is not the same as the set of nodes with which b_i shares a direct link, but rather the set of nodes that influence the threshold of b_i.

Each node b_i has a level of *concern* at time *t*, $c_{i,t}(T_{i,t-1}, L_{t-1}(n(b_i)))$, which is a function of both the previously assessed threat level and of a function $L_t(n(b_i))$ that computes the total impact of the concerns of all nodes within the neighborhood of b_i at time *t*. Our naïve excitation algorithm is defined as follows. Initially, $c_{i,1} = 0$ and $L_1(n(b_i)) = 0$. Let $\bar{c} = \sum_{b_j \in n(b_i)} c_{j,t-1}/|n(b_i)|$ be the mean of the concerns of neighbors, and thus,

$$L_t(n(b_i)) = \begin{cases} 0, & \text{if } \bar{c} < V_0 \\ 1, & \text{if } V_0 \leq \bar{c} < V_1 \\ 2, & \text{otherwise} \end{cases} \quad (3.1)$$

where the current V_0 and V_1 values were informed by the expected traffic flow for each node given the initial network topology and within that expectation, set arbitrarily to 0.34 and 1.34, respectively.

Each node is equipped with a sampling detection algorithm for defeating TCP intrusions such as SYN flooding attacks and port scan activity [30, 31] that we extended to meet the needs of our system. Again, the DU might implement any kind of local intrusion detection intelligence, ideally including a plug-in to dynamically adjust its operational features such as detection thresholds according to $L_{t-1}(n(b_i))$.

We define R_{src} and $R_{\mathrm{dst}} \in \{1, 2, 3, \ldots\}$ as a number of outgoing SYN segments to corresponding incoming ACK segments per source and per destination, respectively, as traffic features corresponding to a_i. At time t_0, we calculate the cdf function of R_{src} based on daily collected packet traces of a trans-Pacific line [32]. We define $st_{i,0} = P(R_{\mathrm{src}} \leq 1.05)$ and $st_{i,\min} = P(R_{\mathrm{src}} \leq 0.99)$ as a minimal value of $st_{i,t}$.

Let $\lambda = L_{t-1}(n(b_i))/\max[T_{i,t-1}(a_i)]$ be called the update rate; thus, the adaptive sensitivity threshold, $st_{i,t}$ could be defined as follows:

$$st_{i,t} = \begin{cases} st_{i,\min}, & \text{if } st_{i,0} * (1 - \lambda) \leq st_{i,\min} \\ st_{i,0} * (1 - \lambda), & \text{otherwise} \end{cases} \tag{3.2}$$

We assign $T_{i,t}(a_i) \in \{0, 1, 2\}$ for each node in each time based on the observed traffic on that node using algorithm a_i with thresholds $st_{i,t}$ informed by the distribution of results from the local intrusion detection algorithm (initially parameterized using "normal network traffic"). Values are defined such that 0 indicates a completely normal classification, 1 indicates that traffic patterns have exceeded some fixed numbers of standard deviations from normal (where this number is itself a component of $st_{i,t}$) but has not yet exceeded threshold to be considered an attack, and 2 indicates classification by the local anomaly detector of a current attack.

Finally, we define a progression of four different test functions $c_{i,t}$ for all nodes to explore the impact of the level of importance being assigned to the concern of neighbors: low: $c_{i,t} = T_{i,t-1}$ unless $L_{t-1} = 2$, in which case $c_{i,t} = 1$; med: $c_{i,t} = T_{i,t-1}$, unless $L_{t-1} = 2$, in which case $c_{i,t} = 1$ if $T_{i,t-1} = 0$, and $c_{i,t} = 2$ otherwise; med+: $c_{i,t} = \max(L_{t-1}, T_{i,t-1})$ unless $T_{i,t-1} = 0$ and $L_{t-1} = 2$, in which case $c_{i,t} = 1$, and high: $c_{i,t} = \max(L_{t-1}, T_{i,t-1})$.

To the best of our knowledge, we are the first to propose to exchange a nonparametric, summarized view of the network state coming from the consensus of both the internal and the external observation.

3.3.4 Communication Protocol

Each of the nodes participates in both periodic and spontaneous message exchange. In trying to keep the communication protocol complexity as low as possible, we define three basic messages: Keep-alive, Update, and Summary [9].

Keep-alive messages are sent every 10 to 30 seconds depending on settings. They include the message type, the TTL value, and the level of concern. Messages are exchanged only between neighbors (cf. Figure 3.4a–c). The reason we consider this type of messages is to make sure that we can always tell the difference between completely normal network state and the state where legitimate traffic is not delivered due to packet losses (possibly caused by intrusive network activity). TTL reflects the number of logical hops between two communicating nodes. Ideally, TTL should reach 0 at the destination node.

Update messages have the same structure as keep-alive messages but are generated spontaneously when the level of concern increases or decreases due to either a change in a threat level or a change in a consensus of level of concern of the neighborhood. As previously, update messages are sent only to neighbors (cf. Figure 3.4a–c). To reduce the potential of exploitation of the system, we limit the number of spontaneous messages possible to send between two successive keep-alive messages.

Summary messages are multicast to many participating nodes to report previously seen attacks and a node position so that the dynamic neighborhood can be potentially redefined according to a strategy, network topology changes, or recently observed attacks. The interval of summary messages could be set to every hour, day, or even week. We limit the lifetime (hop count) of this type of messages by TTL set to 255 (cf. Figure 3.4d). The number of malicious source and destination IP addresses sent in a one message is limited to 20.

3.3.5 Neighborhood Strategies

We investigate the different strategies for creating neighborhoods to maximize the flow of meaningful information while minimizing the number of connections.

The first strategy is based on a hop limit that reflects the geographic or administrative distance between neighbors. In the simplest and yet very effective form, we define a neighborhood of a node by direct physical or logical connection. We also attempt to empirically verify the application of the extended neighborhoods by increasing the TTL value. In other words, nodes

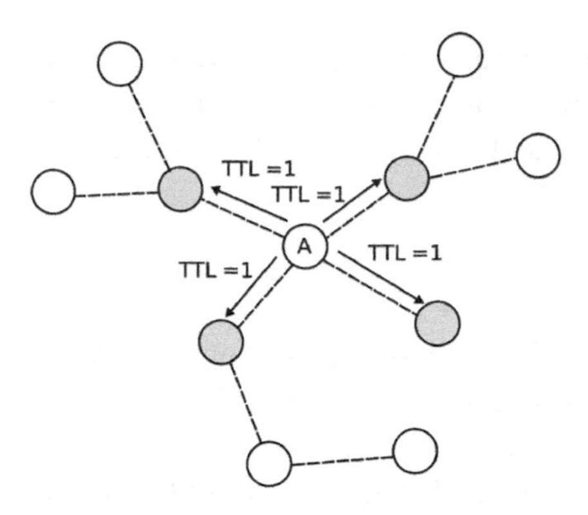

a) Keep-alive and Update message flow; neighborhood based on hop limit TTL = 1

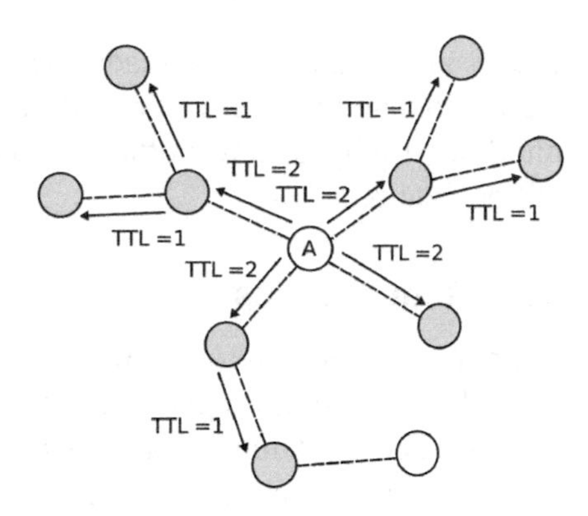

b) Keep-alive and Update message flow; neighborhood based on hop limit TTL = 2

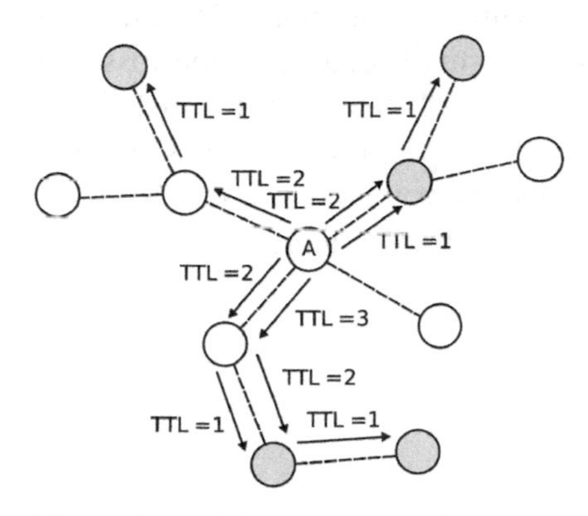

c) Keep-alive and Update message flow; neighborhood based on previously seen attacks;

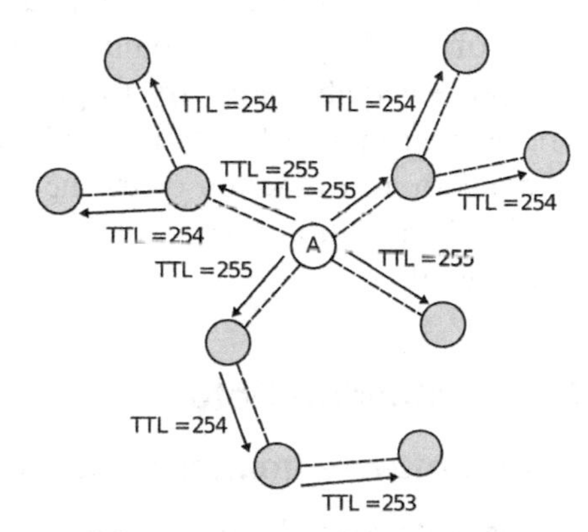

d) Summary message flow; multicast

Figure 3.4 Keep-alive, Update, and Summary message flows based on different communication strategies.

exchange their levels of concerns with their direct neighbors (TTL = 1) and neighbors of their neighbors (TTL = 2) and so on. Another strategy consists of correlating previously observed attacks and constructing neighborhoods based on the assumption that malicious activity may reoccur and be launched from the same set of compromised machines and/or against the same victims (networks, servers). Such a strategy has the following potential advantages: (i) Improve the detection accuracy because the neighborhoods are established between nodes potentially involved in malicious activity, (ii) track botnets because infected machines are often reused by botmasters, and (iii) stop attacks

(DDoS attacks and worm spreads, in particular) close to the source of the attack in an automated way and in real time. This strategy, however, may decrease initial sensitivity to novel attack patterns, and analysis into these trade-offs will be required.

3.3.6 Rogue Nodes

One potential concern for the practicality of any distributed system is that the network may contain rogue nodes that propagate false *levels of concern* for their own benefit. Though not discussed in detail here, a future enhancement to DIAMoND to address this potential risk is the addition of a "weight" feature that reflects the level of trust for each neighbor. Each node will define the weight values for each of their neighbors over time, depending on the received levels of concerns during past interactions with the neighbors. Critically, the slope of weight growth should be reasonably small relative to the loss as a deceptive neighbor may try to gain "weight" and then start to misbehave. The weight update function should vary the slope of growth/loss of level of concern as a function of the neighbor's current weight value. Details of the implementation and efficacy of this method are planned in immediate future work.

3.4 Evaluation Setup

3.4.1 Software Implementation

We have developed our prototype communication protocol as an OpenFlow controller in the POX environment [33]. The main reason for our selection is that leading network equipment manufacturers such as Cisco, Juniper, or HP have integrated the OpenFlow protocol in their products, which opens the way for a seamless transition between our simulations and a real-world operational implementation. We evaluate our switch controller using Mininet 2.0 network emulator [34]. Our initial software system deployment consists of 20 nodes due to computational constraints and up to 20 end-user machines connected to each of the nodes. We use three additional OpenFlow POX components: *forwarding.L2_learning* so that nodes can act as a type of L2 learning switch, *openflow.spanning_tree*, and *openflow.discovery* to build a view of the network in topologies with loops. DIAMoND is added as a security component to the POX controller. Our communication protocol is available to the public [9].

3.4.2 Physical Topologies

In order to explore the impact of network configuration on the success of the distributed detection of DIAMoND, we tested the performance of the algorithm on a variety of network topologies (each with a total network size of 20 nodes due to computational constraints). These included *full mesh*, *mesh*, *bipartite*, *linear*, and *extended star*, defined as follows:

- *Full mesh* is the complete undirected network on 20 nodes, involving $n(n-1)!/n = 190$ edges.
- *Mesh* is a randomly selected subgraph of Full Mesh in which the probability of each edge is 0.3, therefore involving 0.3 $n(n-1)!/n =$ 57 edges for each realization.
- *Bipartite* is the division of the network into two sets of size 10, with the probability of 0.6 for an edge between nodes from different sets and a probability of 0 for an edge between nodes in the same set, therefore also involving 0.6 $n(n-1)!/2n = 57$ edges for each realization.
- *Linear* is a single path of edges among the nodes, i.e., a connected graph in which 2 nodes have exactly 1 incident edge and all others have exactly 2 incident edges, therefore involving $(n - 1) = 19$ edges.
- *Extended star* is a tree, created by initiating the graph with 1 node and then attaching each subsequently created node to one of the already existing nodes with uniform probability until the network size reaches 20, therefore involving $(n-1) = 19$ edges.

3.4.3 Legitimate and Malicious Traffic

In our experimental evaluation, we use traffic captured from the trans-Pacific line (samplepoint-F, 150 Mbps) maintained by the MAWI working group [32].

We first use the traffic labeled by the MAWI group as anomalous or normal using an advanced graph-based method that combines responses from independent anomaly detectors built upon principal component analysis (PCA), the gamma distribution, the Kullback–Leibler divergence and the Hough transform [32]. Then, we develop our method based on *X-means* algorithm.

We implement an algorithm for detecting SYN flooding attacks and port scans [30, 31] using the python API provided by MAWILab-v1. We then calculate the number of outgoing SYN segments to incoming ACKs per source and per destination in 5-second intervals. We cluster the obtained traffic using the Weka's implementation of the *X-means* algorithm based on the Bayesian information criterion (BIC) to approximate the correct number of clusters [35]. We classify all the traffic clustered in the most numerous group with the lowest cluster center as legitimate traffic and classify the remaining ones as anomalous. Finally, we filter all traffic labeled as anomalous by each classification method and use available in our benchmark traffic generator.

We evaluate the capability of our system using two predominant attacks exploiting TCP protocol, namely SYN stealth scans and SYN flooding attacks. We use packet marking to distinguish between legitimate and malicious flows. In every scenario, we compromise a subset of users from randomly selected subnetworks and engage them in malicious activity. Finally, we launch two successive, staggered-over-time attacks to evaluate the potential for self-organizing anomaly detection and dynamic changes of neighborhoods based on repeated attacks.

Our immediate next steps include applying DIAMoND to real-world DNS traffic for botnet detection.

3.5 Emulation Results

3.5.1 Detection Accuracy

In addition to a detailed analysis of sensitivity and specificity defined as follows:

$$Sensitivity = \frac{TP}{TP + FN}, \quad Specificity = \frac{TN}{TN + FP}, \tag{3.3}$$

which is provided in a form of ROC diagrams, and where TP stands for *true positive*, TN for *true negative*, FP for *false positive*, and FN for *false negative*, we calculate the overall system accuracy denoted as follows:

$$Accuracy = \frac{TP + TN}{TP + FP + TN + FN} \tag{3.4}$$

We then quantify the additional information that is gained by deploying our system on top of benchmark local intrusion detectors (BLIDs) (i.e., we ask by how much, if at all, the inclusion of the DIAMoND collaboration among nodes improves their accuracy relative to their use of only the local detection algorithms in isolation). Finally, to evaluate the information gain, we use an information theoretic approach and a Kullback–Leibler (K–L) divergence:

$$D(P||Q) = K\text{–}L(P, Q) = \sum_{x \in X} P(x) \log_2 \frac{P(x)}{Q(x)} \tag{3.5}$$

This provides a way to compare the difference between two probability distributions $P(x)$ and $Q(x)$, where x is the number of accurately classified flows (represented as 5-tuples according to the source and destination IPs and ports, and transport layer protocol) as malicious and legitimate during a single test (which in our case is the entire set), $P(x)$ represents the overall accuracy of our system, and $Q(x)$ is the overall accuracy of BLIDs operating independently. It is important to recall that the potential for improvement in accuracy is scaled by the percent of malicious packets. Since in the case of stealth scans malicious packets constitute a smaller percentage of all network traffic, the increase in accuracy is strictly bounded, meaning that for example, 0.045 represents a substantial improvement relative to the range possible for improvement.

Figure 3.5 (left) depicts a *sensitivity* as a function of 1 – *specificity* for stealth scans in an extended star physical topology and an overlay network where neighborhoods are created on the basis of direct physical connections (TTL = 1). We present results for four test functions corresponding to different excitation levels. They reflect the impact of four levels of importance being assigned to the concern levels of neighbors. In other words, they describe the level of trust that a node places in its neighborhood. The graph indicates a great improvement in sensitivity between approximately 10% and 20% for *low*, *med*, and *med+*, *high* algorithms, respectively, without compromising specificity in comparison with BLID systems operating independently. The reason why $1 - specificity$ does not exceed 3.5% (in the worst case) is attributable to two reasons: (i) precise calibration of the rate-limiting sensitivity thresholds. For example, the consensus of level of concerns of neighbors cannot reduce the sensitivity threshold of a chosen node below $st_{i,\min}$, and (ii) level of concern of a node signals the anomaly, while the decision about assigning particular flows to *legitimate* or *malicious* classes remains with DU. Finally, though not presented in this chapter in consideration of space, the overall information gain of DIAMoND calculated over the accuracy of BLID is approximately twice as large for *med+* and *high* test functions as for *low* and *med* (between 0.022 and 0.047, cf. Table 3.1).

3.5.2 Impact of Physical Topologies

Let us now analyze the impact of physical topologies on the detection accuracy of DIAMoND, which has a direct impact on logical connections between participating nodes.

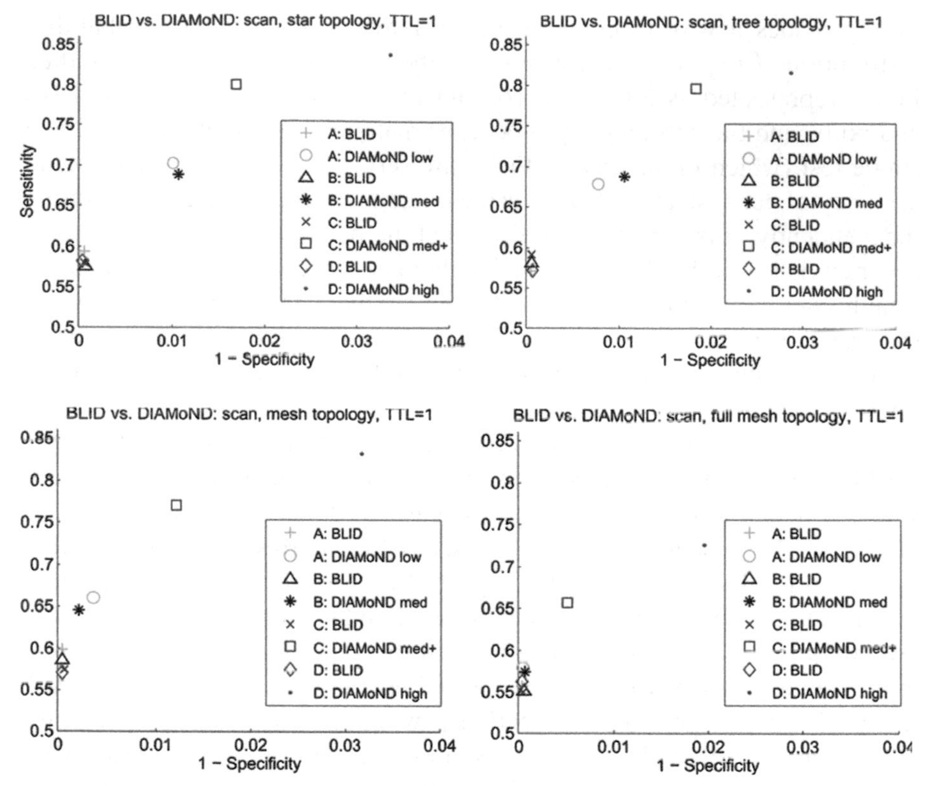

Figure 3.5 *Sensitivity* as a function of $1 - specificity$. Comparison of DIAMoND and BLID for stealth scans in the TTL = 1 neighborhood for different test functions (low, med, med+, and high) and for different topologies: star (left), tree (middle left), mesh (middle right), and full mesh (right).

We find that if there are no transit nodes and we assume a full mesh topology (cf. Section 3.4.2), then the detection accuracy is decreased significantly in comparison with other topologies (cf. Figure 3.5 and Table 3.1), especially in the case of the *low* and *med* excitation functions. This is because stealth scans are not as aggressive as DDoS attacks or worm propagation, so the great majority of nodes within a TTL = 1 neighborhood forming a full mesh topology will not report any suspicious behavior.

Our results also indicate that for aggressive DDoS attacks, the type of topology does not influence the accuracy (see Table 3.1). We observed that the information gain of the overlay detection system is lower (though always positive) in comparison with low-rate malicious activity, but the system can react close to the source of the attack more effectively and thereby reduce the collateral damage to minimum.

Table 3.1 $Sensitivity$, $1 - specificity$ (95% confidence interval), accuracy of BLID and DIAMoND, and the accuracy gain of DIAMoND over BLID for four test functions (A: low, B: med, C: med+, and D: high)

Test	Sensitivity		1 – Specificity		Accuracy		Gain
	BLID	DIAMoND	BLID	DIAMoND	BLID	DIAMoND	
Stealth scan, star topology, TTL = 1 neighborhood							
A	$0.594(\pm0.024)$	$0.702(\pm0.025)$	$6.4e^{-4}(\pm2.2e^{-4})$	$0.01(\pm0.002)$	0.893	0.914	0.022
B	$0.575(\pm0.023)$	$0.688(\pm0.023)$	$7.6e^{-4}(\pm2, 6e^{-4})$	$0.011(\pm0.002)$	0.889	0.911	0.022
C	$0.58(\pm0.02)$	$0.8(\pm0.015)$	$6.2e^{-4}(\pm1, 5e^{-4})$	$0.017(\pm0.003)$	0.889	0.935	0.047
D	$0.583(\pm0.022)$	$0.837(\pm0.013)$	$4.8e^{-4}(\pm6.8e^{-5})$	$0.034(\pm0.003)$	0.887	0.932	0.045
Stealth scan, tree topology, TTL = 1 neighborhood							
A	$0.57(\pm0.022)$	$0.679(\pm0.02)$	$5.7e^{-4}(\pm1.3e^{-4})$	$0.008(\pm0.002)$	0.893	0.915	0.021
B	$0.581(\pm0.021)$	$0.688(\pm0.02)$	$6.7e^{-4}(\pm1.9e^{-4})$	$0.011(\pm0.002)$	0.891	0.911	0.021
C	$0.592(\pm0.027)$	$0.797(\pm0.028)$	$6.7e^{-4}(\pm2e^{-4})$	$0.018(\pm0.003)$	0.891	0.932	0.042
D	$0.571(\pm0.025)$	$0.817(\pm0.024)$	$7.9e^{-4}(\pm3.8e^{-4})$	$0.029(\pm0.003)$	0.89	0.932	0.042
Stealth scan, mesh topology, TTL = 1 neighborhood							
A	$0.599(\pm0.026)$	$0.66(\pm0.028)$	$5.6e^{-4}(\pm1.5e^{-4})$	$0.004(\pm0.001)$	0.889	0.904	0.015
B	$0.586(\pm0.022)$	$0.645(\pm0.024)$	$5.5e^{-4}(\pm1.4e^{-4})$	$0.002(\pm6.6e^{-4})$	0.893	0.907	0.014
C	$0.574(\pm0.019)$	$0.771(\pm0.016)$	$7.2e^{-4}(\pm2.2e^{-4})$	$0.012(\pm0.003)$	0.89	0.932	0.043
D	$0.571(\pm0.02)$	$0.833(\pm0.018)$	$5.5e^{-4}(\pm1.5e^{-4})$	$0.032(\pm0.003)$	0.889	0.939	0.045
Stealth scan, full mesh topology, TTL = 1 neighborhood							
A	$0.557(\pm0.014)$	$0.578(\pm0.016)$	$4.8e^{-4}(\pm1.1e^{-4})$	$5.2e^{-4}(\pm1.3e^{-4})$	0.881	0.887	0.006
B	$0.556(\pm0.019)$	$0.574(\pm0.021)$	$6.9e^{-4}(\pm2.2e^{-4})$	$7.5e^{-4}(\pm2.5e^{-4})$	0.884	0.89	0.006
C	$0.551(\pm0.015)$	$0.657(\pm0.023)$	$5.7e^{-4}(\pm2e^{-4})$	$0.005(\pm0.002)$	0.884	0.908	0.024
D	$0.562(\pm0.016)$	$0.725(\pm0.029)$	$5e^{-4}(\pm1.1e^{-4})$	$0.02(\pm0.004)$	0.882	0.912	0.03
DDoS attack, star topology, TTL = 1 neighborhood							
A	$0.922(\pm0.023)$	$0.938(\pm0.02)$	$0.004(\pm6.6e^{-4})$	$0.018(\pm0.002)$	0.95	0.955	0.004
B	$0.92(\pm0.016)$	$0.937(\pm0.016)$	$0.004(\pm6.5e^{-4})$	$0.018(\pm0.003)$	0.95	0.955	0.005
C	$0.923(\pm0.012)$	$0.962(\pm0.01)$	$0.005(\pm7.3e^{-4})$	$0.032(\pm0.004)$	0.95	0.964	0.014
D	$0.922(\pm0.03)$	$0.967(\pm0.024)$	$0.004(\pm7e^{-4})$	$0.038(\pm0.004)$	0.95	0.965	0.014

(*Continued*)

Table 3.1 Continued

Test	Sensitivity		1 – Specificity		Accuracy		Gain
	BLID	DIAMoND	BLID	DIAMoND	BLID	DIAMoND	
			DDoS attack, full mesh topology, TTL = 1 neighborhood				
A	0.922(±0.023)	0.932(±0.023)	$0.003(\pm 6.2e^{-4})$	0.008(±0.001)	0.949	0.953	0.005
B	0.922(±0.015)	0.931(±0.015)	$0.004(\pm 5.9e^{-4})$	0.007(±0.001)	0.95	0.954	0.004
C	0.921(±0.018)	0.951(±0.018)	$0.004(\pm 4.8e^{-4})$	0.017(±0.002)	0.949	0.963	0.014
D	0.922(±0.019)	0.959(±0.018)	$0.004(\pm 8e^{-4})$	0.028(±0.005)	0.949	0.963	0.015
			Stealth scan, star topology, TTL = 2 neighborhood				
A	0.575(±0.029)	0.684(±0.032)	$6e^{-4}(\pm 1.3e^{-4})$	0.008(±0.002)	0.89	0.912	0.022
B	0.565(±0.022)	0.676(±0.024)	$6.4e^{-4}(\pm 1.3e^{-4})$	0.01(±0.002)	0.889	0.91	0.022
C	0.557(±0.021)	0.787(±0.021)	$7.5e^{-4}(\pm 5.2e^{-4})$	0.019(±0.003)	0.889	0.932	0.045
D	0.561(±0.025)	0.843(±0.014)	$5.3e^{-4}(\pm 1.4e^{-4})$	0.032(±0.003)	0.887	0.936	0.05
			Stealth scan, star topology, TTL = 3 neighborhood				
A	0.599(±0.027)	0.701(±0.028)	$7.3e^{-4}(\pm 1.7e^{-4})$	0.009(±0.002)	0.887	0.909	0.023
B	0.584(±0.024)	0.68(±0.026)	$8.4e^{-4}(\pm 3.1e^{-4})$	0.01(±0.003)	0.89	0.908	0.018
C	0.568(±0.029)	0.793(±0.029)	$6.1e^{-4}(\pm 1.7e^{-4})$	0.02(±0.003)	0.887	0.932	0.045
D	0.578(±0.025)	0.851(±0.015)	$6.7e^{-4}(\pm 1.6e^{-4})$	0.036(±0.003)	0.889	0.935	0.047
			Stealth scan, star topology, attack correlation neighborhood				
A	0.535(±0.022)	0.658(±0.025)	$6.69e^{-4}(\pm 1.7e^{-4})$	0.013(±0.002)	0.889	0.91	0.021
B	0.563(±0.02)	0.702(±0.022)	$9.57e^{-4}(\pm 5.8e^{-4})$	0.016(±0.003)	0.888	0.914	0.027
C	0.528(±0.027)	0.752(±0.027)	$5.55e^{-4}(\pm 1.3e^{-4})$	0.02(±0.003)	0.891	0.931	0.041
D	0.531(±0.023)	0.787(±0.022)	$5.8e^{-4}(\pm 1.6e^{-4})$	0.04(±0.004)	0.891	0.924	0.034
			Stealth scan, full mesh topology, attack correlation neighborhood				
A	0.533(±0.017)	0.586(±0.021)	$4.5e^{-4}(\pm 9.9e^{-5})$	0.007(±0.001)	0.886	0.895	0.01
B	0.528(±0.015)	0.595(±0.018)	$8.2e^{-4}(\pm 3.4e^{-4})$	0.01(±0.002)	0.887	0.896	0.01
C	0.522(±0.015)	0.672(±0.017)	$5.7e^{-4}(\pm 2.6e^{-4})$	0.01(±0.002)	0.881	0.911	0.031
D	0.532(±0.014)	0.739(±0.012)	$5.4e^{-4}(\pm 1.1e^{-4})$	0.03(±0.002)	0.886	0.913	0.027

3.5.3 Influence of Neighborhood Strategies

In this section, we first report on the basic strategy where each node communicates its level of concern with its direct physical or logical neighbors (TTL = 1) and compare it with results for extended neighborhoods (TTL = 2 and TTL = 3) for stealth scans and a star topology (cf. Figure 3.5 left and Figure 3.6, respectively, and Table 3.1). The comparison of three strategies provides quite interesting, though expected results. We observed no major distinction in the detection accuracy and the information gain. This is because nodes announce their levels of concerns instead of threat levels which, in this particular scenario, result in a very similar information flow, regardless of the neighborhood hop limit.

Furthermore, an evaluation of sensitivity, specificity, and information gain for dynamic, self-organizing neighborhoods based on repeated, previously detected intrusions (compare Figure 3.7 with Figure 3.5 left and right and Table 3.1) shows a significant improvement in the case of a full mesh topology. Overall, the primary results highlight the importance of different communication strategies and interesting characteristics of the information flow that is distinctive to DIAMoND.

3.5.4 Minimal and Marginal Deployment Gain

Deployment of networked services across administrative boundaries usually has to take place progressively. In this section, we studied how deployment percentages affect performance of a DIAMoND system. In particular, we tried

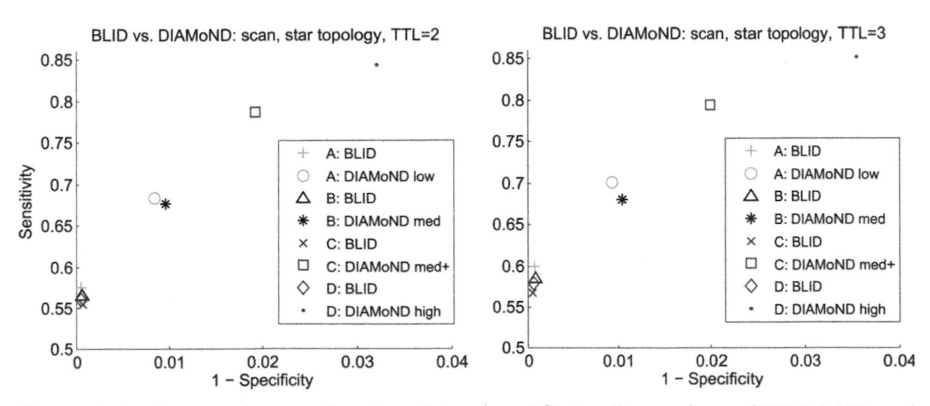

Figure 3.6 *Sensitivity* as a function of $1 - specificity$. Comparison of DIAMoND and BLID for stealth scans in the TTL = 2 and TTL = 3 neighborhoods, for four test functions (low, med, med+, and high) in the star topology.

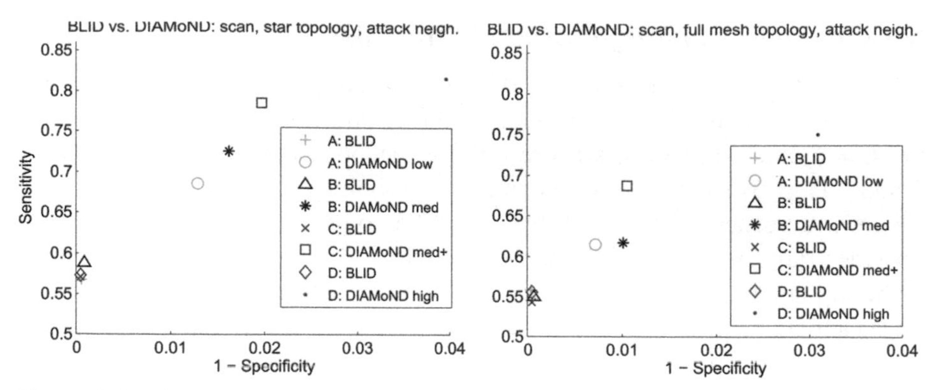

Figure 3.7 *Sensitivity* as a function of $1 - specificity$. Comparison of DIAMoND and BLID for stealth scans in the *attack* neighborhoods for four test functions (low, med, med+, and high) in the star and full mesh topologies.

to understand the minimal deployment percentage needed for DIAMoND to have significant performance impact and the subsequent marginal performance gain achieved with additional deployment.

To quantitatively evaluate the deployment gain, we adapted a calculation of "off-line marginal utility" originally proposed to analyze the impact of additional metrics [36], to instead compute the incremental information gain, U, for each additional node (relative to the information achieved with the BLID system) as follows:

$$U(x^n) = \sum_{x \in X} P(x^n) \log_2 \frac{P(x^n)}{Q(x)}, \quad Q(x) \equiv P(x^m), \quad m = 20$$

Figure 3.8 provides an example analysis of the deployment gain for a 20-node star network under port scan probing. This figure clearly showed a point of diminishing return such that after 30% of the nodes participate in the DIAMoND system, the information gain is close to that achieved when all nodes are participating and the marginal deployment gain from increasing participation is insignificant. On the other side, even when there are only 10% nodes (2 nodes) participating, the information gain is already over 0.01. When 20% nodes are participating, the information gain reached a significant value 0.03. We thus concluded that in this case, (i) minimal effective deployment is 10% of the network nodes participating, (ii) marginal gain is maximized at 20% deployment, and (iii) DIAMoND plateaus after 30% deployment, with a minimal value gained by having additional nodes participating.

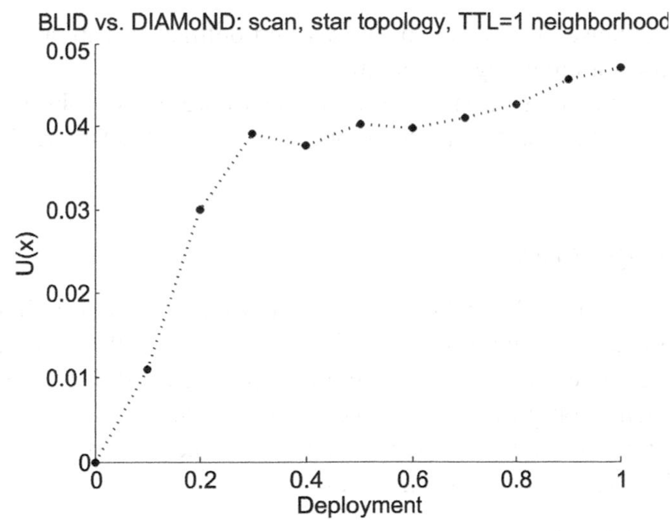

Figure 3.8 Information gain of DIAMoND over BLID as a function of the percentage of system deployment for scan activity in the TTL = 1 neighborhood and the star topology.

3.6 Conclusions

In this chapter, we proposed DIAMoND: a nonparametric distributed coordination framework for network intrusion detection. To illustrate its application, we coupled DIAMoND with local anomaly detection schemes for stealthy port scan and SYN flooding-based DDoS and evaluated its performance on an emulation test bed. DIAMoND consistently demonstrated up to 20% sensitivity enhancement without sacrificing specificity over a wide range of coordination network topologies. In this chapter, we also systematically investigated several automated coordination neighborhood construction strategies and found that DIAMoND exhibits stable performance gain over neighborhoods of TTL = 1, 2, 3. This leads us to conclude that DIAMoND is robust to neighborhood size. We further conclude from our experiments that DIAMoND has no strong correlation with network topologies either. Deployment impact showed that DIAMoND quickly reaches information gain plateau after 30% of network nodes are participating in coordination, which enhances the deployability of DIAMoND.

DIAMoND allows multiple entities, who may be functionally and/or legally prohibited from sharing cyber data, to leverage each other's insight and increase their effectiveness at cyber defense. Further, DIAMoND enables real-time adaptation, eliminating the identification-designed response delay inherent in defenses that react to known and defined threats and allowing active defense for emerging, novel attacks.

As our immediate next step, we plan to explore the scalability of DIAMoND coordination protocol and apply it to a broad set of real network topologies.

Acknowledgments

The US Department of Homeland Security sponsored this research under the Air Force Research Laboratory (AFRL) agreement number FA8750-12-2-0232. The US government is authorized to reproduce and distribute reprints for governmental purposes. This work represents the views of the authors and does not represent official policies or endorsements, either expressed or implied, of AFRL or the US government.

References

[1] Wang, P., Wu, L., Aslam, B., and Zou, C. C. (2009). A Systematic Study on Peer-to-Peer Botnets. In *Proceedings of ICCCN*. IEEE Computer Society, pp. 1–8.

[2] Gallagher, S. (2013). Microsoft Disrupts Botnet That Generated $2.7 M per Month for Operators, http://arstechnica.com, December.

[3] MITRE, Structured Threat Information eXpression: A Structured Language for Cyber Threat Intelligence Information. [Online]. Available: https://stix.mitre.org

[4] MITRE, Trusted Automated eXchange of Indicator Information: Enabling Cyber Threat Information Exchange. [Online]. Available: https://taxii.mitre.org

[5] Melancon, D. (2014). ISACs: Let the Sharing Begin, [Online]. Available: http://www.tripwire.com

[6] Fonash, P. (2014). Using Automated Cyber Threat Exchange to Turn the Tide against DDoS, [Online]. Available: http://rsaconference.com

[7] Serrano, O., Dandurand, L., and Brown, S. (2014). On the Design of a Cyber Security Data Sharing System. In *Proceedings of the ACM Workshop on Information Sharing & Collaborative Security*. ACM, pp. 61–69.

[8] Belmega, E., Sankar, L., Poor, H., and Debbah, M. (2012). Pricing Mechanisms for Cooperative state Estimation. In *Proceedings of IEEE ISCCSP*, May, pp. 1–4.

[9] Korczyński, M. (2015). Simulation Testbed of DIAMoND, http://mkorczynski.com/diamond.html, January.

[10] Ullrich, J. (2000). DShield: Internet Storm Center – Internet Security, http://www.dshield.org, November.

[11] Vigna, G., and Kemmerer, R. A. (1998). NetSTAT: A Network-Based Intrusion Detection Approach. In *Proceedings of ACSAC*. IEEE Computer Society, p. 25.

[12] Porras, P. A., and Neumann, P. G. (1997). EMERALD: Event Monitoring Enabling Responses to Anomalous Live Disturbances. In *Proceedings of NISSC*, pp. 353–365.

[13] Li, J., Lim, D.-Y., and Sollins, K. (2007). Dependency-based Distributed Intrusion Detection. In *Proceedings of DETER*. USENIX Association.

[14] Zhou, C., Karunasekera, S., and Leckie, C. (2005). A Peer-to-Peer Collaborative Intrusion Detection System. In *Proceedings of the IEEE 7th Malaysia International Conference on Communication*, vol. 1, November.

[15] Chhabra, P., Scott, C., Kolaczyk, E. D., and Crovella, M. (2008). Distributed Spatial Anomaly Detection. In *Proceedings of INFOCOM*. IEEE, pp. 1705–1713.

[16] Locasto, M., Parekh, J. J., Keromytis, A. D., and Stolfo, S. J. (2005). Towards Collaborative Security and P2P Intrusion Detection. In *Proceedings of the IEEE Information Assurance Workshop (IAW)*, pp. 333–339.

[17] Zhou, C. V., Karunasekera, S., and Leckie, C. (2005). A Peer-to-peer Collaborative Intrusion Detection System. In *Proceedings of IEEE ICON*, vol. 1.

[18] Yegneswaran, V., Barford, P., and Jha, S. (2004). Global Intrusion Detection in the DOMINO Overlay System. In *Proceedings of NDSS*.

[19] Dash, D., Kveton, B., Agosta, J. M., Schooler, E., Chandrashekar, J., Bachrach, A., and Newman, A. (2006). When Gossip is Good: Distributed Probabilistic Inference for Detection of Slow Network Intrusions. In *Proceedings of NISSC*, pp. 1115–1122.

[20] Fung, C. J., Zhu, Q., Boutaba, R., and Basar, T. (2011). Poster: SMURFEN: A Rule Sharing Collaborative Intrusion Detection Network, in *ACM CCS*. ACM, pp. 761–764.

[21] Boggs, N., Hiremagalore, S., Stavrou, A., and Stolfo, S. J. (2011). Cross-Domain Collaborative Anomaly Detection: So Far Yet So Close. In *Proceedings of RAID*. Springer-Verlag, pp. 142–160.

[22] Zhu, Q., Fung, C., Boutaba, R., and Basar, T. (2010). A Distributed Sequential Algorithm for Collaborative Intrusion Detection Networks. In *Proceedings of IEEE ICC*, May, pp. 1–6.

[23] Korczyński, M., Hamieh, A., Huh, J. H., Holm, H., Rajagopalan, S. R., and Fefferman, N. H. (2016). Hive oversight for network intrusion early warning using DIAMoND: a bee-inspired method for fully distributed cyber defense, *Commun. Mag., IEEE.*

[24] Grochocki, D., Huh, J. H., Berthier, R., Bobba, R., Sanders, W. H., Cardenas, A. A., and Jetcheva, J. G. (2012). AMI Threats, Intrusion Detection Requirements and Deployment Recommendations. In *Proceedings of IEEE SmartGridComm*. IEEE, pp. 395–400.

[25] Robinson, M., Mirkovic, J., Michel, S., Schnaider, M., and Reiher, P. (2003). DefCOM: Defensive Cooperative Overlay Mesh. In *Proceedings of the DARPA Information Survivability Conference and Exposition*, vol. 2, pp. 101–102.

[26] Zhang, Y., Wang, L., Sun, W., Green II, R. C., and Alam, M. (2011). Distributed intrusion detection system in a multi-layer network architecture of smart grids, *IEEE Trans. Smart Grid*, 2(4), 796–808.

[27] Bremler-bar, A. and Levy, H. (2005). Spoofing Prevention Method. In *Proceedings of IEEE INFOCOM*, March, pp. 536–547.

[28] Snoeren, A. C., Partridge, C., Sanchez, L. A., Jones, C. E., Tchakountio, F., Kent, S. T., and Strayer, W. T. (2001). Hash-Based IP Traceback. In *Proceedings of ACM SIGCOMM*, August.

[29] Ferguson, P., and Senie, D. (2000). Network Ingress Filtering: Defeating Denial of Service Attacks Which Employ IP Source Address Spoofing, RFC 2267.

[30] Korczyński, M., Janowski, L., and Duda, A. (2011). An Accurate Sampling Scheme for Detecting SYN Flooding Attacks and Portscans. In *Proceedings of IEEE ICC*, June, pp. 1–5.

[31] Korczyński, M. (2012). Classifying Application Flows and Intrusion Detection in the Internet Traffic, Ph.D. Dissertation, École Doctorale Mathématiques, Sciences et Technologies de l'Information, Informatique (EDMSTII), Grenoble, France, November.

[32] Fontugne, R., Borgnat, P., Abry, P., and Fukuda, K. (2010). MAWILab: Combining Diverse Anomaly Detectors for Automated Anomaly Labeling and Performance Benchmarking. In *Proceedings of CoNEXT*. ACM, pp. 1–12.

[33] Open Networking Lab, "POX," https://openflow.stanford.edu, 2014.

[34] Lantz, B., Heller, B., and McKeown, N. (2010). A Network in a Laptop: Rapid Prototyping for Software-defined Networks. In *Proceedings of the 9th ACM SIGCOMM Workshop on Hot Topics in Networks*. ACM.

[35] Hall, M., Frank, E., Holmes, G., Pfahringer, B., Reutemann, P., and Witten, I. H. (2009). The WEKA data mining software: an update, *SIGKDD Explor. Newsl.*, 11(1), 10–18.

[36] Barford, P., Bestavros, A., Byers, J., and Crovella, M. (2001). On the Marginal Utility of Network Topology Measurements. In *Proceedings of the 1st ACM SIGCOMM Workshop on Internet Measurement*. ACM, pp. 5–17.

4

Detection of Service Level Agreement (SLA) Violations in Memory Management in Virtual Machines

Xiongwei Xie[1], Weichao Wang[1] and Tuanfa Qin[2]

[1]Dept. of Software and Information System, UNC Charlotte, Charlotte, NC 28223, USA
[2]School of Computer and Electronic Information, Guangxi University, Nanning, Guangxi, China

Abstract

In cloud computing, quality of services is often enforced through service level agreement (SLA) between end users and cloud providers. While SLAs on hardware resources such as CPU cycles or network bandwidth can be monitored by low layer sensors, the enforcement of security SLAs at memory management level stays a very challenging problem. Several high-level architectures for such SLAs have been proposed. However, details of implementation still need to be filled before they can be deployed.

In this chapter, we propose to design mechanisms to detect violations of SLAs at the memory management level. Specifically, we propose to measure the changes in memory page access orders and data access delay in order to detect three types of SLA violations in virtual machines: (1) unauthorized accesses to memory pages of a virtual machine, (2) violation of the memory deduplication policies, and (3) under-allocation of memory to virtual machines. Through measuring the accumulated memory access latency, we try to derive out whether or not the memory pages have been swapped out and the order of accesses to them has been changed. These events will then be compared to access commands issued by the local VM. In this way, violations of the SLA in memory management can be detected. Compared to existing approaches, our mechanisms do not need explicit help from the hypervisor or third parties. Therefore, it can detect SLA violations even when they are initiated by the hypervisor. We implement our approaches under VMWare with Windows virtual machines. Our experimental results show that the VM can effectively detect the violations with small increases in overhead.

Keywords: Security service level agreement, Memory management, Virtual machines, Attack detection.

4.1 Introduction

In the past few years, cloud computing has attracted a lot of research efforts. At the same time, more and more companies start to move their data and operations to public or private clouds. For example, out of 572 business and technology executives that were surveyed in [1], 57% believed that cloud capability could improve business competitive and cost advantages, and 51% relied on cloud computing for business model innovation. These demands also become a driving force for the development of cloud security, which ranges from very theoretical efforts such as homomorphic encryption to very engineering mechanisms defending against side-channel attacks through memory and cache sharing.

In parallel with the active research in cloud security, enforcement of service level agreement (SLA) also becomes a very hot topic. In cloud computing, customers usually need to outsource their data processing or storage to service providers. To guarantee the system performance and data safety, many customers rely on service level agreement (SLA) with the providers to enforce such properties. The resources that are monitored under SLAs include CPU time [2–5], network downtime [5, 6], and bandwidth [2, 5]. Several multi-layer monitoring structures [4, 6–10] have been proposed to link low-level resources with high-level SLA requirements.

Compared to the research in SLA enforcement for QoS parameters, investigation into security SLA validation falls behind in many aspects. For example, in [11] the authors define the concept of an accountable cloud and propose an approach to differentiate the responsibility of a user from that of the service provider when some security breach happens. An infrastructure to enforce security SLA is described in [12]. However, the high-level discussion often lacks implementation details. It is very hard to generate concrete defense mechanisms for detecting SLA violations based on these descriptions.

To bridge this gap, in this chapter, we study mechanisms to detect violations of security SLA for memory management in virtual machines. Under many conditions, end users of a cloud environment may sign some agreement with the cloud provider on the usage and monitoring of the memory of their virtual machines. For example, many prominent virtual machine hypervisors such as VMware ESX and ESXi [13], Extended Xen [14], and KSM (Kernel Samepage Merging) [15] of the Linux kernel use the technique of memory deduplication to reduce memory footprint size of virtual machines. Since previous research has shown that page-level memory sharing could create a side channel for information leakage [16–22], many end users ask the hypervisor to disable memory deduplication for their VMs. However, there exists no solution for end users to verify the execution of this agreement other than trusting the words of the cloud provider.

As another example, cloud providers usually have the privilege to take a sneak peek at the memory pages of the VMs under their management and reconstruct their internal views [23–26]. To protect their own privacy, end users could sign an SLA with the provider that prevents it from peeking at the memory pages without their permission. However, if no technical mechanism can detect violations of such an SLA, it cannot be enforced.

Last but not least, because of the sharing property of cloud computing, memory overcommitment is a widely used technique by hypervisors [27–30]. During the initiation of a virtual machine, a user can request the size of RAM and also identify the minimum physical memory that the hypervisor needs to guarantee. However, if no violation detection mechanism is implemented, a greedy hypervisor may cut the allocated memory to the VM. Previous research [31] has shown that when a VM gets too little physical memory, its performance can be severely impacted since the CPU will either be busy in handling page swapping or use most of time waiting for data to be loaded into the system.

To detect such violations, we propose to design mechanisms based on memory access latency. When we revisit the three types of violations described above, we find out that all attacks described above would lead to changes in access orders to the memory pages [18, 32, 33]. For the attack on memory deduplication, although the other VM is accessing only its own pages, the victim VM is impacted because of the shared memory. For the attack of unauthorized peek, the attacker violates the security aspect of SLA and reads the memory of the victim VM. For the memory under-allocation violations, the reduced physical memory resources to a VM will lead to extra swapping. Under all cases, the order of accesses to the VM's memory pages changes. Our detection mechanisms will try to capture such changes.

According to the documents released by major hypervisor companies such as VMWare and research results of other investigators [34–36], Least Recently Used (LRU) memory pages are still the best choices during memory reclaiming. Therefore, unauthorized accesses to the memory pages of the victim VM will lower their priority of being swapped out. We propose to introduce a group of reference pages into the virtual machine memory and access them with different time intervals. In this way, we can set up a series of reference points in time for memory swapping operations. Through comparing the access latency to these reference pages with that to the pages we try to monitor, we can determine which pages are still in the memory and which pages have been swapped out. Since the reference pages are hidden within the real data and program pages, it is very hard for the attacker to identify them and treat them differently. We have implemented the approaches in virtual machines under VMWare and tested them. Our experimental results show that the approaches can effectively detect the violations to SLAs in memory management with small increases in overhead.

The contributions of the chapter are as follows. First, existing SLA enforcement mechanisms usually focus on hardware resources such as CPU cycles and network bandwidth. Our approaches study this problem from a different aspect and try to enforce security SLAs. Second, in this research we choose SLA violations to memory management in virtual machines as an example. We design mechanisms to detect such attacks based on changes in memory access latency. Last but not least, we implement the approaches and evaluate them in real systems. The experimental results show the effectiveness of the approaches.

The remainder of the chapter is organized as follows. In Section 4.2, we introduce the related work. Specifically we focus on the enforcement of SLAs and information leakage through changes in memory and cache access latency. In Section 4.3, we present the details of our approaches. In Section 4.4, we present the experimental results in real systems. In Section 4.5, we discuss the safety and efficiency of the approaches. Finally, Section 4.6 concludes the chapter.

4.2 Related Work

4.2.1 Information Leakage among Virtual Machines

With the proliferation of cloud computing, more and more companies start to use it. One big security concern in cloud is the coresidence of multiple virtual machines belonging to different owners in the same physical box. Existing investigation into this problem can be classified into two groups. In the first group, side-channel attacks through shared cache have been carefully studied. The shared cache enables timing-based attacks [20, 21, 37, 38] to steal information about key stroke or Internet surfing histories. Research in [39] shows that in a practical environment such as Amazon EC2, a cache-based covert channel can reach the effective bandwidth of tens of bits per second. An implementation of the key extraction attack was presented in [40]. More recently, researchers have shown that even when multiple virtual machines are put on different cores of a CPU, cache-level information stealing is still a viable attack [41]. The delay caused by separation of deduplicated memory pages in virtual machines has been used to identify guest OS types [16] or derive out memory page contents [42].

In the second group, researchers have designed security mechanisms to prevent information leakage among the virtual machines. In hardware-based approaches, special components have been embedded into the architecture to manage information flow. For example, in [43] the processor is responsible for updating the memory page mapping and the page table. Szefer et al. [44] proposed to use memory-level isolation or encryption to protect guest VMs from a compromised hypervisor.

The software-based approaches adopt more diverse mechanisms. For example, in [45] the cache that is used by security-related processes is labeled with different colors to prevent side-channel attacks. In both [46, 47], the hypervisor monitors the memory access procedures to prevent cross-VM information flow. In [48], researchers tried to establish a very small compartment to allow VMs to run in an isolated state. Using lightweight run-time introspection, Baig et al. identify side channels which could potentially be used to violate a security policy, and reactively migrate virtual machines to eliminate node-level side channels [49].

4.2.2 Service Level Agreement Enforcement

In cloud computing environments, service level agreements are often described at the business level. Their enforcement, however, is often at the technical level. Such discrepancy creates a challenge for researchers. Two groups of approaches have been designed to solve this problem. In group one, a middleware layer is implemented to bridge the high-level service requirements with low-level hardware resources [2, 3, 6]. Group two pushes the approaches one step further through formalization of both service capabilities and business process requirements [7]. In this way, a language can be used for direct communication between the two layers. Dastjerdi et al. [50] designed an ontology-based approach that can capture not only changes in individual resources, but also their dependency.

Compared to research in SLA enforcement for QoS parameters, investigation into security SLA validation deserves more efforts. As a pioneer, Henning [51] raised the question of whether or not security can be adequately expressed in an SLA. Casola et al. [52] proposed a methodology to evaluate and compare security SLAs in Web services. Chaves et al. [53] explored security management through SLAs in cloud computing. An infrastructure to enforce security SLA in cloud services is described in [12]. Haeberlen [11] proposed an approach to differentiate the responsibility of a user from that of the service provider when some security breach happens. Efforts have also been made to quantify the security properties in cloud computing environments so that automated and continuous certification of security properties could be implemented [54, 55].

4.3 The Proposed Approaches

In this section, we will present the details of the proposed approaches. We first introduce the techniques of memory overcommitment and deduplication and how they impact memory access in virtualization environments. We will then discuss the assumptions of the environments to which our approaches can be applied. Finally, the details of the approaches and mechanisms to turn the idea into practical solutions are presented.

4.3.1 Memory Overcommitment in Virtualization Environments

Virtualization enables service providers to consolidate virtual hardware on less physical resources. The consolidation ratio is an important measure of the virtualization efficiency. To boost up system performance, overcommitment has been adopted to enable the allocation of more virtual resources than available physical resources. For example, two virtual machines with 4-GB RAM each could be powered on in a VMWare ESXi server with only 4-GB physical memory. To support smooth operations of the two virtual machines, various techniques such as page sharing, memory ballooning, data compression, and hot swapping have been designed [27]. Almost all prominent hypervisors use some type of memory overcommitment [56]. For example, both DiffEngine [14] and Singleton [57] have tried to use page deduplication to reduce memory footprint of virtual machines. Ginkgo [58] implements a hypervisor-independent overcommitment framework to adjust CPU and memory resource allocations in virtual machines jointly.

Among different techniques for memory overcommitment, memory ballooning is an active method for reclaiming idle memory from virtual machines. If a VM has consumed some memory pages, but is not subsequently using them in an active manner, VMWare ESXi attempts to reclaim them from the VM using ballooning. Here an OS-specific balloon driver inside the virtual machine will first request memory from the OS kernel. Once granted, it will transfer the control of the memory to ESXi, which is then free to reallocate it to another VM. The whole procedure includes both inflation (getting idle memory from the first VM) and deflation (giving memory to the second VM) of the balloon. An example is shown in Figure 4.1.

Although memory overcommitment can improve the consolidation ratio in virtualization, it also has the potential to severely impact system performance. For example, the set of memory pages that are being actively accessed by a VM is called the working set. Previous research [27] shows that when the total working set of VMs remains within the physical memory limit, the VMs will not experience significant performance loss. On the contrary, when the overcommitment factor becomes larger than 2, the operations per minute (OPM) of a VM can reduce 17% to 200%, depending on the properties of the applications.

4.3.2 Memory Deduplication in VM Hypervisors

The memory deduplication technique takes advantage of the similarity among memory pages so that only a single copy and multiple handles need to be preserved in the physical memory, as shown in Figure 4.2. Here each of the two virtual machines *VM1* and *VM2* needs to use three memory pages. Under the normal condition, six physical pages will be occupied by the VMs. If memory deduplication is enabled, we need to store only one copy of multiple identical pages. Therefore, the two VMs can be fit into four physical pages

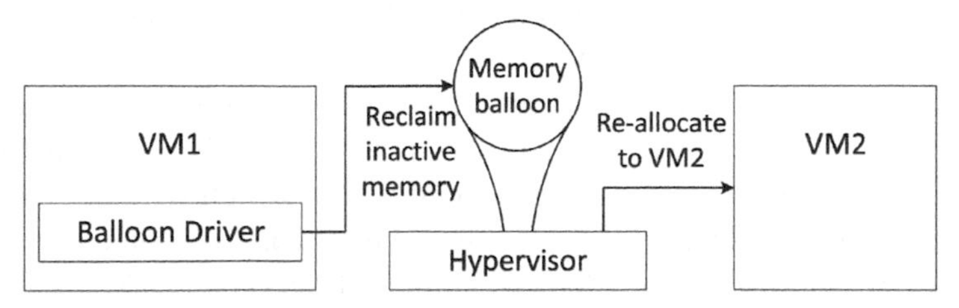

Figure 4.1 One technique of overcommitment: memory ballooning.

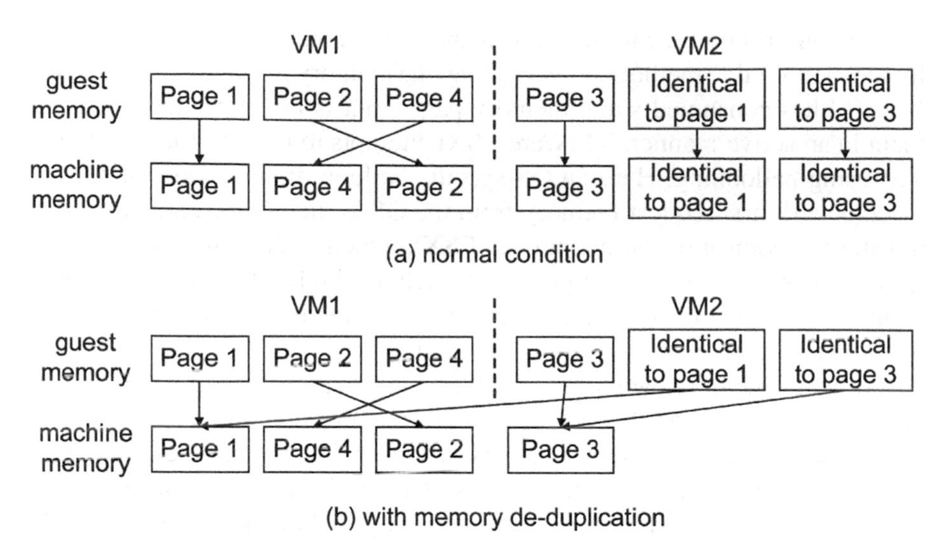

Figure 4.2 Memory deduplication reduces the memory footprint size.

(note that we illustrate both inter- and intra-VM memory deduplication in the figure). This technique can reduce the memory footprint size of VMs and the performance penalty caused by memory access misses.

Although the implementation of memory deduplication in different hypervisors may be different, the basic idea is similar. To avoid unnecessary delay during page loading, whenever a new memory page is read from the hard disk, the hypervisor will allocate a new physical page for it. Later, the hypervisor will use idle CPU cycles to locate the identical memory pages in physical RAM and remove duplicates by leaving pointers for each VM to access the same memory block. Hash results of the memory page contents are used as index values to locate identical pages. To avoid false deduplication caused by hash collisions, a byte-by-byte comparison between the pages will be conducted. While the reading operations to the deduplicated pages will access the same copy, copy-on-write is used to prevent one VM from changing another VM's memory pages. Specifically, on writing operations a new page will first be allocated and copied. This procedure will incur extra overhead compared to writing to not-shared pages, which will lead to a measurable delay when a large number of shared pages are allocated and copied. This delay will allow us to detect violations of security SLAs on memory management.

4.3.3 System Assumptions

In the investigated scenarios, we assume that the security SLA signed between the cloud provider and customers includes the following three requirements: (1) The provider cannot apply memory deduplication technique to the memory pages of the guest VMs; (2) without a customer's permission, the provider cannot peek at the memory pages of her VM, and (3) the provider needs to guarantee the minimum size of allocated physical memory to the VM. We assume that a customer has total control over the memory usage of her virtual machine. For example, she can initiate application programs and load data files into the memory of the VM. She can also measure time durations accurately so that they can be used for attack detection (more discussion in subsequent sections). We do not assume the customer can decide how much physical RAM her VM can consume. Without losing generality, we assume that VMs use 4-KB memory pages.

We assume that the cloud service provider is not malicious but very curious. At the same time, it wants to squeeze as many virtual machines as possible into a single physical box to maximize its profits. From this point of view, it has the motivation to enable memory deduplication or allocate less memory for VMs. The cloud service provider may not be the developer of the hypervisor software. For example, a private cloud provider runs VMWare software to manage the cloud. It does not have the capability to alter the source code of the hypervisor. However, it may configure the software to enable memory deduplication. It may also read the memory page contents of other VMs through the virtual machine manager. All such operations can be conducted without the permission of end users.

4.3.4 Basic Ideas of the Proposed Approaches

In this part, we will first introduce the basic ideas of the approaches. The difficulties to turn these ideas into practical mechanisms and our solutions to these problems will then be discussed.

The basic idea of the proposed detection mechanisms is to map violations of security SLAs on memory management to changes in memory access latency. If a memory page is located in physical memory, its access delay is at the level of several to tens of micro-seconds. On the contrary, if a memory page has been swapped out, its access delay is partially determined by the hard disk performance. The waiting time is usually at the level of milliseconds. From this point of view, we can easily differentiate a page in physical memory from that on hard disk.

Now let us re-examine the three violations of interest. When the hypervisor or an attacker takes a sneak peek at memory pages of a guest VM, it will change the sequence of access operations, thus impacting the priority of page swapping. We can choose a group of memory pages to serve as references and keep records of access operations to them. If we detect that the order of memory swapping is different from that of the memory access commands initiated by the guest VM, we can confirm that some unauthorized access has happened. Since it is often difficult to predict the access order to real data pages, we propose to insert guardian pages into data files to serve as the comparison landmark. More details will be described in Sections 4.3.5 and 4.3.6.

Similar technique can be applied to detect the violation of memory deduplication policies. A user can load two files (we call them *F1* and *F2*) with the same contents into her VM memory. If the hypervisor does not enable memory deduplication, the files will occupy different chunks of memory. Otherwise, their memory pages will be merged. To differentiate between these two cases, we need to conduct the following operations. We will access the pages of *F1* and *F2* regularly to keep them in the main memory. We can estimate the progress of deduplication based on our previous research [16]. When this procedure is finished, we can initiate "writing" operations to the pages. If the memory pages of the two files are not merged, the writing delay will be relatively short. On the contrary, if their pages are deduplicated, "copy-on-write" must be conducted for every page. A measurable increase in delay can be detected. Based on the measurement results, we can figure out whether or not the SLA for memory deduplication has been violated.

Determination of the allocated physical memory for a virtual machine deserves more discussion. Here we consider two scenarios. In the first scenario, the working set of a VM is larger than the allocated physical memory. Under this condition, the VM performance will deteriorate because of the increases in page misses. Previous research [58] has shown that through continuous sampling of application metrics under a variety of memory configurations and loads, we can derive out a performance-to-memory correlation model. In this way, a virtual machine can use the size of its active working set and measured system performance to estimate the allocated physical memory. It can then compare the estimation result to the promised memory by the SLA to detect possible violations.

In the second scenario, the size of the working set is smaller than the allocated physical memory. Under this condition, the VM's performance is not restricted by the physical memory size. If the VM determines that its working set is already larger than the physical memory size promised by the SLA, no further detection needs to be conducted. On the contrary, if the working set is smaller than the promised memory size, we can request new memory allocation to boost the usage to the promised value. We can then locate the pages from the working set that has the least recent access records and read them. If the hypervisor has allocated less than SLA promised memory to this VM, these pages would have been swapped out. Therefore, through measuring access latency to these memory pages, we can derive out whether or not there exists a violation in memory allocation for the VM.

4.3.5 Details of Implementation

Although the basic ideas of the detection mechanisms are straightforward, we face several difficulties in implementation. For example, we need to carefully select the memory pages that we access to reduce false alarm rates. We also need to consider the time measurement accuracy and the order of memory page accesses. Below we discuss our solutions to these problems.

4.3.5.1 Choice of memory pages

The first problem that we need to solve is to choose the memory pages that will be used for the detection of SLA violations. There are several criteria that we need to follow when we choose these pages. First, the selected pages must incur very small performance penalty on the system. If the proposed approaches impact the system efficiency to a large extent, it will become extremely hard to promote their wide adoption. Second, these pages should not be easily identified by the hypervisor or attackers. Otherwise, they may handle these pages differently to avoid detection.

We design different methods to choose memory pages for the detection of the investigated violations. The selection of memory pages for the detection of unauthorized access is very tricky. Theoretically, attackers or the hypervisor could choose any memory pages of the guest VM to read. It is impossible for the guest VM to know beforehand which pages to examine. In real world, however, the selection range is much smaller. An end user usually cares most about the data files that he/she is processing with sensitive information. Therefore, we propose to insert guardian pages into these sensitive data files. Since these pages are integrated into the real data files, the hypervisor or attackers will not be able to identify them. Therefore, when they conduct unauthorized memory access to the guest VM, these pages have a high probability to be touched.

To detect whether or not the cloud provider secretly enables memory deduplication without notifying end users, we need to make sure that the following two requirements are satisfied: (1) There exist memory pages that can be merged, and (2) more importantly, the guest VM can read from/write to these pages to measure the access delay. We propose to construct data files and actively load them into our VM's memory. Since deduplication is conducted at the page level, the offsets of the pages in data files will not impact the final result. Therefore, we can construct different data files through reorganizing the order of the pages. This scheme will also introduce randomness into the data files so that it is difficult for the cloud provider to discover the detection activities. After constructing these files, we can initiate different application softwares to load them into memory. Since memory deduplication can happen in both intra-VM and inter-VM modes, we can read different files in different VMs. Under this case, we can use methods in [59] to make sure that these VMs are located in the same physical box so that deduplication can be conducted.

To detect under-allocation of physical memory to a VM, we need to identify the pages with the least recent access records. Here a malicious hypervisor also knows these pages. However, there is very little that it can do to hide the fact that there is not enough memory allocation for the VM. For example, the SLA may have promised that at least 1-GB physical memory will be guaranteed for the victim VM while in real life only 800 MB is provided. Under this case, when our detection mechanism boosts the memory usage to 1 GB and the hypervisor refuses to provide new memory, at least 200-MB data needs to be swapped out. Since the victim VM can choose any memory pages in the working set as the detection sensors and measure access delay to them, it is very hard for the hypervisor to hide the increased loading latency when we consider the time difference between reading from memory and reading from hard disk.

The detection mechanisms described above often involve the usage of extra memory pages for attack detection. This usage could cause performance penalty if too much overhead is added. We have conducted a group of experiments to measure the impacts when different amounts of memory is used, the results of which will be presented in Section 4.4.

4.3.5.2 Measurement of access time

To successfully detect the SLA violations, we must accurately measure the data access time. Traditionally a computer provides three schemes to measure the length of a time duration: time of the day, CPU cycle counter, and tickless timekeeping. The first method provides the measurement granularity of seconds which is too coarse for our application. The second method will be a good candidate for time measurement if the operating system completely owns the hardware platform. In a VM-based system, however, it cannot accurately measure the time duration. For example, if a page fault happens during our reading operation, the hypervisor may pause the CPU cycle counter while it fetches the memory page. Therefore, the delay caused by hard disk reading will not be measured. Based on the analysis in [60], tickless timekeeping can keep time at a finer granularity. Therefore, we choose the Windows API $QueryPerformanceCounter$ to measure the duration. Previous research [61] has also shown that the time measurement accuracy may be impacted by the workload on the physical box. We can use the lightweight toolset $TiMeAcE.KOM$ [61] to assess and correct the measurement results.

4.3.5.3 Verification of memory access order

As explained in Section 4.3.4, the detection of SLA violations in memory management depends on the verification of some facts: Some memory pages that should have been swapped out are still in memory, while some other pages that should have stayed in memory are swapped out. The reason that these discrepancies happen is because some access operations change the order of swapping. To verify this fact, we need to set up a group of memory pages to serve with references to time. Through controlling access to these reference pages, we can derive out whether or not the data pages should have been swapped out. While the basic idea is straightforward, we need to consider several issues when we choose these reference pages. First, the reference pages should not belong to frequently used OS or application software. In this way, they will not be merged by the deduplication algorithm. Second, we want these reference pages to be randomly distributed in the memory. In this way, if the cloud provider or attackers access the guest VM memory stealthily, they have a very low probability to read many reference pages.

To satisfy these requirements, we propose to use the memory pages that are unique in the Windows 95 system as the reference pages. Our previous research [16] has successfully identified these pages. Since almost no users are still using Windows 95, these pages will not be deduplicated. We will allocate space for each individual page and chain them together with pointers to form a linked list. Since we do not allocate a big chunk of continuous memory for these pages, they may distribute all over the memory. Therefore, it is very hard for the attacker to touch many reference pages if he randomly selects guest VM pages to read. In this way, the reference pages can effectively serve their purposes.

4.3.6 Detection Procedures of the SLA Violations

With all the building blocks, we need to construct the detection algorithms; below we describe the details of the detection procedures. We introduce the algorithms, respectively, for the three SLA violations.

Figure 4.3a illustrates the detection of unauthorized memory accesses. The guest VM will load both the reference pages and the guardian pages into

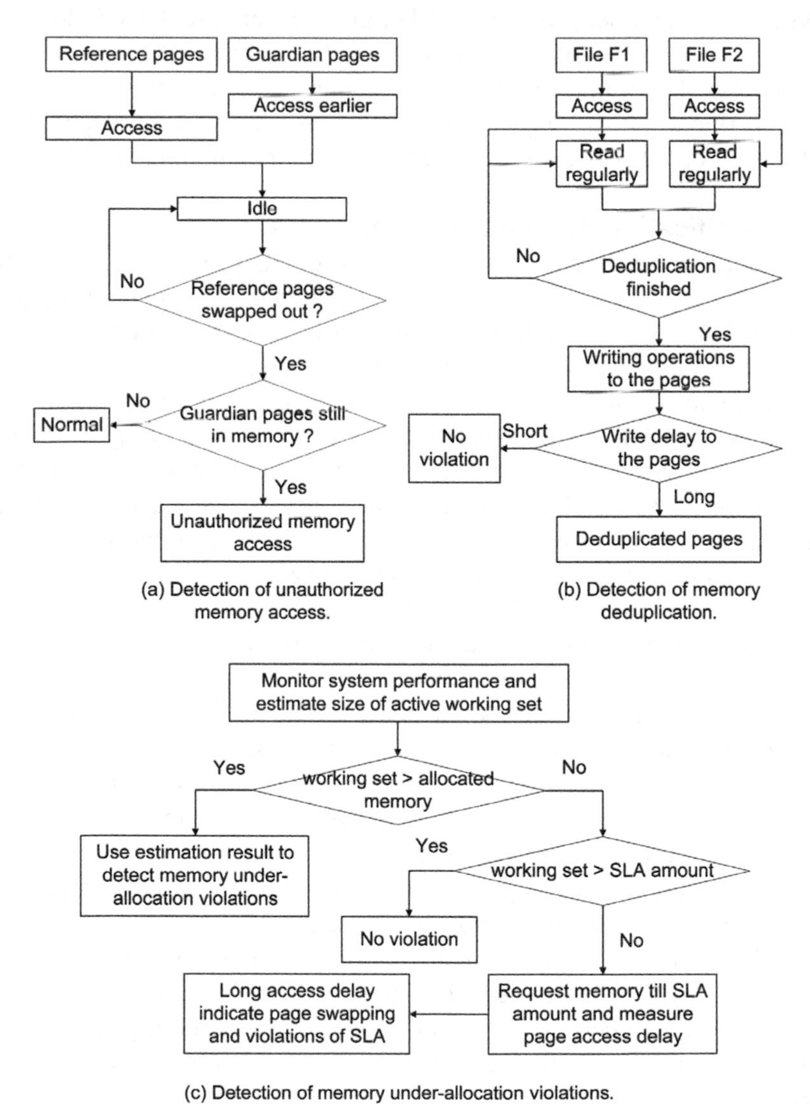

Figure 4.3 Detection procedures of the SLA violations.

its memory. Since the swapping operations heavily depend on the memory usage, we propose to divide the reference pages into multiple groups and access them at different intervals. As illustrated in Figure 4.4.*top*, we will first read all the guardian pages before the reference pages. In this way, the guardian pages would have been swapped out before the reference pages if no one else

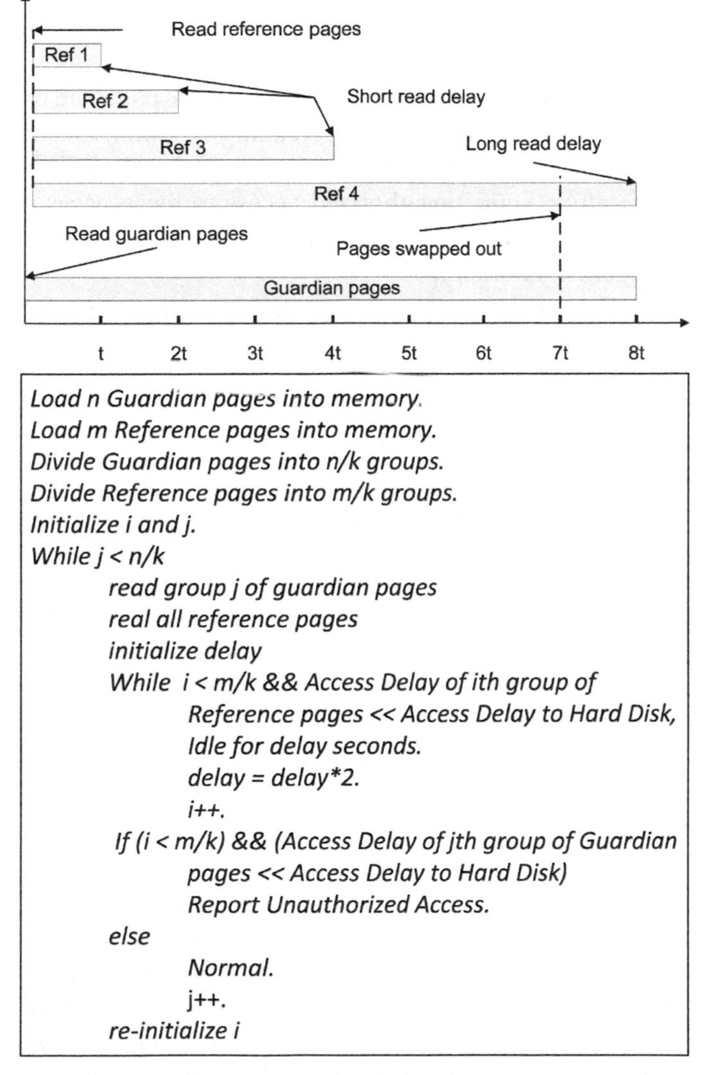

Figure 4.4 Using reference pages to detect memory swap operations. *Top:* the read operations on guardian and reference pages. *Bottom:* the algorithm to detect unauthorized memory access.

touches them afterward. These pages will then be left idle. We will access the reference pages with different intervals. For example, the intervals shown in Figure 4.4.*top* for different groups of reference pages increase exponentially. Assuming that at time $7t$ the VM is under pressure for more memory and has to swap many pages out. Since the guardian pages are accessed before the 4th group of reference pages, they will be swapped out first. Then the 4th group of reference pages is also swapped out. At time $8t$, when the predetermined interval for group 4 expires, we will read this group of reference pages. Since they have been swapped out, the reading delay will be long. As soon as we detect the long reading delay, we can derive out that these pages are no longer in memory. We will then immediately conduct reading operations on the guardian pages. If the reading delay is short, we can derive out that these pages are still in memory. Therefore, some unauthorized access to these pages must have happened after our first reading. The algorithm to implement the procedure is illustrated in Figure 4.4.*bottom*.

Figure 4.3b illustrates detection of the violations of deduplication policies. Here two applications in the guest VM will read the files $F1$ and $F2$ into its memory, respectively. The two files contain identical memory pages. Should deduplication is enabled in the guest VM, the pages of the two files will be merged. We will read *F1* and *F2* regularly so that their pages will not be swapped out. Using our previous research in [16], we will estimate the time that is needed to accomplish memory deduplication. When the time expires, we will conduct a group of writing operations to these pages. If the pages of *F1* and *F2* have very short writing delay, they have their own copies. On the contrary, if they are merged, the "copy-on-write" operations will introduce a measurable delay. We can use the results to determine whether or not the SLA has been violated.

Figure 4.3c illustrates the detection of memory under-allocation violations. Here the VM will first estimate the size of its active working set. It will then monitor the system performance and use the results in [58] to estimate the physical memory that is allocated to the VM. If the working set is larger than the allocated memory, we can compare the estimation result to the minimum memory size promised by the SLA and detect any violations. On the contrary, the monitoring results may show that the allocated physical memory is larger than the working set. Under this condition, if the working set is already larger than the promised minimum memory size, the hypervisor is keeping the SLA. If the working set is smaller than the promised value, we will request extra memory from the hypervisor and boost the usage to the SLA amount.

We will then measure the access delay to the least recently accessed pages to detect any violations.

4.4 Experimental Results

In this section, we present our experimental results. We first introduce the experimental setup. The detection capabilities and overhead of our approaches will then be assessed.

4.4.1 Experimental Environment Setup

The experimental environment setup is as follows. The physical machine has a Dual Core 2.4 GHz Intel CPU, 2-GB RAM, and SATA hard drives. We choose a machine with relatively small memory size so that it is easy for applications to exhaust the memory and trigger swapping. The hypervisor that we use is VMWare Workstation 6.0.5. We choose this version since it provides explicit interfaces for memory sharing and access between virtual machines. All user's virtual machines are using Windows XP SP3 as the operating system. Each virtual machine will occupy one Dual Core CPU and 8-GB hard disk. The amount of virtual memory that we allocate for each virtual machine will be determined by the experiment. Our experiments show that when there are more than twenty (20) pages that need to be read from the hard disk or separated from the merged memory, the accumulated delay can be accurately detected. Therefore, in our experiments we choose the size of each group of reference pages and guardian pages to be 20.

4.4.2 Experiments and Results

We conduct three groups of experiments to evaluate the detection of unauthorized memory access, violation of deduplication policies, and memory under-allocation violations, respectively. Below we will present the results of the baseline experiments and the detection capabilities and overhead in more complicated scenarios.

The first group of experiments tries to evaluate the mechanism for detecting unauthorized memory access. To simplify the experimental setup and examine the practicability of our approach, we initiate only two virtual machines in the physical box: the guest VM that runs our detection algorithms, and an attacker's VM that stealthily accesses the memory of the victim. Instead of locating some malware to penetrate VMWare and get access to the guest VM's memory, we propose to use the Virtual Machine Communication Interface (VMCI) [62] to simulate such an attack. Specifically, in the guest VM, we create a block of shared memory so that the attacker's VM can access the guardian pages remotely. The guest VM will load both reference pages and guardian pages into its memory, as described in Section 4.3.6. After initial access to these memory pages, we would launch some applications to consume all of the memory. In this way, the system will choose memory pages to swap out. Since the attacker's VM remotely accesses the guardian pages, they will be kept in the memory. On the contrary, the reference pages will be swapped out first. When the guest VM measures the access delay to the guardian pages, it will figure out that they are still in the memory and detect the unauthorized access.

Figure 4.5a illustrates the detection results. We conduct five reading operations to the memory pages. On the Y-axis, we show the average access latency to every page. Since the delays span across multiple degrees of magnitude, we use log-scale Y-axis. Reading Operation 1 has long delay for both groups of pages since they are loaded from hard disk. Reading Operation 2 is conducted immediately after Operation 1 to verify the contents. The interval between Reading Operations 2 and 3 represents the idle time. We can see that the access latency to reference pages at Reading Operation 3 is much longer than that to the guardian pages since they have been swapped out. After that, we conduct another two rounds of reading operations to measure the delay. From this figure, we could infer that the guardian pages must have been touched by someone after the initial access. Since the access command is not issued by our VM, it is unauthorized access.

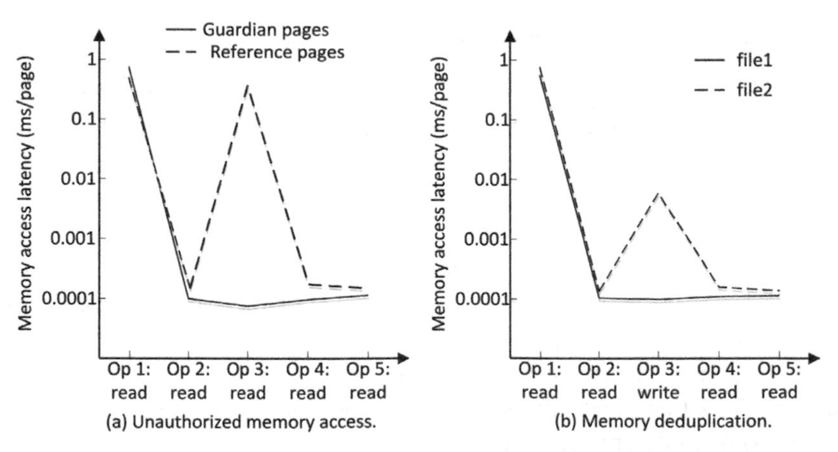

Figure 4.5 Detection capabilities of the approaches in baseline scenarios.

The second group of experiments assesses the detection of the violations to deduplication policies. We configure the corresponding parameters in VMWare so that the page sharing process will scan the memory and merge the pages with identical contents. As illustrated in Figure 4.3b, the constructed files *F1* and *F2* contain many such pages. We will access the two files regularly so that they will not be swapped out from the memory. These reading operations will not impact the deduplication procedures. Using the experimental results in [16, 17, 42], we can estimate the time that the algorithm needs to merge the pages. When the estimated delay expires, we will issue a "write" command to the pages of *F2*. If the pages have been merged, the "copy-on-write" operations will introduce a measurable increase in delay. On the contrary, if each page has its own memory, the write delay will be much shorter.

Figure 4.5b illustrates the detection results. Reading Operation 1 has long delay for both files since they are loaded from hard disk. The interval between Operations 2 and 3 represents the deduplication procedure. At the Writing Operation 3, we first write to pages of *F2*. We can see that the delay is very long because of the separation of the pages. After that, we write to the pages of *F1*. Since the merged pages have been separated, the writing delay is short. Using this result, we can figure out that the deduplication function is enabled.

We conduct another group of experiments to evaluate the detection capabilities of the proposed approaches in more complicated scenarios. In this experiment, the guest VM is running the software package *Prime* to generate prime numbers. This application demands a lot of CPU resources. We run the detection algorithms for the two violations. The results are shown in Figure 4.6.

From the figure, we can see that the proposed mechanisms can still effectively detect the violations. Since our approaches will read from/write to memory pages at a sparse interval, they do not incur heavy CPU overhead. Therefore, the execution of CPU-intensive applications does not impact our approaches to a large extent.

In the third group of experiments, we assess the detection of the memory under-allocation violations. Two experiments with different memory demands are conducted. In experiment one, we initiate one virtual machine and assume that the hypervisor promises to allocate at least 4-GB RAM to the VM. However, the total physical memory in the machine is only 2 GB. More memory resources are promised than available memory. We have three applications running in the guest VM. Application 1 reads file *F1* periodically so that it will be kept in memory. Application 2 reads file *F2* only once so that its memory pages become the least recently accessed ones. Finally, Application 3

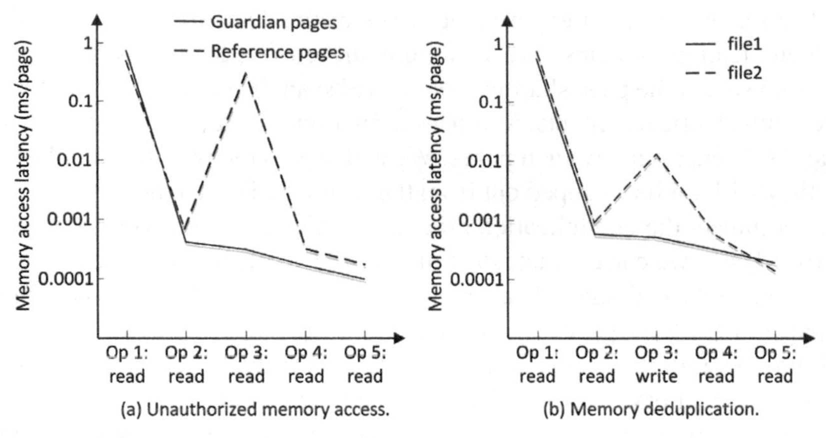

Figure 4.6 Detection capabilities of the approaches under intense CPU demand.

consumes more than 2 GB but less than 3.5-GB memory. From the guest VM's point of view, if it actually has 4-GB RAM, the file *F2* would have stayed in memory. However, Figure 4.7a shows that the memory access latency to *F2* is much longer than that to *F1*. This indicates that the actual RAM is less than 4 GB. The memory under-allocation violation is detected.

In experiment two, we initiate two virtual machines. The total physical memory in the machine is still only 2 GB. The hypervisor promises 3-GB

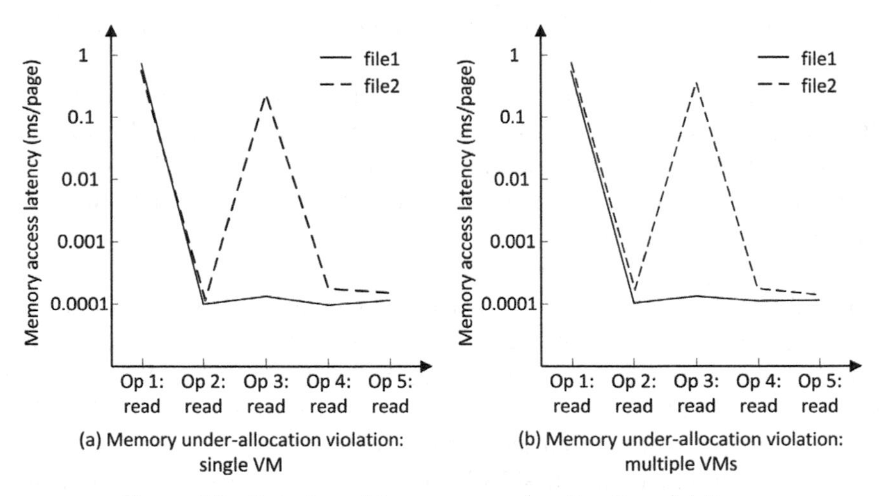

Figure 4.7 Detection of the memory under-allocation violations.

physical memory to each of the VMs. Here *VM2* is running a memory-intensive application that demands about 2-GB memory. *VM1* needs about 1-GB memory to read the files *F1* and *F2* as in experiment one. The monitoring operations in *VM1* find out that the active working set is way lower than 3 GB. Therefore, extra memory is demanded from the hypervisor to reach the promised amount. After that, reading operations to *F1* and *F2* are conducted. Since we read *F1* periodically, only the pages of *F2* are swapped out, as shown in Figure 4.7b. Again the memory under-allocation violation is detected.

4.4.3 Impacts on System Performance

The proposed violation detection mechanisms demand resources such as memory and CPU time during their execution. Therefore, they may impact the overall system performance. In the following group of experiments, we try to assess such impacts. Since the detection algorithm for unauthorized memory access demands more CPU and memory resources than the mechanisms for the other two types of violations, we choose it as the subject of evaluation. During our experiments, we want to test an extreme scenario. When the detection algorithm is launched, it will allocate many groups of reference pages and try to exhaust the main memory of the guest VM as soon as possible in order to force swapping operations. Other application software will have to compete with our detection algorithm for resources for their execution.

We are especially interested in the impacts on two groups of applications. The first group includes the CPU-intensive applications. We choose three examples: (1) the $Fibonacci$ benchmark that computes the Fibonacci sequence; (2) the $Prime$ benchmark that generates prime numbers; and (3) the $Narcissistic$ benchmark that generates the narcissistic numbers. Each of these software packages is running in parallel with the detection algorithm for unauthorized memory access. The measured CPU usage is very close to 100%. We measure the execution time of the software since this is the most intuitive parameter that end users adopt to evaluate the system performance.

The second group includes those CPU/memory-intensive applications. We also choose three examples: (1) the N-$Queens$ benchmark that computes solutions to the N-Queens problem in chess and stores the results in memory; (2) the $Combination$ benchmark that computes all possible combinations of the input numbers; and (3) the $Permutation$ benchmark that computes all possible permutations of the input numbers. We measure their execution time when each of them is running in parallel with the proposed mechanism.

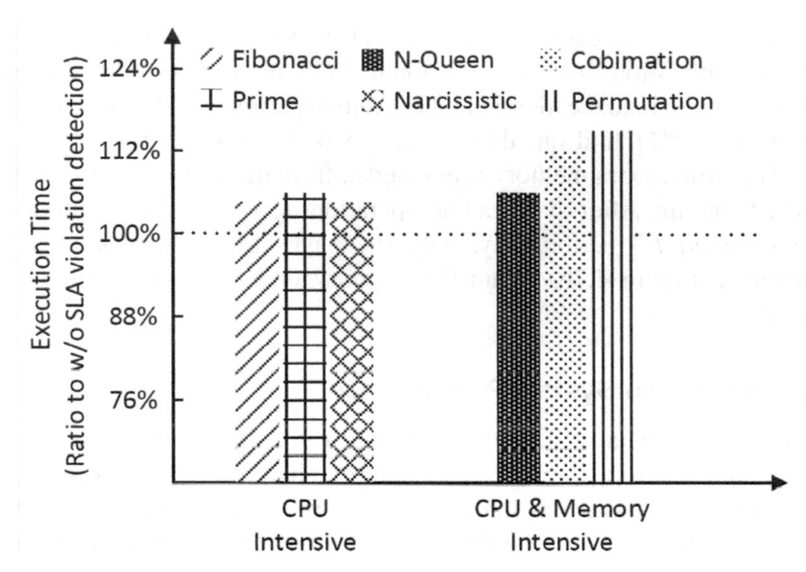

Figure 4.8 Impacts on system performance.

From Figure 4.8, we can see that for CPU-intensive applications, the increase in execution time is less than 6%. The increase in execution time for CPU/memory-intensive applications is smaller than 15%. Please note that this is an extreme case since the detection algorithm runs continuously and tries to force memory swapping by grabbing as much RAM as possible. In real world, end users can reduce the detection frequency (e.g., once every 5 min). Under that condition, the increased execution time is smaller than 1% on average for both groups of applications.

To better understand the impacts of the size of guardian pages and reference pages on the system performance, we have conducted another group of experiments. In these experiments, the size of the data file that we want to protect is 100 MB. We also use the extreme scenario in which we consume as much memory as possible to force guest VM to swap out pages. We run both CPU-intensive application (Prime Number Benchmark) and CPU/memory-intensive application (Permutation Benchmark) to evaluate the performance overhead. In Figure 4.9.*top*, we keep the size of the reference pages unchanged and increase the volume of guardian pages from 20% to 120% of the data file size. In Figure 4.9.*bottom*, we keep the size of the guardian pages unchanged and increase the volume of reference pages. From the figures, we can see that under both conditions their impacts on system performance do not change largely.

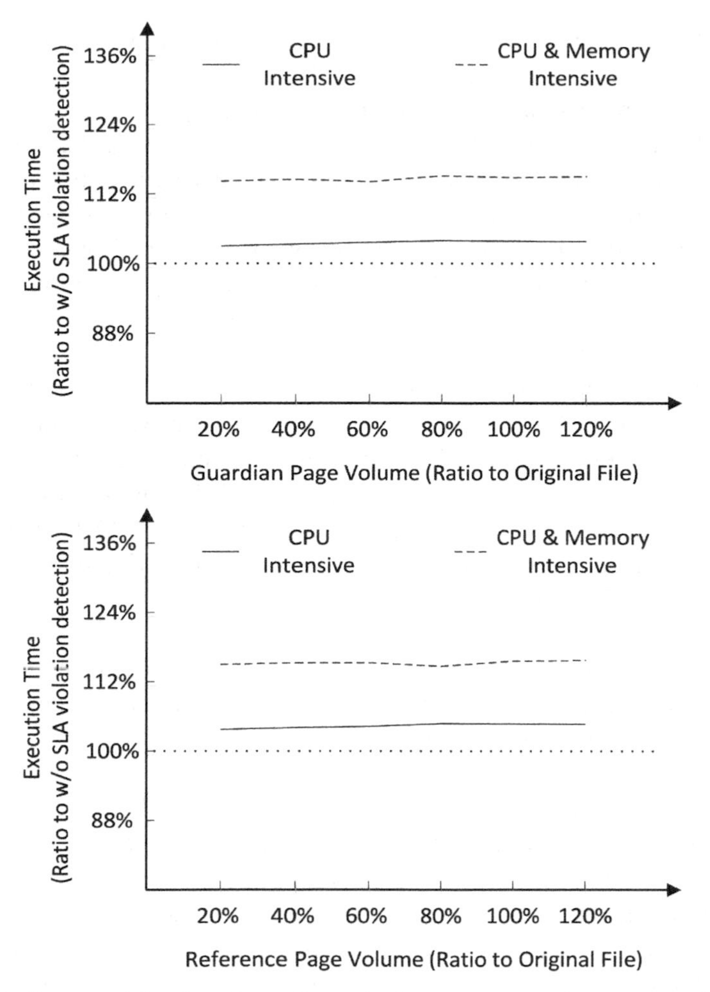

Figure 4.9 Impacts of the size of guardian and reference pages on the system performance.

4.5 Discussion

4.5.1 Reducing False Alarms

The proposed approaches use memory access latency in guest virtual machines to detect violations of SLA on memory management. Different from many interactive security mechanisms that involve third parties, our approaches do not need collaborations from other virtual machines. In this way, it reduces the attack surfaces of the approaches. Below we discuss the schemes that attackers can adopt to avoid detection and our mitigation mechanisms. We will also discuss the schemes to reduce false alarms.

Since the access latency to a single memory page is too short to be accurately measured, the detection of the three types of SLA violations depends on the accumulated delay. Therefore, the hypervisor or attackers can reduce the memory access frequency and volume to the guest VM to reduce the chance of detection. For example, the hypervisor can randomly select memory pages of the guest VM to read. In this way, when our detection algorithm measures the access delay, only a small percentage of the guardian pages would still be in memory. Attackers can hide the short delay of these pages within other swapping operations. This scheme, however, will also impede the information-stealing procedures. For example, our experiments show that if more than twenty pages under monitoring are impacted by an attack, the accumulated change in access time can be detected. This will restrict the speed at which an attacker reads the victim VM' memory. For a 1-MB data file, if the end user conducts a round of detection every 15 min and in every round the attacker can read only twenty pages to avoid detection, it may take the attacker three hours to read all pages of the file. As another example, the hypervisor may adjust the memory deduplication parameters to reduce the merging speed. Under this condition, the virtual machines will keep a relatively large memory footprint size, which will diminish the purpose of deduplication.

If the data file is very valuable, a dedicated attacker may want to spend a lot of time to access the memory pages in a stealthy way. Under this condition, two methods with higher costs to the legitimate user could be adopted. In the first approach, we can increase the ratio of guardian pages to the data file. For example, if the volume of guardian pages and the size of data file have the ratio of 1:1, the amount of time that the attacker needs to get the whole copy of the file will double. In the second method, the user can benefit from the latest advances in homomorphic encryption and process only encrypted memory pages in the cloud. In this way, it will become very difficult for either the cloud provider or an attacker to steal sensitive information directly from the user.

One factor that may impact the detection accuracy of the proposed approaches is the prefetching technique. Prefetching tries to predict the information that the OS will need in the near future and loads the data in before the actual instructions are issued. This technique may introduce false-positive alarms into our system since some guardian pages may be read into memory even though no unauthorized access has been conducted. To mitigate such problems, we can adopt two schemes. First, prefetching uses the property of locality and is usually applied to the subsequent pages of current contents. If we know the number of pages that a prefetching operation will load, we can use the guardian pages beyond the prefetching range to detect violations. Another scheme that we can use is to chain the guardian pages together to form a linked list. This technique can also diminish the impacts of prefetching.

4.5.2 Impacts of Extra Memory Demand

Some users may worry that the extra memory demand for the detection of under-allocation violations will impact the system performance. We will justify the approach from the following aspects. First and most importantly, the extra memory demand will happen only when the active working set is smaller than the minimum memory amount promised by the SLA. At the same time, if the hypervisor is keeping its SLA, this memory would have belonged to the VM any way. Under this case, the demand will not impact the system performance. Second, the memory demand is incurred only during the detection procedures, which would happen infrequently. Last but not least, our experimental results in Section 4.4.3 have shown that the detection algorithms will cause very small impacts on the system performance.

4.5.3 Building A Unified Detection Algorithm

For the clarity of the chapter, we have presented the detection algorithms for the three types of violations separately. In real life, we can build a unified detection algorithm. As shown in Figure 4.10, the unified detection mechanism will consist of five components. The memory management component will communicate with the virtual machine to request and return memory pages. A timer will issue commands to the memory reading/writing component based

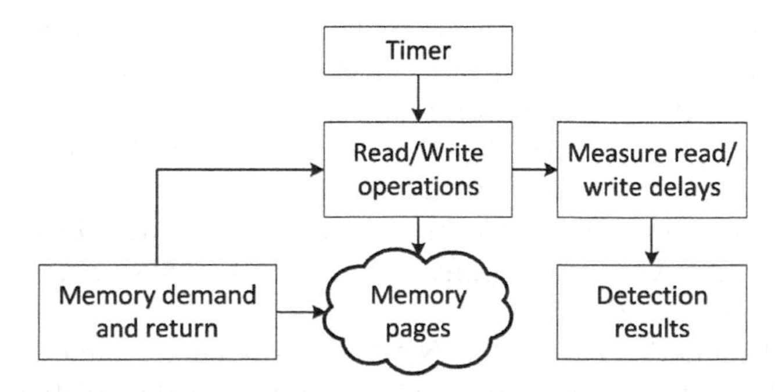

Figure 4.10 Architecture of a unified detection algorithm.

on predetermined clock intervals so that certain pages will be kept in memory. The reading/writing component will also work with the time measurement component to get the accurate access latency. The measurement results of these operations will be provided to the detection component. Based upon different target violations, the algorithm will return detection results to the end user.

4.6 Conclusion

In this chapter, we propose mechanisms to detect violations of the SLA on memory management in virtual machines. Instead of proposing a generic security SLA enforcement architecture, we design mechanisms to detect three types of memory management violations in guest VMs. We have implemented the detection approaches under VMWare and tested them. The results show that they can effectively detect the violations with a small increase in overhead. We also discuss the techniques to improve our approaches.

Immediate extensions to our approaches consist of the following aspects. First, we plan to explore other types of security SLA violations in memory management and design a generic approach for their detection. We will also experiment with other hypervisors such as extended Xen and Linux KSM to generalize the mechanisms. Second, we want to study the relationship between our approaches and existing security SLA enforcement architectures. If we can integrate them into the existing architecture, we will have a solid platform for future extension. The research will provide new information for strengthening protection to virtual machines and end users of cloud computing.

References

[1] Berman, S. J., Kesterson Townes, L., Marshall, A., and Srivathsa, R. (2012). How cloud computing enables process and business model innovation, *Strategy & Leadership*, 40(4), 27–35.

[2] Brandic, I., Emeakaroha, V. C., Maurer, M., Dustdar, S., Acs, S., Kertesz, A., and Kecskemeti, G. (2010). Laysi: A layered approach for sla-violation propagation in self-manageable cloud infrastructures. In *Proceedings of the IEEE Annual Computer Software and Applications Conference Workshops*, pp. 365–370.

[3] Emeakaroha, V., Ferreto, T., Netto, M., Brandic, I., and De Rose, C. (2012). Casvid: Application level monitoring for sla violation detection

in clouds. In *IEEE Annual Computer Software and Applications Conference (COMPSAC)*, pp. 499–508.

[4] Kyriazis, D., Menychtas, A., Kousiouris, G., Boniface, M., Cucinotta, T., Oberle, K., Voith, T., Oliveros, E., and Berger, S. (2014). A real-time service oriented infrastructure. *GSTF Int. J. Comput. (JoC)*, 1(2).

[5] Li, C.-C. and Wang, K. (2014). An sla-aware load balancing scheme for cloud datacenters. In *International Conference on Information Networking (ICOIN)*, pp. 58–63.

[6] Emeakaroha, V., Brandic, I., Maurer, M., and Dustdar, S. (2010). Low level metrics to high level slas-lom2his framework: Bridging the gap between monitored metrics and sla parameters in cloud environments. In *International Conference on High Performance Computing and Simulation (HPCS)*, pp. 48–54.

[7] Comuzzi, M., Kotsokalis, C., Spanoudakis, G., and Yahyapour, R. (2009). Establishing and monitoring slas in complex service based systems. In *IEEE International Conference on Web Services (ICWS)*, pp. 783–790.

[8] Emeakaroha, V. C., Netto, M. A. S., Calheiros, R. N., Brandic, I., Buyya, R., and De Rose, C. A. F. (2012). Towards autonomic detection of sla violations in cloud infrastructures, *Future Gener. Comput. Syst.*, 28(7), 1017–1029.

[9] Haq, I. U., Brandic, I., and Schikuta, E. (2010). Sla validation in layered cloud infrastructures. In *Proceedings of the 7th International Conference on Economics of Grids, Clouds, Systems, and Services*, pp. 153–164.

[10] Muller, C., Oriol, M., Franch, X., Marco, J., Resinas, M., Ruiz-Cortes, A., and Rodrguez, M. (2014). Comprehensive explanation of sla violations at runtime, *IEEE Trans. Serv. Comput.*, 7(2), 168–183.

[11] Haeberlen, A. (2010). A case for the accountable cloud, *SIGOPS Oper. Syst. Rev.*, 44(2), 52–57.

[12] Bernsmed, K., Jaatun, M., Meland, P., and Undheim, A. (2011). Security slas for federated cloud services. In *Sixth International Conference on Availability, Reliability and Security (ARES)*, Aug, pp. 202–209.

[13] VMWare. "esxi configuration guide," VMware vSphere 4.1 Documentation, 2010.

[14] Gupta, D., Lee, S., Vrable, M., Savage, S., Snoeren, A., Varghese, G., Voelker, G., and Vahdat, A. (2010). Difference engine: harnessing memory redundancy in virtual machines, *Commun. ACM*, 53(10), 85–93.

[15] Rong, H., Wang, H., Liu, J., Zhang, X., and Xian, M. (2015). Windtalker: An efficient and robust protocol of cloud covert channel based on memory

deduplication. In *IEEE International Conference on Big Data and Cloud Computing (BDCloud)*, pp. 68–75.

[16] Owens, R., and Wang, W. (2011). Non-interactive os fingerprinting through memory de-duplication technique in virtual machines. In *IEEE International Performance Computing and Communications Conference (IPCCC)*.

[17] Suzaki, K., Lijima, K., Yagi, T., and Artho, C. (2011). Memory deduplication as a threat to the guest os. In *Proceedings of the Fourth European Workshop on System Security*, pp. 1–6.

[18] Xiao, J., Xu, Z., Huang, H., and Wang, H. (2013). Security implications of memory deduplication in a virtualized environment. In *IEEE/IFIP International Conference on Dependable Systems and Networks (DSN)*, pp. 1–12.

[19] Yarom, Y., and Falkner, K. (2014). Flush+ reload: a high resolution, low noise, l3 cache side-channel attack. In *USENIX Security Symposium (USENIX Security)*, pp. 719–732.

[20] Pratiba, D., Shobha, G., Tandon, S., and Srushti, S. (2015). Cache based side channel attack on aes in cloud computing environment. *Int. J. Comp. Appl.*, 119(13).

[21] Younis, Y. A., Kifayat, K., Shi, Q., and Askwith, B. (2015). A new prime and probe cache side-channel attack for cloud computing. In *IEEE International Conference on Dependable, Autonomic and Secure Computing (DASC)*, pp. 1718–1724.

[22] Liu, F., Yarom, Y., Ge, Q., Heiser, G., and Lee, R. B. (2015). Last-level cache side-channel attacks are practical. In *IEEE Symposium on Security and Privacy*, pp. 605–622.

[23] Jiang, X., Wang, X., and Xu, D. (2007). Stealthy malware detection through vmm-based 'out-of-the-box' semantic view reconstruction. In *Proceedings of the ACM Conference on Computer and Communications Security*, pp. 128–138.

[24] Li, M., Zang, W., Bai, K., Yu, M., and Liu, P. (2013). Mycloud – supporting user-configured privacy protection in cloud computing. In *Annual Computer Security Applications Conference*.

[25] Fattori, A., Lanzi, A., Balzarotti, D., and Kirda, E. (2015). Hypervisor-based malware protection with accessminer, *Comput. Secur.*, 52, 33–50.

[26] Hwang, T., Shin, Y., Son, K., and Park, H. (2014). Virtual machine introspection based rootkit detection in virtualized environments. In *International Conference on Advanced Computing and Services (ACS)*.

[27] Banerjee, I., Guo, F., Tati, K., and Venkatasubramanian, R. (2013). Memory overcommitment in the esx server, VMWare Inc., VMWare Technical Journal, Tech. Rep.

[28] Gordon, A., Hines, M. R., da, D. S., Ben-Yehuda, M., Silva, M., and Lizarraga, G. (2011). Ginkgo: Automated, application-driven memory overcommitment for cloud computing. In *RESoLVE: Runtime Environments/Systems, Layering, and Virtualized Environments Workshop (co-located with ASPLOS)*.

[29] Amit, N., Tsafrir, D., and Schuster, A. (2014). Vswapper: A memory swapper for virtualized environments, *ACM SIGPLAN Notices*, 49(4), 349–366.

[30] Shaikh, F., Yao, F., Gupta, I., and Campbell, R. H. (2014). Vmdedup: Memory de-duplication in hypervisor. In *IEEE International Conference on Cloud Engineering (IC2E)*, pp. 379–384.

[31] Heo, J., Zhu, X., Padala, P., and Wang, Z. (2009). Memory overbooking and dynamic control of xen virtual machines in consolidated environments. In *IFIP/IEEE International Symposium on Integrated Network Management (IM)*, pp. 630–637.

[32] Moon, S.-J., Sekar, V., and Reiter, M. K. (2015). Nomad: Mitigating arbitrary cloud side channels via provider-assisted migration. In *Proceedings of the ACM SIGSAC Conference on Computer and Communications Security*, pp. 1595–1606.

[33] Zhang, Y., Juels, A., Reiter, M. K., and Ristenpart, T. (2014). Cross-tenant side-channel attacks in paas clouds. In *Proceedings of ACM SIGSAC Conference on Computer and Communications Security*, pp. 990–1003.

[34] Lee, C., Hong, C.-H., Yoo, S., and Yoo, C. (2014). Compressed and shared swap to extend available memory in virtualized consumer electronics, *IEEE Trans. Consum. Electron.*, 60(4), 628–635.

[35] VMWare. "Understanding memory resource management in vmware esx 4.1," VMware ESX 4.1 Documentation, EN-000411-00, 2010.

[36] Jian, W., Du Wei, J. C.-F., and Xiang-Hua, X. (2014). Overcommitting memory by initiative share from kvm guests. *Int. J. Grid Distr. Comput.*, 7(4).

[37] Ristenpart, T., Tromer, E., Shacham, H., and Savage, S. (2009). Hey, you, get off of my cloud: exploring information leakage in third-party compute clouds. In *Proceedings of the 16th ACM Conference on Computer and Communications Security (CCS)*, pp. 199–212.

[38] Tromer, E., Osvik, D. A., and Shamir, A. (2010). Efficient cache attacks on aes, and countermeasures, *J. Cryptol.*, 23(1), 37–71.

[39] Xu, Y., Bailey, M., Jahanian, F., Joshi, K., Hiltunen, M., and Schlichting, R. (2011). An exploration of L2 cache covert channels in virtualized environments. In *Proceedings of ACM Workshop on Cloud Computing Security Workshop*, pp. 29–40.

[40] Zhang, Y., Juels, A., Reiter, M. K., and Ristenpart, T. (2012). Cross-vm side channels and their use to extract private keys. In *Proceedings of the ACM Conference on Computer and Communications Security*, pp. 305–316.

[41] Irazoqui, G., Eisenbarth, T., and Sunar, B. (2015). S$a: A shared cache attack that works across cores and defies vm sandboxing – and its application to aes. In *IEEE Symposium on Security and Privacy (SP)*, pp. 591–604.

[42] Owens, R., and Wang, W. (2011). Fingerprinting large data sets through memory de-duplication technique in virtual machines. In *IEEE Military Communications Conference (MILCOM)*.

[43] Jin, S., Ahn, J., Cha, S., and Huh, J. (2011). Architectural support for secure virtualization under a vulnerable hypervisor. In *Proceedings of the IEEE/ACM International Symposium on Microarchitecture*, pp. 272–283.

[44] Szefer, J., and Lee, R. B. (2011). A case for hardware protection of guest vms from compromised hypervisors in cloud computing. In *Proceedings of the International Conference on Distributed Computing Systems Workshops*, pp. 248–252.

[45] Shi, J., Song, X., Chen, H., and Zang, B. (2011). Limiting cache-based side-channel in multi-tenant cloud using dynamic page coloring. In *Proceedings of the IEEE/IFIP International Conference on Dependable Systems and Networks Workshops*, pp. 194–199.

[46] Li, C., Raghunathan, A., and Jha, N. K. (2010). Secure virtual machine execution under an untrusted management os. In *Proceedings of the IEEE International Conference on Cloud Computing*, pp. 172–179.

[47] Mundada, Y., Ramachandran, A., and Feamster, N. (2011). Silverline: Data and network isolation for cloud services. In *USENIX Workshop on Hot Topics in Cloud Computing*.

[48] Raj, H., Robinson, D., Tariq, T. B., England, P., Saroiu, S., and Wolman, A. (2011). Credo: Trusted computing for guest vms with a commodity hypervisor, Microsoft Research, MSR-TR-2011-130, Redmond, WA, USA, Tech. Rep.

[49] Baig, M., Fitzsimons, C., Balasubramanian, S., Sion, R., and Porter, D. (2014). Cloudflow: Cloud-wide policy enforcement using fast vm

introspection. In *IEEE International Conference on Cloud Engineering (IC2E)*, pp. 159–164.

[50] Dastjerdi, A. V., Tabatabaei, S. G. H., and Buyya, R. (2012). A dependency-aware ontology-based approach for deploying service level agreement monitoring services in cloud, *Softw. Pract. Exper.*, 42(4), 501–518.

[51] Henning, R. R. (1999). Security service level agreements: Quantifiable security for the enterprise? in *Proceedings of the Workshop on New Security Paradigms*, pp. 54–60.

[52] Casola, V., Mazzeo, A., Mazzocca, N., and Rak, M. (2006). A sla evaluation methodology in service oriented architectures. In *Quality of Protection*, ser. Advances in Information Security, Gollmann, D., Massacci, F., and Yautsiukhin, A., Eds. Springer US, 23, 119–130.

[53] de Chaves, S., Westphall, C., and Lamin, F. (2010). Sla perspective in security management for cloud computing. In *International Conference on Networking and Services (ICNS)*, March, pp. 212–217.

[54] Katopodis, S., and Spanoudakis, G. (2014). Towards hybrid cloud service certification models. In *IEEE International Conference on Services Computing*.

[55] Rahulamathavan, Y., Pawar, P. S., Burnap, P., Rajarajan, M., Rana, O. F., and Spanoudakis, G. (2014). Analysing security requirements in cloud-based service level agreements. In *Proceedings of the International Conference on Security of Information and Networks*, pp. 73–76.

[56] Moniruzzaman, A. B. M. (2014). Analysis of memory ballooning technique for dynamic memory management of virtual machines (vms), arxiv.org document 1411.7344.

[57] Sharma, P., and Kulkarni, P. (2012). Singleton: system-wide page deduplication in virtual environments. In *Proceedings of International Symposium on High-Performance Parallel and Distributed Computing (HPDC)*, pp. 15–26.

[58] Hines, M., Gordon, A., Silva, M., Silva, D. D., Ryu, K. D., and Ben-Yehuda, M. (2011). Applications know best: Performance-driven memory overcommit with ginkgo. In *IEEE International Conference on Cloud Computing Technology and Science (CloudCom)*, pp. 130–137.

[59] Varadarajan, V., Zhang, Y., Ristenpart, T., and Swift, M. (2015). A placement vulnerability study in multi-tenant public clouds. In *USENIX Security Symposium*, pp. 913–928.

[60] VMware. "Timekeeping in vmware virtual machines," http://www.vmware.com/files/pdf/techpaper/Timekeeping-In-VirtualMachines.pdf, 2011.

[61] Lampe, U., Kieselmann, M., Miede, A., Zller, S., and Steinmetz, R. (2013). A tale of millis and nanos: Time measurements in virtual and physical machines. In Lau, K.-K., Lamersdorf, W., and Pimentel, E., (Eds.) *Service-Oriented and Cloud Computing*. Lecture Notes in Computer Science, 8135. Springer: Berlin, pp. 172–179.
[62] VMWare. (2012). Configure the virtual machine communication interface in the vsphere web client. In *vSphere Virtual Machine Administration Guide for ESXi 5.X*.

5

Analysis of Mobile Threats and Security Vulnerabilities for Mobile Platforms and Devices

Syed Rizvi, Gabriel Labrador, Whitney Hernandez and Kelsey Karpinski

Department of Information Sciences and Technology,
Pennsylvania State University, Altoona-PA, USA

Abstract

Mobile devices, such as smartphones, tablets, smart TVs, and others have become increasingly popular since they provide essential functionality in our everyday life. However, mobile devices present greater security and privacy issues to users in terms of both mobile platform vulnerabilities and the vast advancements in sophistication of malicious software. By installing malicious software, mobile devices can be infected with worms, Trojan horses or other virus families, which can compromise the user's security, privacy, or even gain complete control over the device. In this research, we present an analysis of contemporary mobile platform threats and give an in-depth overview of threat environment and security mechanisms built into state-of-the-art mobile operating systems. Specifically, we have the following three research objectives to achieve: Our first research objective is to present a comprehensive discussion on the three competing modern mobile platforms: iOS, Android, and BlackBerry. Moreover, we discuss the common vulnerabilities of the mobile platforms and the existing security models of these operating systems to protect mobile devices. Our second research objective is to present a threat model of each mobile platform. Specifically, we first discuss the factors that motivate attackers to breach mobile security. We also present the attack vectors and the modern exploitation techniques used by attackers to breach the security of a mobile device by inserting malicious codes. In addition, we discuss some common types of malwares particularly designed for mobile devices such as mobile Trojans/worms and other viruses. Our third research objective is to present state-of-the-art security defense mechanisms to protect mobile platforms and devices along with a brief discussion on future trends in mobile security and its applications.

Keywords: Mobile platforms, Mobile devices, Threat environment, Mobile security.

5.1 Introduction

Mobile devices are rapidly emerging as popular appliances with increasingly powerful computing, networking, and sensing capabilities. The number of mobile users has drastically increased over the past few years. For example, the number of mobile users is forecast to reach 4.77 billion by the year 2017 [1]. According to one recent report [2], mobile data traffic grew 69% in 2014. Specifically, mobile data traffic reached 2.5 exa-bytes per month at the end of 2014, up from 1.5 exa-bytes per month at the end of 2013. These vast growing mobile device platforms are supported by three competing operating systems: Android, iPhone operating system (iOS), and BlackBerry. The numbers of Android and iOS users are also increasing at a faster rate. The global mobile operating system (OS) market share shows that Android OS reached 69.7% at the end of 2012, racing past Symbian OS, BlackBerry OS, and iOS [3].

Although we have seen significant advancements in mobile technology in recent years, it comes with a similar increase in the number and sophistication of malicious software targeting popular platforms. Currently, mobile devices present greater security and privacy issues to users than in the previous decade. This is mainly due to the fact that mobile devices incorporate several ways to leak highly sensitive information about a user's location, pictures/videos, and so on [3]. One major source of getting malware on mobile devices is the inherited ability of devices to download third-party applications (apps) from online markets. These online app markets are further divided into two types: open market versus closed/controlled market. In an open online app market, mobile users are free to install any app from any available market, whereas in the closed/controlled market, app markets are restricted to access by certain legitimate users. For example, the open market is typically used in the Android security model which mainly relies on user's judgment to install applications from reliable sources or to evaluate whether the application requests reasonable permissions for its intended operation [4]. Since these markets contain hundreds to thousands of apps, determining which are malicious and which are not is still a formidable challenge.

Even though there have been many schemes/architectures recently proposed in literature to analyze and detect mobile malware, adopting these schemes/architectures for one specific type of mobile platform is still an open research question. For example, some of these malware analysis/detection schemes work for Android OS, whereas the others are completely biased toward iOS. From the mobile stakeholder's perspective, which includes mobile users, industry, and the research community, it is critical to have more understanding of these mobile platforms, their threat environment, and the capabilities of the existing security defense systems.

One of the objectives of our research is to advance the current research on the analysis of mobile platforms. Specifically, we present a comprehensive discussion on the three competing mobile OS: iOS, Android, and BlackBerry. Moreover, we discuss the common vulnerabilities of the mobile platforms and the existing security models of these OS to protect mobile devices. In addition, we present a threat model of each mobile platform. Specifically, we first discuss the factors that make attackers breach mobile security. Secondly, we present the attack vectors and the modern exploitation techniques used by the attackers to breach the security of a mobile device by inserting malicious code. These modern exploitation techniques include (a) Web connections/browser, (b) social engineering, (c) network services, and (d) applications. Thirdly, we discuss some of the common types of malwares particularly designed for mobile devices such as mobile Trojans/worms and viruses. In this research work, we also present a comprehensive discussion on the modern defense schemes designed to protect the three major mobile platforms. Specifically, we identified seven unique defense methods to protect mobile platforms from variety of different existing security threats. Finally, we present our own analysis on the existing mobile threats and the future trends in mobile security and its different applications.

5.2 Analysis of Mobile Platforms

In this section, mobile platforms are introduced and explained. To begin, a mobile platform includes both hardware and software that create an environment for mobile devices [5]. Some mobile platforms that will be discussed in this section include Apple, Android, and BlackBerry. The number of mobile smart devices continues to grow from the recorded 2.6 billion users in 2014 [6]. In fact, it is predicted by the year 2020, the number will jump to 6.1 billion users. This number represents 70% of the global population. In addition, by the year 2020, around 90% of the world's population will at least have access to data coverage [6]. To put this into perspective, there are around twenty new smartphone subscribers every second of the day [7]. The objective of this section is to understand the operating system of each mobile platform. In addition, the security features and flaws will also be discussed in the following sections. Lastly, the different types of mobile malware, attacker motives, and potential vulnerabilities will be reviewed. By the end of this section, smartphone users will have a clear understanding on how mobile platforms work and the general concepts that relate to the term. In the next section, an analysis of three major mobile platforms will be discussed in detail.

5.2.1 Dominating Mobile Platforms

When it comes to mobile platforms, there are two major operating systems that rule the market: iOS and Android. The BlackBerry OS, however, is attempting to make a comeback in the mobile device world. In the following subsections, the fundamentals of these three platforms will be explained in detail starting with the iOS.

5.2.1.1 iPhone Operating System (iOS)

The iOS platform is an operating system which is being used on all mobile devices produced and distributed by the Apple. These devices include iPhones, iPads, iPods, and the Apple watch. The most recent version of the operating system is iOS 9, and its capabilities are cutting edge. For example, it allows devices to hold a charge longer [8] by not allowing notifications to light up the device when it is facing in a downward direction. A new low-power mode is also made available when the devices reach 20% and 10% [8]. This option stops background processes, darkens the screen, and goes into sleep mode more quickly. Also, in iOS 9, the system is using predictive information to understand the user. This is now as a proactive assistant feature that now suggests things such as the recipient of an email or when you should leave to avoid traffic jams based of your typical locations [8]. Lastly, another update includes the *note* application qualities. Its new qualities include a variety of formatting options such as drawing or including an image from your gallery. Users can also save links or pictures directly to the note application instead of copying and pasting [8]. It is important to note that these are just a few of the main updates that represent the potential of iOS 9.

The capabilities of iOS are what makes it a major player in the mobile device world. Many users of iOS remain loyal to the operating system. For example, 1.1% of users each month upgrade to the newest model, while only 0.4% switch to the Android operating system [7]. However, during the new iOS launch, the switching rate from Android to iOS increases from 73% to 93% in the weeks following the release. As a result, the customer loyalty drops for Android users [7].

5.2.1.2 Android operating system (Android)

The Android operating system's main purpose, like iOS, is to grant the hardware of the mobile device the ability to function [9]. The operating system is written in the computer coding language Java and is run through a virtual machine. This machine is known as the Dalvik virtual machine which expresses Android's byte code. This machine is the runtime environment for the Android platform [10]. The application code, however, is coded in Java and expressed or executed in the virtual machine Dalvik. To elaborate, the overall architecture of the android application begins with built-in applications or third-party applications [10]. Next, the application framework layer is established, which deals with the content provider, package management, location management, etc. [10]. The native libraries follow behind the application framework. Within the libraries, there is SQLite, WebKit, Secure Socket layer, etc. [10]. Running at the same time as the native libraries is the Android runtime machine Dalvik and the core library [10]. Lastly, the Linux kernel includes the Wi-Fi driver, a flash drive, keyboard driver, and more [10]. This operating system was established by Android Inc. and would later be bought by Google. Google would make Android an open-source project in 2007, allowing companies to use the software freely with altered terms to the license. As a result of the open nature of the Android platform, security is a major concern. The Android platform faces a variety of different threats due to types of applications available to Android customers. The applications provided on the Android official and non-official markets often contain malicious code, because of the freedom the platform offers for application development [10].

Updates in the Android operating systems happen at a much quicker pace than to iOS [9]. In fact, the updates for this operating system are released every few months. The rapid updates allow Android to remain a dominate player in mobile platforms. Every month, around 1.7% of Android users update to the newest smart device, while only 0.3% switch to their competitor Apple (iOS) [8]. Overall, 82% of Android users stay loyal to their operating system.

The only time that Android sees a major decrease in customer loyalty is after the release of new iOS devices. At this time, the percentage drops from 82% to 76% [8]. While Android and iOS are the most popular mobile platform, the BlackBerry platform cannot be counted completely out of the race.

5.2.1.3 BlackBerry operating system

The BlackBerry operating system is only used on BlackBerry devices, such as the BlackBerry Bold, Curve, Pearl, and Storm series [11]. This system is known for its ability to push emails using the BlackBerry Enterprise Server (BES). While Android has the capacity to run on different types of mobile devices, BlackBerry cannot and must only be run on BlackBerry devices [11]. Creating applications for BlackBerry can be done using the Java Micro Edition or the Web development platform. The Web development option utilizes the widgets software development kits which allow applications to be made up of HTML, CSS, and JavaScript code. BlackBerry had once been the smartphone to have in 2009 when about 40% of the United States smartphones were the BlackBerry brand [12]. However, that number has now decreased dramatically to less than 1% of smartphones [12]. This is because the company failed to turn their email-based devices into complete smartphones.

By the year 2014, the company had lost just under six billion dollars after making a last ditch effort and producing its first touchscreen mobile devices. However, recently, the company has been making money off of their operating system with new devices such as the Passport, which is aimed at enterprises. This new device has the highest rating on Amazon [12]. BlackBerry had once been the number one smartphone platform and while it is taking great strides to earn the title back it will likely be a challenging process.

5.2.2 Security Models for Mobile Platforms

Every mobile platform has a security model. This implies that each operating system has security measures put in place to protect the mobile platform. The purpose of the security measures is to protect the hardware, software, and sensitive information that makes up the mobile device environment. In this section, iOS's, Android's, and BlackBerry's security models will be analyzed and compared. Table 5.1 illustrates common security features for mobile platforms such as system security, which protects the hardware and software of the device. Each feature is fully explained through the following subsections. This comparison framework depicts which major security features that the common mobile platform actively incorporates into their operating systems.

Table 5.1 General security features that each platform provides or lacks

Mobile Platform	System Security	Encryption/Data Protection	Application Protection	Network Security	Device Controls (Protects Against Authorized Users)	Privacy Controls	Security Management System IOS
iOS	Yes	Yes	Yes	Yes	Yes	Yes	Yes
Android	No	Yes	No	No	Yes	Yes	No
BlackBerry	Yes	Yes	No	Yes	No	No	Yes

5.2.2.1 iOS security model

The iOS has a number of security options to keep its mobile platform safe. The newest version of iOS is a completely new architecture in regard to security. The security methods in iOS include protecting and working with software, hardware, and services. According to Apple, the model is divided into eight sections [13]. The first section includes system security which focuses on securing the software and hardware for the iOS platform. As seen in Table 5.1, BlackBerry also focuses on the software and the hardware protection. Next, the encryption and data protection secures the user's information from unauthorized users. App security permits applications to run, while the integrity of the platform remains intact. Network security protects the transportation of data through authentication and encryption. Apple Pay's security addresses the protection of mobile payments. The Internet services secure and protect messaging through the platform. Device controls permit the device to be wiped clean if it were to get into the hand of an unauthorized user [13]. Lastly, the privacy controls allow the users to decide whether location and certain data can be accessed. The security model of iOS 9 has been successful and positivity reviewed; however, even though Apple users say that their devices are the most secure, iOS 9 has still faces security vulnerabilities [14]. For example, iPhone users found a way to bypass any passcode and access a device through Apple's assistant Siri. At this time, the only way to avoid this issue is to disable Siri in the settings of the device [14].

5.2.2.2 Android security model

Currently, Android has focused its security efforts on encryption, restricted profiles, and authentication, as seen in Table 5.1. The new full disk encryption method in Android 5.0 allows the user's password to be protected from brute force attacks [15]. Android 5.0 also allows for a guest mode to be created on a device so that personal data are not given to unauthorized users. In addition, trustless authentication allows access to devices when it is near a trusted device or by a trusted face [15]. The Google Play Store also has some forms of protection to verify that an application is not going to be harmful to a device.

The major issue with Android is the overall lack of security. In fact, around 87% of Android devices are vulnerable to different attacks [16]. One type of attack on Android devices is Ransomware [17], an attack that locks down the device until a ransom is paid to the hostage takers. Another issue is the malicious links that are within the Android applications. If a user clicks on the link and it opens in Webview, it can then install an application without the user's consent. There are also applications that do not activate the malware until a later time. This allows the user to think the application is safe until later, when a popup or malicious Websites are presented to the user. Android Installer Hijacking affects almost 50% of Android smart devices [17], although the newer versions of Android claim to be secure from the issue at this time. The hijacking occurs when a user downloads an application and another one is downloaded onto the device instead. This all happens in the background of the device so users are left unaware. After reviewing all three platforms, it becomes clear that there are significant malware issues with Android. Many believe these issues are due to the poor security in third-party Android markets. The issues mentioned previously with the Android platform are far from the true number of malware problem that are out there.

5.2.2.3 BlackBerry security model

BlackBerry security has recently been noteworthy for its success to keep information secure. The release of the BlackBerry Enterprise Server 12 (BES 12) has not had any serious security breaches in the last twelve months [12]. The purpose of this sever is to have a formal mobility management solution for smartphones used in the business world. In addition, BlackBerry is the only platform shining a light on Internet of things (IoT) security [12]. This means that BlackBerry offers protection of data being transferred through a network, as shown in Table 5.1. This seems a huge competitive advantage, especially for their target audiences (e.g., businesses and large organizations). BlackBerry has a very organized security model that includes a number of features. Below is the outline of how BlackBerry organizes their security abilities. To review the meaning of each term, you can refer to the BlackBerry 10 Security Overview [18].

Even with all of the forms of security, BlackBerry still affected by different types of vulnerabilities. The first issue that BlackBerry users may face is with the Adobe Flash, as there are some Flash Player remote code issues that can affect several different types of BlackBerry devices [19]. For an attacker to successfully manipulate this code, a customer would have to access it on a Website or download a version of the Adobe AIR application that was altered [19]. The WebKit remote code (BRST-2013-008) also has a number of vulnerabilities [19]. For an attacker to use the issue to his or her benefit, they would need to create a malicious Website or alter a Website for malicious use. The user would have to use that Website to obtain the malicious JavaScript. In addition, the WebKit code (BSRT-2013-010) also requires an attacker to make a malicious Website. Lastly, the libexf flaw (BSRT-2013-009) requires an attacker to create a malicious image that the user must download and save. Overall, there are issues with all mobile platforms; however, BlackBerry has been able to make great strides in the field of security.

5.2.3 Existing Security Vulnerabilities in Mobile Platforms

In this section, the motives and purposes of mobile attackers will be discussed in detail. Three main motives are information theft, espionage, and sabotage. Information theft occurs when the attacker attempts to steal information owned by an organization or target [20]. Typically, this is accomplished by targeted attacks. This implies that the attackers know exactly what information they need to steal and go after that information. Espionage is the monitoring of activities that may compromise a particular target [20]. Sabotage uses blackmail to get the target of the attack to accept certain demands. In Table 5.2, the amount of malware in popular mobile platforms is depicted. It is clear that the Android makes up the majority of malwares, with 79% of total malware issues. BlackBerry makes up only 0.3% and iOS is only at 0.7% [21]. The other lesser-known mobile platforms are still significantly less than the Android market. Table 5.2 represents the amount of malware each platform is responsible for creating due to their own respective vulnerabilities.

5.2.3.1 Potential vulnerabilities

One of the major vulnerabilities that mobile platforms face is social engineering. Social engineering is a concept that hackers use to trick humans into breaking security protocols [22]. This is typically done to steal sensitive

Table 5.2 This table depicts the amount of malware mobile platform are responsible

Mobile Platform	Malware Occurrence
Android	79%
Symbian	19%
iOS	0.7%
Other	0.7%
Windows	0.3%
BlackBerry	0.3%

information from a user that can be used to benefit that hacker's agenda. This can be done in a number of ways including phishing, pretexting, baiting, and spamming. In addition to social engineering applications, downloads can also cause major issues for mobile platforms. This is a major flaw that Android must overcome due to its third-party markets. While there is protection for the Google Play Store, additional markets are not regulated and caused a number of malware issues for Android users. As mentioned previously, network services are on the security radar for all of the mentioned platforms. This means that the data being transferred over a network is protected in some form. For iOS, this is through authentication and encryption [13]. If the network was to be breached, this could cause major issues for any mobile platform. Bluetooth is also a potential vulnerability that can spread malware to devices in its proximity.

5.2.3.2 Mobile device malware

There is a numerous amount of different malwares in existence today. A virus is a certain type of malware that infects software once it is executed [23]. This typically occurs when information or data are being shared between devices. A Trojan is a type of virus that appears to be a safe application or Webpage that tricks the user into accessing the malware unknowingly. A worm is a type of malware different from a virus that can replicate itself and does not require user intervention for execution. Adware is a type of malware that can be on a mobile device with no replication capabilities. For example, certain ads may appear in an application based on the user's Internet searches, making the user more likely to click the link. While not all Adware is malware, it can be which cause major concerns for users [23]. The total amount of malware Adware accounts for 17% of all malware, as illustrated in Table 5.3 [24]. Spyware is a type of malware that spies on a user by tracking activities by using tools such as keyloggers.

Table 5.3 This graph illustrates the distribution of the malware for mobile devices

Malware Distribution Method	Occurrence
Sending SMS	24%
Aggressively displaying Ads	17%
Malicious software	15%
Remote access of the device	13%
User data stolen	05%
Malware download	04%
Monetary theft	04%
Other	16%

This type of malware accounts for monetary theft and user data being compromised as depicted in Table 5.3. Organizations often use keyloggers to view what their employees are doing throughout their work day [23]. This type of malware was once only a concern for computer users; however, mobile devices have become handheld computers. This means that every type of malware a desktop can face, a mobile device has the potential to be vulnerable as well.

5.3 Threat Model for Mobile Platforms

In this section, we present the threat model for mobile platforms. Our threat model is composed of three main components. First, we discuss the factors that motivate an attacker to breach mobile device security. Second, we discuss the modern exploitation techniques used by an attacker to accomplish malicious activities such as inserting a malicious content into a mobile device. Third, we discuss some of the common types of malwares that are particularly designed to target mobile devices. Figure 5.1 illustrates these three components as well as their sub-components.

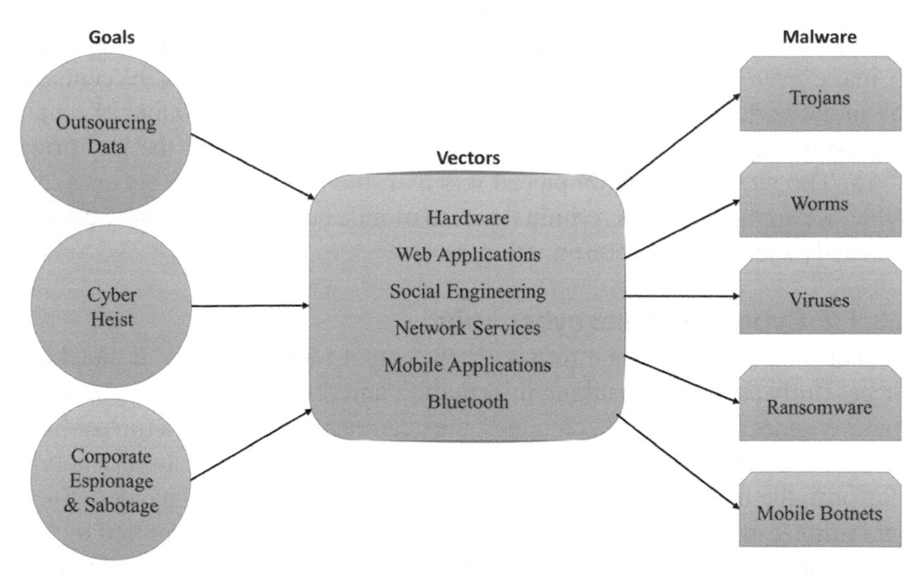

Figure 5.1 An illustration of an attack tree showing the typical relationship among attacker goals, targeting mobile devices, attack vector, and a choice of a malware.

5.3.1 Goals and Motives for an Attacker

This section presents some of the common goals that attackers try to achieve when breaching the security of mobile devices by exploiting common vulnerabilities. Identifying a malware's motivation on smart devices is paramount to gain a better understanding of its behavior and can be used to develop targeted detection strategies [3]. Such goals range from fraud and service misuse that is driven by economic incentives, to spamming, espionage, data theft, and sabotage.

5.3.1.1 Cybercriminals: outsourcing sensitive data

Cybercriminals who target mobile devices often do so in order to access personal information. Personal information can range from local mailing addresses, credit or debit card numbers, banking information, email addresses, phone numbers, and much more. Cybercriminals can then do one of two things: Sell their vast information stash to a potential buyer or keep it for their own use. The former choice is most desirable. In the underworld filled with criminals of all sorts, information is highly valuable for making money or other malicious deeds. In the black market that keeps the criminal world thriving, hackers that obtain personal data sell it to a variety of buyers, including identity thieves, organized crime rings, spammers, and botnet operators, who use the data to make even more money [25]. A simple name that is worth 99 cents in the underworld may be the target the local mafia is after. A couple of email addresses may be the best debut for a new spammer, only for the low price of $5. The more information pieced together, the more damage occurs [25]. When we apply these cybercriminal acts to mobile devices, many more people are easily robbed of their information.

5.3.1.2 Cybercriminals: cyber heist

Hackers can also keep the information they steal for themselves. If attackers successfully acquire all banking information and other personal data through social engineering or other techniques, they can easily act as the victims online. Identity theft was already a problem on users with computers. In the mobile platform, the methods of extracting user's information have expanded. Such data mining methods include intercepting data as they are transmitted to and from mobile devices and inserting malicious code into software applications to gain access to users' sensitive information [26]. The attackers can then use the victim's credit card to buy whatever they want, or withdraw funds from their victim's bank account. Other than stealing cash from the victim through stolen information, more innovative techniques have surfaced. On the mobile platform, many apps exist that a user can easily download as he or she pleases. A user downloads a seemingly harmless app, only to be locked out of their phone unless they pay the app a fee. An attacker can use less conniving methods to earn cash. For example, Adware is a popular form of malware used by mobile attackers. They can create an app, useless or not, and populate it with advertisements, generating revenue per specific amount of views [27].

5.3.1.3 Cybercriminals: corporate espionage and sabotage

Big companies such as Amazon, Netflix, and Apple possess personal information about each and every customer they have, in addition to their own unique trade and technology secrets. All it takes is a couple of cybercriminals to hack into the database system or user's device in order to gain the required credentials and they will have access to the wealth of information a company or business has. Take the 2012 incident as an example regarding the data breach of the online retailer Zappos [25]. Cybercriminals gained access to 24 million customers' Zappos profile information with the exception of customer's credit card and sensitive payment data [25]. What this demonstrates is that even if a company promises that its Website security is solid, its defenses are not impenetrable. Gaining access to corporate networks is often accomplished via external attackers or internal workers. Company employees form the latter.

In our digital age, almost every person has a phone or smartphone. At a company such as Zappos, it is safe to assume that most, if not all, employees had a phone. An attacker can simply acquire login credentials for an employee and simply login to the company database without a hitch. Using the data mining techniques stated earlier, an external attacker can easily act as an internal user without anyone knowing. The company will not discover the breach because the attack was done on an employee's personal phone, not through the company's networks. In this scenario, attackers do not have to be external or underworld criminals at all. An employee who wishes to exact revenge on his cold-hearted boss may access the database at work and outsource personal information or trade secrets to others. A rival business may want to acquire the fiscal earnings' report or secretly attack through the mobile platform. In essence, the mobile platform is an easy target for malicious attackers to enter the secure environment of a company.

5.3.2 Attack Vectors or Modern Exploitation Techniques for Mobile Devices

In this section, we discuss the modern exploitation techniques commonly used by attackers to accomplish malicious activities such as inserting a malicious content into a mobile device.

5.3.2.1 Susceptibility on the mobile through hardware

One of the most basic methods an attacker can use to steal data or wealth from a user's phone is by a direct access. These days, it is easy to entrust your smartphone to a friend or a store employee, or a stranger for few seconds. Less common is the theft of personal mobile devices, but it is relevant to the hardware-related attacking vector. Inside most smartphones, there exists a subscriber identity module (SIM) card which is owned by a mobile network operator or MNO. Such operators include Verizon, T-Mobile, Sprint, and Virgin Mobile. An outside source that was able to acquire a smartphone for malicious motives can either target the SIM card or the mobile device.

An example of targeting the SIM cards themselves is the TurboSIM technique. Within a mobile phone, the device must communicate with the SIM card in order to accomplish variety of tasks. This communication channel between phone and SIM card is not protected. Therefore, an attacker may insert a TurboSIM card into the device itself which interferes with the communication process [28]. The attacker may then eavesdrop on messages sent from the phone, send messages to the phone itself as a trusted source, or simply remove the SIM lock placed on the SIM card.

An attacker may also target the phone itself through forensic analysis or joint test action group (JTAG) card method [29]. An attacker uses forensic analysis by directly inspecting the mobile device's hardware and targeting stored data [8]. Such a method can be mitigated by a multitude of encryption schemes and other security methods on the SIM card, memory cards, etc. The JTAG is a standard for testing and debugging hardware on a variety of devices including the mobile platform [28]. JTAG, if available on phones, presents a vulnerability risk to users. JTAG works by using a boundary scan, which when matched with the correct port in the device, can access processor and memory [29]. JTAG, originally designed for the developers of the mobile device, however, can end up in the wrong hands if the mobile device is lost. With the right skillset, an attacker can do anything to the mobile device and its stored data using only the JTAG method.

5.3.2.2 Attacking through the Web

The browsers of mobile devices, such as Google's Chrome or Apple's Safari, are similar to their PC counterparts except these browsers must cater to the main functionalities of the mobile device. For example, the phone's Web browser must be required to allow the user to answer or receive voice and video calls at no delay, but the requirements often depend on which browser app is being used. This requirement brings forward the fact that Web browsers for mobile devices are also applications in themselves, and their security often conflicts with the security of the device and other apps [28]. Think of it this way, it is wise to download and utilize protection software for your computer, such as Symantec-related products. However, there will often be advice or suggestion stating that it should not run alongside another virus- or malware-related software, because each software is different and may interfere with one another. Running Symantec alongside McAfee virus software may seem like double the protection, but often it is hindering the performance of the other. Like the example just given, the difference in security between a mobile Web browser and the mobile device itself can be exploited by any attacker.

Another threat that is related to mobile Web browsers includes what is called geo-inference attacks. In each mobile device, the user may turn on the location feature to use for apps. Such apps include Google Maps, SnapChat, and even the Web browsers themselves. An attacker can set up their own Website, which, when accessed by a user either on their PC or mobile device, will start to "sniff" through the browsing history and browser cache of any location-oriented Website they visited via timing-based side channels [30]. Web and mobile developers alike will need to increase security measures when it comes to the Web browsers and mobile features such as location to combat the emergence of innovative threats.

5.3.2.3 Mobile intrusion and deception through social engineering

Today's society is steeped in social media more than ever. People communicate with each other through venues such as email, Facebook, Instagram, and other sites. Social engineering does not require skills of a top-class hacker nor the skills of an amateur. One must, however, possess a knack for deceiving people online. We have all been victimized by social engineering techniques. An attacker may have acquired a large stash of email addresses to spam their devious ploy: "Your rich grandmother has died and willed her entire fortune to you! All you have to do is send the Nigerian caretakers a few hundred dollars in order to have the fortune shipped to your address in no time." On social media, an attacker can comment on public posts about suspicious job offers or bare links which will redirect victims to Websites they created to phish information. Another method most people have experienced is the numerous advertisements seen on social media sites placed by both the Website itself and adware placed by other outside sources.

This technique is part of a multitude of schemes to fish out cash, personal information, or other sensitive data from a user. Web applications created in different countries also add to the issue. Chinese-based Web applications such as social media Renren, or QQ, may have Chinese personnel committing cyber espionage on corporate or government files through employees' mobile devices [31]. Given that China had a couple of incidents on record for alleged cyber-attacks on U.S. government entities and other valuable targets throughout the years, security on foreign-based Web applications is in focus. However, with the unlimited growth of the Internet, it is a difficult task to control the vast number of Websites and applications that emerge.

5.3.2.4 Attacking through the mobile network

A mobile phone is either a GSM or CDMA network-reliant device. These two radio networks make it possible for one to call or message another at anytime and anyplace. GSM stands for global system for mobiles while CDMA stands for code division multiple access [32]. Let us take a look at the global common variety—GSM. GSM is a radio network, relying on connecting mobile devices to the supporting network operator's base via air link [28]. This channel is encrypted to keep it secure. The SIM chip of all GSM phones encases all sensitive data a user has, and is accessed by the mobile device by using cryptographic algorithms A3, A8, and A5 for authentication, key derivation, and encryption, respectively [28]. How can outside sources have the opportunity to bypass this secure "handshake" between SIM and a phone? To understand what requires accomplishing these malicious tasks, let us take a look at the weaknesses of the GSM network.

First, one weakness is that the GSM system encodes a message first and then encrypts it [28]. This particular order of generating the message leaves much uptime on the data being sent. Second, GSM phones in developing countries often use weaker states of encryption such as A5/2, which combined with the first weakness, provides the window of opportunity for attackers to strike [28]. Stronger encryption schemes such as A5/1 are much more secure, but weak versions such as A5/2 suits countries less advanced than the developed countries. By using their own cryptographic methods against the weaknesses of the network, they can trick the handshake system implemented by the GSM which results in conversation eavesdropping by the mobile device. This allows the attacker to accomplish his or her goals listed earlier in Section 5.3.1.

5.3.2.5 Cyber Arson through common mobile applications

Every smartphone in the world today contains variety of applications. Ranging from the business-related TD-Ameritrade app to games such as Angry Birds, mobile applications help pass the time or, to the opposite effect, increase productivity. Yet, some developers or attackers create/mix malicious code with certain applications that can drain battery life or spy on the user's actions [33]. With every app downloaded, there are always requested permissions such as access to photos or other data files, location, and personal information. In order to use some apps, Facebook for example, these permissions are required. Yet for games such as Candy Crush Saga, one must really connect its Facebook profile with the app. Malware-afflicted apps can be used for any purpose including eavesdropping, data mining, or stealing money. An app may lock a user out of his or her phone entirely until a ransom fee is paid (see ransomware above) or could be secretly sending short message service (SMS) to an expensive service number to accrue profit for the attacker [28]. An example of this SMS method is illustrated in Figure 5.2. The victim of the latter will only notice the malware in action when the phone bill arrives. For mobile applications, one must be vigilant for which apps are trustworthy and never entrust sensitive information to third-party applications unless necessary.

5.3.2.6 Attacking via Bluetooth connection

Another common method of communication between digital devices these days is Bluetooth technology. Bluetooth allows short-ranged communication via short wavelengths to connect with "enslaved" electronic devices [34].

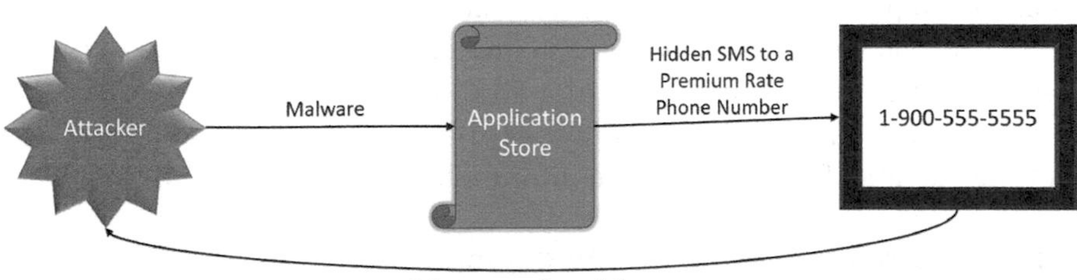

Figure 5.2 The SMS trick through mobile applications. The attacker uses a malware of choosing and sends it to app store or market. Victim unsuspectingly downloads the app and attacker then sends covert SMS to a premium-rate number they own to have the victim pay.

Bluetooth devices range from Bluetooth-enabled speakers, laptops, mobile devices, etc. While it is short range, which means a lower chance of an intruder interception, it does not mean it is impossible to breach. An attacker who successfully penetrated the Bluetooth communication can inject its malicious malware numerous times, complicating security measures of the victimized mobile device [28]. Bluetooth is a double-edged sword. Each time it is activated, it broadcasts the information of the phone such as its unique name, list of services, and other technicalities [34]. One can connect to their beloved Bluetooth speaker, but at the same time, a nearby attacker can easily note the vulnerable mobile device and take actions. Some examples of Bluetooth attack methods include BlueSmack, BlueBug, BlueSpoof, and BlueBump [34]. These attacks vary from causing massive data overflows, resulting in a distributed denial of service (DDoS) attack, or utilizing software to completely hijack a victim's phone.

5.3.3 Types of Malwares in Mobile Devices

In this section, we discuss different types of malwares commonly designed to target mobile devices which include Trojan horses, worms, and a family of viruses.

5.3.3.1 Trojan-related malware

Trojans are common malware that cause misfortune to PC users. However, with the dawn of the smartphone age, especially the Android market, Trojans are beginning to cause trouble for mobile devices. A Trojan is essentially any software that appears to have normal functions but covertly contains malicious coding [35]. Apps that may seem like some harmless games could contain a malicious Trojan program: takes Zeus, for example, a Trojan mostly used by attackers for targeting banking applications. Messages sent from a trusted banking company can be intercepted by attackers using Mitmo, or man in the mobile, techniques. Then, attackers can implement the Zeus Trojan to fake a certificate from the banking application [36]. People who are not tech savvy or cautious can easily fall for this fake certificate and proceed to an SMS or Website that asks for more personal information, only to be faked.

5.3.3.2 Worms targeting mobile devices

Worms are another set of malware that can infiltrate and harm an electronic device. Unlike Trojans, worms are programs that make copies of themselves, infecting devices through each other and use different transportation mechanisms through the network each time, all under the nose of the victimized user [35]. Worms were prevalent in computers, but are just as easily implemented in mobile devices. Take the first worm ever observed in mobile devices, the Cabir. Cabir was masterminded by an eastern bloc hacker group, first discovered in 2004 when it used Bluetooth technology to spread to multiple mobile devices using Symbian OS [35]. Worms are not as harmful as viruses or Trojans, but can considerably worsen the efficiency of an infected mobile device.

5.3.3.3 Viruses on the mobile

Viruses are like worms, yet slightly different. Even though they also make copies of themselves, viruses are set pieces of malicious code that can attach themselves to virtually anything in a PC or mobile device [35]. Viruses can infect programs, data files, pictures, games, movies, etc. The mosquito virus that infected a computer game illegally pirated to a person's mobile device allowed the attacker to hijack the phone covertly and use it to their pleasure [37]. Viruses can give way to something even more malicious than a single colony of viruses: the botnet. A botnet can be likened to a zombie army being controlled by one master. An attacker can send viruses to multiple mobile devices, infect them, and control them as a group to enact further schemes such as sending mass spam emails or a coordinated DDoS attack [35]. It is important for users of mobile devices to think about investing in legitimate antivirus software to add protection against viruses and also increase their overall awareness of malwares.

5.3.3.4 Ransomware: a mobile kidnapping

Ransomware is a type of malware related to the mobile applications vector. An attacker who utilizes ransomware aims to gain mostly cash or some form of imbursement from the victim. Ransomware is mostly a Trojan or a virus that infiltrated the mobile device and encrypts user data, requiring the victim to pay up in order to receive the appropriate encryption key [28]. This is why it is very important for each user, especially those who own the more "open" Android-based mobile device, to be cautious what data they store on their phones. Some people keep important data such as banking information and passwords on a note widget or took a couple of photos of their SSN for emergency purposes. Even on the less vulnerable iPhone device, it is still possible to be infiltrated. People will often jailbreak their iPhones in order to have the sort of privileges Android phones do. However, when users jailbreak, they end up creating a vulnerability for attackers to exploit. An attacker can scan via Internet IP addresses of jailbroken iPhones and upload its ransomware app for them to unsuspectedly download [35]. The ransomed iPhones can then be used by the attacker to secretly install a worm to spread the ransomware to other iPhones and hold the phones ransom.

5.3.3.5 Mobile botnets

As stated earlier in the virus section, a botnet is essentially a swarm of infected devices used together for malicious intentions. In canon, a botnet is defined as a "set of devices infected by a virus that gives an attacker the ability to remotely control them [35]." In mobile devices, a botnet could be more personal and closer than ever before. The diverse communication channels of the mobile world along with the fact that they are almost always turned on for use means, the threat of a botnet attack can come at anytime, anywhere [28]. Botnets can be used for activities that require a large-scale operation such as DDoS attacks or mass spamming.

5.4 Defense Mechanisms for Securing Mobile Platforms

Our mobile devices are always subject to malicious attacks such as from suspicious third-party applications or through Bluetooth interceptions. However, security software and measures evolve parallel to the rise of outside threats. The three mobile platforms of today increasingly demand social society to be protected and those security measures can come from phone developers or third-party companies that seek to secure mobile devices. In this section, we outline various common security methods and present tables detailing which mobile platform offers which security measure and a summary of how the defense techniques work. An illustration of mobile platform securities is present in Figure 5.3.

5.4.1 Keychain Authentication and Encryption

Mobile platforms these days will often contain default encryption settings already installed in supported devices. They also share what is called a "keychain," which is a secure storage for protecting the sensitive files. They often contain a device key paired with a user passcode that protects all applications and iOS software-sensitive data from external attacks [38]. This code prevents a stranger from accessing your iOS's valuable contents, whether the phone was lost or stolen. The keychain API also protects the information even when backed up. When a phone is backed up, the keychain password is not included in the backups, meaning that a perpetrator who was able to acquire the phone user's backup data cannot access the keychain's secured

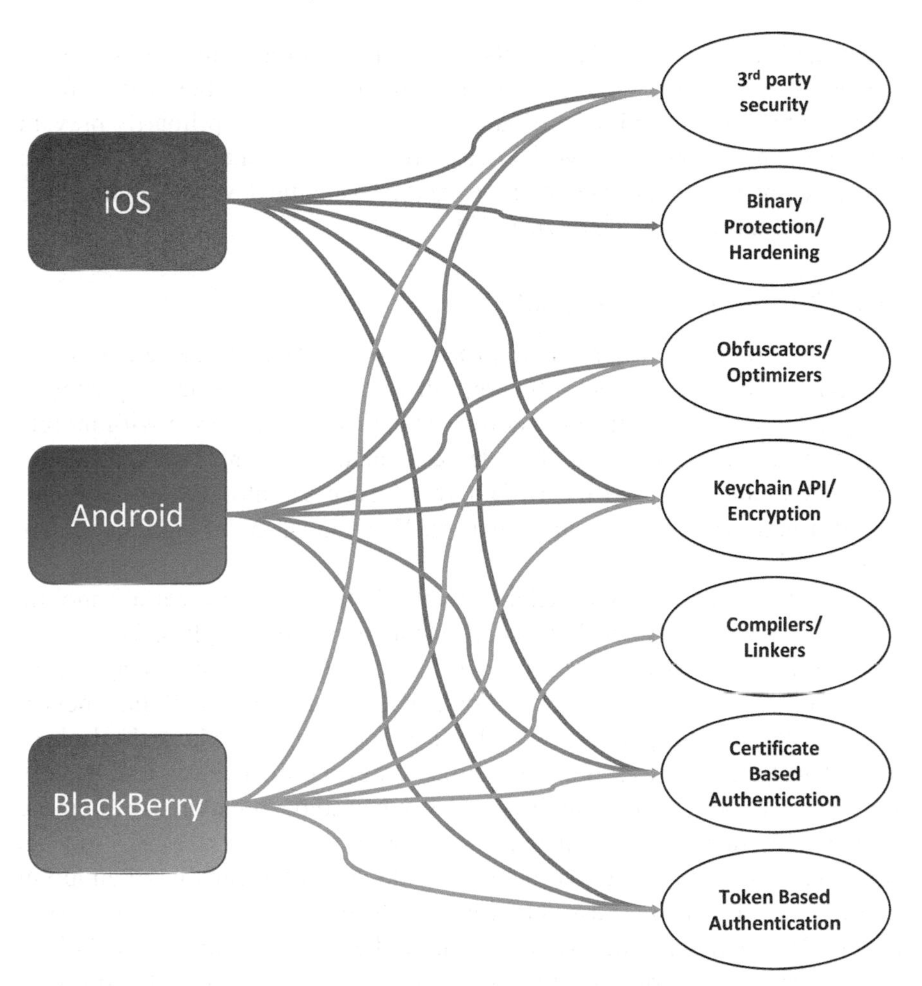

Figure 5.3 Mobile OS security implementation chart.

contents [39]. The downside for this defense mechanism is the possibility of brute force access and advanced hacking schemes that can easily undo common encryption schemes on the mobile device.

5.4.2 Binary Protection and Hardening

This method of security has seen a trending rise in the iOS platform [40]. What is the meaning behind protecting binary? The art of binary hardening is modifying the binary of applications and software of a mobile platform to strengthen security and resistance to attacks such as reverse engineering, debugging, and patching [41]. Unlike keychain APIs, binary protection focuses more on the structure of applications and software, making it less vulnerable to brute force login attempts and techniques. Binary protections serve to primarily prevent apps operating in unsafe network environments, harden memory encryption, obfuscate reverse engineering attempts, and prevent the device from malware threats originating from the device [42].

5.4.3 Third-Party OS Products

A common and often easier way to protect a user's mobile device is through third-party applications and software, either during phone development or through the device's application store. These products designed with mobile security in mind include 1) mobile device management (MDM) and mobile application management (MAM), 2) network access control (NAC), 3) security information and event management (SIEM), and 4) point mobile security (PMS) [43].

Each product type offers defense methods that focus on certain mobile vulnerabilities. MDMs and MAMs are what their name implies. They track how the device is used, can utilize remote location tracking, manage apps, register the device, and even support data loss prevention (DLP) that encrypt secure containers [44]. The downside of MDMs and MAMs is the lack of threat detection and prevention capabilities. They require the user to manually control how to protect their mobile devices, when some threats can be far more advanced for the average consumer to successfully apprehend. NAC products serve to govern who can access the mobile device, often capable of endpoint security checking, and access enforcement. While NAC products can ensure that the OS is updated with endpoint security checking, it cannot detect hidden attacks that were inserted in an OS's patch version, installed applications, or network activity.

SIEMS functions the same as the intrusion detection systems for a computer unit. SIEMS allows a user to monitor and manage network activity, analyze and log information, and detect threats to the mobile device [44]. SIEMS is effective in that they have the potential to allow users to make the decisions to prevent threats before their devices are infected. However, SIEMS is not effective in the security manner, since it can only detect threats but cannot stop and quarantine a threat alone. PMS applications are what the general public would think of when it comes to mobile security. PMS software for mobile devices includes mobile antivirus, mobile antimalware, ad-blocks, and mobile authentication and network security [44]. These third-party applications are often more intensive than mere MDMs or MAMs but the problems remain. A user who downloads a simple antivirus-secured app may not be protected from malicious malware as effectively, and vice versa. A top of the line third-party PMS product that completely encompasses a mobile device's vulnerability scope is still far from the market. Even now, mobile products that combine two or more of the services offered above end up being costly and taxing on the device hardware.

5.4.4 Obfuscators and Optimizers

One of the reasons why Android is so popular to launch an attack on is its diversity in API and the numerous ways a developer can code. A reverse engineer can undo even the most complex code structures and turn what was once a secured application into a vulnerable one. Obfuscators offer a service that can increase the time and possibly deter a reverse engineer from hacking into a user's mobile device, while the optimizer simply optimizes the obfuscated files. The Android developing kit comes with obfuscating and optimizing software, such as Proguard [45]. However, one can purchase more advanced obfuscator and optimizer software for added security and variation of methods. Functions that support mobile security include string encryption, class encryption, reflection to hide API calls, tamper detection, removal of dead code, and renaming class variables [42]. While they are a viable choice of protection for developers, it is important to note that attackers may also utilize obfuscators to damage a mobile device and its contents.

5.4.5 Compiler and Linker Defense Mechanisms

A defense option that is available to developers lies right in their own equipment. Code compilers and their accompanying linkers can provide in-depth security, depending on mobile platform. Some of the features this defense method includes are as follows [42]:

1. Stack Cookies—Protect against stack-based overflows.
2. Relocations Read Only—Protects against overwrite of the relocation section which contains function pointers.
3. Bind Now—Loads all library dependencies at load time and resolves them, allowing the Global Offset Table to be set to read only and protect from direct overwriting.
4. Position-Independent Code/Executable—Allows libraries and program executable to benefit from address space layout randomization by not assuming it will load at a particular memory address.

Developers can implement this style of security starting at the moment they get their hands on the BlackBerry's native software development kit (SDK). If the developer secures the device with the correct configuration, one can remove vulnerabilities that would allow attackers to take control of a system, and instead only cause an application crash [46]. BlackBerry achieves this method on a common basis more than their Android and iOS counterparts.

5.4.6 Certificate-based Mobile Authentication

Most IT devices today are protected by authentication certificates or digital certificates. They are the gateways into one's network, device, application, etc. Certificate authorities (CAs) are able to secure many devices and software and easily support all current mobile OS platforms [47]. Much like its name implies, this type of authentication relies on a series of trust certificates and would synergize well if paired alongside a MDM product.

Certificate authentication works by utilizing a secure and trusted CA, and communication of the encryption keys—known privately by the owner (private key) and the encryption key that is known to all public entities (public key) the owner is in contact with [48]. This method of defense has been widely implemented on the PC side but also works well with mobile devices, improving the experience of the mobile users. Certificate authentication allows a mobile user to access vulnerable applications and software without the need of typing tedious usernames and passwords and allow companies to use freeware CA software or collaborate with a third party [47]. While CAs have been proven to ensure device security for years, the attackers are finding more ways to crack this method of defense, launching the need for users to use more than one method of mobile defense.

5.4.7 Token-based Mobile Authentication

Despite CAs still holding popularity in securing PCs and mobile devices, its vulnerabilities are ever-changing and dangerous. Token-based authentication serves to enhance the performance of other security implementation services, or even work as a standalone product. Token-based authentication revolves around the mobile's hardware, relying on user interaction to generate a password created with a timestamp synchronized with a server application [49]. This method ensures that the user alone will have to access a token-based protected device rather than two users: one the victim and one the imposter. Rather than a hacker who may employ techniques such as brute forcing or deceiving the victim with spam email, the mobile device will not break unless the required hardware usage process is completed [49]. Token-based authentication may prove costlier than the usual CAs or less adequate than biometrics but they are quite formidable for mobile authentication security.

5.4.8 Summary

Each mobile OS is built differently and thus may not have the ability to defend a certain way another OS does. Understandably, it is paramount that developers for a mobile OS need to implement more than one security method to ensure a secure experience for customers. Customers will also need to understand the significance of utilizing third-party apps and common knowledge of applications and social media that may contain malicious threats. Table 5.4 provides an overview of what each OS offers in terms of security.

5.5 Related Work

Authors in [50] discuss the vulnerabilities and security solutions that were from 2004 to 2011 on high-level attacks. The solutions are grouped into categories according to the detection principles, architectures, collected data, and operating systems, with a special focus on intrusion detection system (IDS)-based models. Also, the authors have found that mobile malware is still small compared to the PC malware but growing at a faster rate. The solutions founded by the authors indicate differences in the PC environment since smartphones have limited resources available with a large number of features that can be exploited by some attackers. La Polla et al. [50] state that there are four different operating systems that can be exploited: Symbian, Android, Windows Mobile, and iPhone OS. Symbian is an open-source OS that is used by Nokia's smartphones. The interface components are based on the

Table 5.4 Security services of mobile platforms

	iOS	Android	BlackBerry
Keychain API	✓	✓	✓
Third-Party Security	✓	✓	✓
Binary Protection	✓		
Obfuscators/Optimizers		✓	✓
Compilers/Linkers			✓
Certificate-based Authentication	✓	✓	✓
Token-based Authentication	✓	✓	✓

S60 5th Edition. This operating system is on its third version. A monitoring approach can be used to extract features that could be exploited. In addition, they found that certain vulnerabilities that are exploited can render this type of operating system unusable. The Android operating system is based on a modified version of Linux kernel. The applications used on this type of device are developed by a community of developers to allow the user to find new functions to add to the phone. To check for security flaws in this operating system, evaluations were done to analyze the Android framework and Linux kernel. Security solutions have been proposed for mitigating the Android threats. Windows mobile OS was developed by Microsoft. This operating system is based on Windows CE 5.2 kernel. This was developed to be similar to the desktop version of the Windows. The first version has been replaced by Windows Phone 7. There is a security system called Window mobile malware detection system that detects malware for Windows-based platforms. The iPhone iOS is a mobile platform that was developed by Apple. Apple uses a vetting process that ensures that all applications conform to Apple's rules before they can even enter the Apple Store. This is a good approach but the vetting process is not well documented so some malicious applications have been able to gain access to the Apple Store.

Lopes et al. [51] explain that 96% of all smartphones do not have preinstalled security software in place. The current security solution software that is used in PCs is not available to use on mobile devices. Since smartphones are small and convenient, many people do personal tasks on them which allow attackers to exploit and gain access to their personal data. The authors have identified four areas that are susceptible to threats. That includes security of mobile devices, operating systems, mobile databases, and mobile networks. A solution to device security is to install anti-phone-theft software. The software allows the user to remotely access the device. In addition, the use of a security PIN will make the phone less desirable to a thief because they cannot gain access to personal data. To stop threats to the operating system, vendors should ensure that security is built into the core of the OS. Solutions proposed by the authors in [4, 51] suggest that the data execution protection and layout randomization are critical for a comprehensive security mechanism. Encryption is also a good way to protect data, add antivirus software, and ensure secure password storage. In addition to encryption, secure boot functions and antimalware defenses are needed. Solutions to ensure the security of mobile database include adding virus scanning, implementing firewalls, utilizing some form of user authentication, using industry standard cryptography, adding firmware/software patches, implementing Router/Firewall Configuration Changes, applying auditing and logging of everything but sensitive information, and, finally, making sure that sensitive data are not saved in plain text. To protect against mobile network threats, a password should be changed every 60 days. Browsing history, system caches, picture caches, network caches, installation log, viewed SMS, and viewed email should be deleted from the phone.

The Apple iOS security model makes sure that third-party applications are in an isolated environment so that the application is only able to access its own data and permitted system resources. This being said all third-party applications are granted the same data and capability access (i.e., location and notifications). Application distribution is mandated to go through a manual review process with restriction based on policies regarding data collection API usage, content appropriateness, and user interface guidelines compliance. The assumption is that the device is not "jailbroken" and only apps from the Apple App Store will be downloaded to ensure that the phone is safe. Android OS has a security model that supports open application distribution. In this model, applications are gated access by permissions that are accepted when downloaded options cannot be changed. After reading the permissions, user may choose not to install the application for various reasons. In addition to this, if users find the app suspicious, they can report it. Since there is no formal review of applications before being placed in the market, this model heavily relies on the user being able to evaluate the permission and make their own decision on whether to install or not [51].

Delac et al. [4] found that smartphones are becoming, in many ways, more and more like PCs so mobile security should be assessed similarly as if you were trying to secure a computer. There is a potential to download worms, Trojan horses, or other types of viruses that can compromise a user's security and privacy. If the phone becomes compromised, the attackers could have access to voice-recording, photo captures, and SMS messages, or even gain the user's location. Authors in [4] present an attacker-centric threat model. First, the motive is stated to gauge what interests that attacker and what will be targeted. Next, the model takes attack vectors to present where the attacker may enter into the device. Lastly, the threat types are taken into consideration. Some of the threats that an attacker may use are collecting private data; the use of Bluetooth to spread malware from device to device; compromising the USB and other peripherals; and use of mobile malware, Trojan horses, Botnet, worms, and Rootkit. To prevent these attacks, Android relies on user judgment to install an application or decide not to since it could harm the device. Delac et al. [4] suggest that Android permission-based security model needs improvements. Also, for iOS devices, isolation should be applied. However, if it were to become compromised, a malicious application would have access to critical system and private data. This is especially a problem if the device is jailbroken.

Shabtai et al. [52] focus on evaluating the security solutions for Android devices. They proposed a criterion for evaluating the security mechanisms that are just for Android devices. Attacks could range from social engineering to hardware where cybercriminals exploit vulnerabilities that exist in human nature, software/hardware, and network. The exploitation of such vulnerabilities results in many security concerns such as privilege escalation, third-party library vulnerability, Dalvik vulnerabilities, man-in-the-middle attack, return to libc attack, JIT-spraying attack, Android debug bridge, and kernel attacks. To minimize these vulnerabilities, the security solutions that were proposed include antimalware, firewalls, access control mechanisms, spam filters, automated application analysis, data leakage prevention (DLP), security extensions, and mobile device management (MDM).

Sujithra et al. [53] proposed a new way to secure mobile platforms in a remote cloud using cryptographic techniques with minimal performance degradation. Mobile cloud computing (MCC) services allow the user to adjust processing and storage capabilities transparently and offloading intense and storage demanding jobs. The problem with MCC is that there is a high risk of unauthorized access to the data. Using a three-tier encryption method to overcome this problem, a hybrid model of cloud architecture was proposed. The first tier encryption is done using the message digest (MD5) algorithm with the key. Second tier is then encrypted using AES algorithm. Final tier further encrypts using either elliptic curve cryptography (ECC) or RSA algorithm. Zhang et al. [54] indicate that current protection for Android-based phones has many shortages in security. To address these security weaknesses, the authors in [54] proposed a browser-free multilevel smartphone privacy protection system. Protection is ensured by SMS. This is done by the user sending a SMS to the phone remotely with operating instructions. The sensors on the remote phone then execute the instructions and return the information. Also, the sensor on the daemon process mechanism is used so that the sensors do not become closed and uninstalled. This proposed system adopts a SIM-detecting mechanism to indicate whether or not there is a SIM card or if it has been changed out. If it has been deemed that there is a change, the phone is locked by the inside sensor.

5.6 Threats Analysis and Future Trends

Threat trends are on the rise in every aspect, from cyber-attacks to physical attacks. Richard Piggin [55] recently published an article that outlines what cyber security trends are most likely to keep CEOs up at night. He states that according to recent reports from the U.S. Department of Homeland Security, cyber attackers are becoming more sophisticated which will lead to more destruction and loss. These attackers are using control systems to penetrate and wreak havoc [55]. Table 5.5 illustrates several security applications for exploited mobile operating systems.

Gajjar and Parmar [56] conducted research to provide proactive and reactive solutions for information security to prevent mistakes that users ultimately make. These mistakes range from granting permission without understanding the consequences, avoiding security features built into the device, and storing information without encrypting it. The authors state that attackers are becoming more equipped and consumers are not protecting themselves well enough against attacks. Some solutions that were proposed were to look at the security installed in the phone over other features when purchasing, using screen locks, avoiding downloading unauthorized apps, updating and backing up the phone regularly, checking permissions, encrypting data, turning off Bluetooth, Wi-Fi, and location services when they are not in use, and many others. Also, they proposed reactive steps for security such as reporting stolen devices, erasing lost device if conceivable, formatting devices to detect malware, and reporting incidents of impersonations [10].

After extensive research, we have found that mobile security is a problem that has been going on for years and will not be solved anytime soon [4, 50, 54]. With the increased sophistication of attackers, the field is ever-changing and consumers are completely ignoring the safety measures that have been put into place to somewhat protect them. In this chapter, we have presented comparisons of some of the vulnerabilities and threats that are known to mobile devices as of now; however, it is unclear what is to come with how fast

Table 5.5 Security applications for exploited mobile operating platforms

Exploited Operating Systems	Interface	Security for Applications
Android	Linux kernel	User based judgment
Symbian	S60 5th Edition	Isolation
Windows Mobile	Windows CE 5.2 kernel	Window mobile malware detection
iPhone OS	BSD UNIX kernel	Vetting process

technology changes even from day to day. We have found that 95% of mobile devices do not have any form of security software installed [53]. This would be an easy fix if someone were to develop an universal security software that could protect mobile devices. Apple has the most security control in place when it comes to mobile apps. They keep them in an isolated environment, so that the application is only able to access its own data and permitted system resources. Android OS, on the other hand, has a security model that supports open application distribution. In this model, applications are gated access by permissions that cannot be changed and are accepted upon installation of the applications [51]. Mobile devices are becoming more and more like PCs, although the security of them is not top priority like it should be. Consumers want more features, so that is what companies tend to focus on, rather than trying to protect consumers who are ignorant of the potential damages, and who potentially leave themselves open to attacks.

5.7 Conclusion

In this chapter, we presented some key concepts that surround mobile platforms. This includes the definition, major players, security features, and types of attacks. At this point in time, it is important to understand the variety of mobile platforms made available because of the increase in smart device use. Users of mobile platforms (e.g., Android, BlackBerry, and iOS) should be aware of the security of their platforms, as well as how to protect themselves against potential flaws. By understanding the potential threats and the types of malwares, users will be able to make informed/rational decisions when it comes to choosing an operating system. Moreover, understanding the motivations behind malware can lead to a better identification of its behavior. Our analysis of mobile platforms and their security models shows that malware is becoming increasingly complex and adaptive with constantly changing goals and using multiple distribution and infection strategies.

References

[1] Mobile phone users worldwide 2013–2019 (fee-based). Available at http://www.statista.com/statistics/274774/forecast-of-mobile-phone-users-worldwide/

[2] Cisco Visual Networking Index: Global Mobile Data Traffic Forecast Update 2014–2019 White Paper. http://www.cisco.com/c/en/us/solutions/collateral/service-provider/visual-networking-index-vni/white_paper_c11-520862.html

[3] Suarez-Tangil, G., Tapiador, J., Peris-Lopez, P., Ribagorda, A. (2014). Evolution, detection and analysis of malware for smart devices. In *Communications Surveys & Tutorials, IEEE*, vol. 16, no. 2, pp. 961–987, Second Quarter.

[4] Delac, G., Silic, M., and Krolo, J. (2011). Emerging security threats for mobile platforms. In *Proceedings of the 34th International Convention MIPRO,* Opatija, Croatia, pp. 1468–1473.

[5] Xanthopoulos, S., and Xinogalos, S. (2013). A comparative analysis of cross-platform development approaches for mobile applications. In *Proceedings of the 6th Balkan Conference in Informatics (BCI '13).* ACM, New York, NY, USA, pp. 213–220.

[6] Chin, E., Felt, A. P., Sekar, V., and Wagner, D. (2012). Measuring user confidence in smartphone security and privacy. In *Proceedings of the Eighth Symposium on Usable Privacy and Security (SOUPS '12). ACM,* New York, NY, USA, Article 1, 16 pages.

[7] Ericsson Mobility Report, 2015. [Online]. Available: http://www.ericsson.com/res/docs/2015/mobility-report/ericsson-mobility-report-nov-2015.pdf. [Accessed: 08-Dec-2015].

[8] Kelly, H., Apple iOS9: The pros and cons, *CNNMoney*, 2015. [Online]. Available: http://money.cnn.com/2015/09/16/technology/apple-ios-9-review/. [Accessed: 08-Dec-2015].

[9] Brahler, S., Analysis of the Android Architecture, 2015. [Online]. Available: https://os.itec.kit.edu/downloads/sa_2010_braehler-stefan_android-architecture.pdf. [Accessed: 01-Dec-2015].

[10] Zou, F., Zhang, S., Wan, T., and Pan, L. (2014). A survey of android mobile platform security. In *10th International Conference on Wireless Communications, Networking and Mobile Computing (WiCOM 2014).* Beijing, pp. 520–527.

[11] Techopedia.com, What is BlackBerry OS? – Definition from Techopedia, 2015. [Online]. Available: https://www.techopedia.com/definition/25196/blackberry-os. [Accessed: 01-Dec-2015].

[12] Wheatley, M., BlackBerry is poised to make a stunning comeback in 2015, *SiliconANGLE*, 2015. [Online]. Available: http://siliconangle.com/blog/2015/01/15/blackberry-is-poised-to-make-a-stunning-comeback-in-2015/. [Accessed: 01-Dec-2015].

[13] apple.com, iOS Security, 2015. [Online]. Available: https://www.apple.com/business/docs/iOS_Security_Guide.pdf. [Accessed: 01-Dec-2015].

[14] AppleInsider, iOS 9 security flaw grants unrestricted access to Photos and Contacts, 2015. [Online]. Available: http://appleinsider.com/articles/

15/09/23/ios-9-security-flaw-grants-unrestricted-access-to-photos-and-contacts. [Accessed: 01-Dec-2015].

[15] static.googleusercontent.com, Google Report Android Security 2014 Year in Review, 2015. [Online]. Available: https://static.googleusercontent.com/media/source.android.com/en//security/reports/Google_Android_Security_2014_Report_Final.pdf. [Accessed: 01-Dec-2015].

[16] Engadget, Most Android phones are vulnerable due to lack of security patches, 2015. [Online]. Available: http://www.engadget.com/2015/10/14/android-vulnerabilities/. [Accessed: 01-Dec-2015].

[17] Coccimiglio, J., Malware on Android: The 5 Types You Really Need to Know About, *MakeUseOf*, 2015. [Online]. Available: http://www.makeuseof.com/tag/malware-android-5-types-really-need-know/. [Accessed: 01-Dec-2015].

[18] Help.blackberry.com, BlackBerry Protect-BlackBerry 10 Security Overview – latest, 2015. [Online]. Available: https://help.blackberry.com/en/blackberry-security-overview/latest/blackberry-security-overview-html/ada1406165579198.html. [Accessed: 01-Dec-2015].

[19] Infosecurity Magazine, BlackBerry Issues Four Security Advisories for BB 10 Devices, 2013. [Online]. Available: http://www.infosecurity-magazine.com/news/blackberry-issues-four-security-advisories-for-bb/. [Accessed: 01-Dec-2015].

[20] Trendmicro.com, Understanding Targeted Attacks: Goals and Motives – Security News – Trend Micro USA, 2015. [Online]. Available: http://www.trendmicro.com/vinfo/us/security/news/cyber-attacks/understanding-targeted-attacks-goals-and-motives. [Accessed: 01-Dec-2015].

[21] Sutherland, E., FBI and DHS label Android primary malware target, *idownloadblog.com*, 2016. [Online]. Available: [40] http://www.idownloadblog.com/2013/08/26/android-fbi-dhs-security/. [Accessed: 16-Apr-2016].

[22] Rouse, M., What is social engineering? – Definition from WhatIs.com, *Search Security*, 2015. [Online]. Available: http://searchsecurity.techtarget.com/definition/social-engineering. [Accessed: 01-Dec-2015].

[23] Devotta, N., Malware vs Viruses: What's the Difference?, *Comodo Antivirus Blogs | Anti-Virus Software Updates*, 2014. [Online]. Available: https://antivirus.comodo.com/blog/computer-safety/malware-vs-viruses-whats-difference/. [Accessed: 01-Dec-2015].

[24] Chebyshev, V., and Uncheck, R., The Enemy on your Phone – Securelist, *Securelist.com*, 2016. [Online]. Available: https://securelist.com/analysis/publications/68916/the-enemy-on-your-phone/. [Accessed: 16-Apr-2016].

[25] Levinson, M., Are You at Risk? What Cybercriminals Do With Your Personal Data, CIO: Security, 2012. [Online]. Available at: http://www.cio.com/article/2400064/security0/are-you-at-risk–what-cybercriminals-do-with-your-personal-data.html. [Accessed: 2015].

[26] Wilshusen, G., and Barkakati, N. Information Security: Better Implementation of Controls for Mobile Devices Should Be Encouraged, U.S. GAO – U.S. Government Accountability Office, 2012. [Online]. Available at: http://www.gao.gov/products/gao-12–757. [Accessed: 2015].

[27] Ballano, M., Mobile Attacks: Cybercriminals' New Cash Cow, Mobile Attacks: Cybercriminals' New Cash Cow, Nov-2014. [Online]. Available at: http://www.symantec.com/connect/blogs/mobile-attacks-cybercriminals-new-cash-cow. [Accessed: 2015].

[28] Becher, M., et al. (2011). Mobile Security Catching Up? Revealing the Nuts and Bolts of the Security of Mobile Devices. In *IEEE Symposium on Security and Privacy*, IEEE. doi: 10.1109/SP.2011.29.

[29] Park, K., Yoo, S., Kim, T., and Kim, J. (2010). JTAG Security system based on credentials. *J. Elect. Test.*, 26(5), 549–557.

[30] Jia, Y., Dong, X., Liang, Z., and Saxena, P. (2014). I know where you've been: geo-inference attacks via the browser cache. *IEEE Int. Comput.*, 9(1), 44–53.

[31] Colin, A. (2013). Mapping Social Media Insider Threat Attack Vectors. In *46th Hawaii International Conference System Sciences (HICSS),* IEEE. doi: 10.1109/HICSS.2013.392.

[32] Segan, S. (2015). CDMA vs. GSM: What's the Difference? PCMAG, Jun-2015. [Online]. Available at: http://www.pcmag.com/article2/0,2817,2407896,00.asp. [Accessed: 2015].

[33] Dhaya, R., and Poongodi, M. (2014). Source code analysis for software vulnerabilities in android based mobile devices. *Int. J. Comput. Appl.*, 93(17), 11–14.

[34] Nasim, R. (2012). Security threats analysis in bluetooth-enabled mobile devices. *Int. J. Netw. Secur. Appl.*, 4(3), 41–56.

[35] La Polla, M., delle, N., Consiglio, R., and Martinelli, F. (2012). A survey on security for mobile devices. *Commun. Surv. Tutor. IEEE*, 15(1), 446–471.

[36] Valero, G. (2011). The resurgence of Zeus and other banking Trojans. *Netw. Secur.*, 2011(3), 2–2.

[37] Ketari, L., and Khanum, M. A. (2012). A review of malicious code detection techniques for mobile devices. *Int. J. Comput. Theory Eng.*, 4(2), 212–216.

[38] Thiel, D. (2016). iOS Application Security: The Definitive Guide for Hackers and Developers, 1st ed. San Francisco, California: No Starch Press, Inc.

[39] Keychain Services Programming Guide, Keychain Services Concepts, 2014. [Online]. Available at: https://developer.apple.com/library/mac/documentation/security/conceptual/keychainservconcepts/02concepts/concepts.html. [Accessed: 30-Mar-2016].

[40] Tipton, S., White, D., Sershon, C., and Choi, Y. B. (2014). iOS security and privacy: authentication methods, permissions, and potential pitfalls with touch ID. *Int. J. Comput. Inform. Technol.*, 3(3), 482–489.

[41] Griffiths, A. Binary Protection Schemes, Binary Protection Schemes. [Online]. Available at: http://www.bitlackeys.org/resources/binary_protection_schemes.pdf. [Accessed: 11-Apr-2016].

[42] Chell, D. (2015). *The Mobile Application Hacker's Handbook*, 1st ed. Indianapolis, IN: John Wiley & Sons.

[43] Steiner, P. (2014). Going beyond mobile device management. *Comput. Fraud Secur.*, 2014(4), 19–20.

[44] Shaulov, M. (2016). *Bridging mobile security gaps, Network Security*, 2016(1), 5–8, ISSN 1353–4858.

[45] Strazzere, T., and Sawyer, J. (2014). Android Hacker Protection Level 0, www.defcon.org, 10-Aug-2014. [Online]. [Accessed: 13-Apr-2016].

[46] Blackberry. Using compiler and linker defenses – Native SDK for PlayBook. [Online]. Available at: http://developer.blackberry.com/playbook/native/documentation/com.qnx.doc.native_sdk.security/topic/using_compiler_linker_defenses.html. [Accessed: 13-Apr-2016].

[47] Symantec. Why Digital Certificates Are Essential for Managing Mobile Devices, *White Paper*, 2012. [Online]. Available at: https://www.symantec.com/content/en/us/enterprise/ white_papers/b-why-certs-mobile-devices-wp-21259170-en.us.pdf. [Accessed: 20-Apr-2016].

[48] Bradbury, D. (2012). Digital certificates: worth the paper they're written on? *Comput. Fraud Secur.*, 2012(10), 12–16.

[49] Tanvi, P., Sonal, G., and Kumar, S. M. (2011). Communication Systems and Network Technologies. In *Communication Systems and Network Technologies International Conference*, pp. 85–88.

[50] La Polla, M., Martinelli, F., and Sgandurra, D. (2013). A survey on security for mobile devices. *IEEE Commun. Surv. Tutor.*, 15(1), 446–471.

[51] Lopes, H., and Lopes, R. (2013). Comparative analysis of mobile security threats and solution. *Int. J. Eng. Res. Appl.*, 3(5), 499–502.

[52] Shabtai, A., Mimran, D., and Elovici, Y. (2015). Evaluation of Security Solutions for Android Systems. arXiv preprint arXiv:1502.04870.

[53] Sujithra, M., Padmavathi, G., and Narayanan, S. (2015). Mobile device data security: a cryptographic approach by outsourcing mobile data to cloud. *Procedia Comput. Sci.*, 47, 480–485.

[54] Zhang, W., He, H., Zhang, Q., Kim, T. (2014). PhoneProtector: protecting user privacy on the android-based mobile platform. *Int. J. Distrib. Sens. Netw.*, 1–10.

[55] Piggin, R. (2016). Cyber security trends: What should keep CEOs awake at night. *Int. J. Crit. Infrastruct. Prot.*, http://dx.doi.org/10.1016/j.ijcip.2016.02.001.

[56] Gajjar, K., and Parmar, A. (2016). A Study of Challenges and Solutions for Smart Phone Security. Emerging Research in Computing, Information, Communication and Applications. Springer India. 325–334

PART III

Cryptographic Algorithms

6

Quasigroup-Based Encryption for Low-Powered Devices

Abhishek Parakh, William Mahoney, Leonora Gerlock and Matthew Battey

University of Nebraska at Omaha, Omaha, NE 68182, USA

Abstract

The first part of this chapter discusses recently proposed quasigroup-based block cipher with applications in low-powered computationally constrained environments. We present some preliminary analysis of the block cipher using NIST Statistical Analysis Tool (second half discusses the linear cryptanalysis). We also present our results on hardware implementation of quasigroup-based block cipher.

In the second part of the chapter, we determine whether any key material can be found by conducting a linear cryptanalysis of the cipher matrix lookup transformations on the input blocks using the key bytes. Linear cryptanalysis involves a known-plaintext attack such that a set of plaintexts is known to have a specific statistical relationship to a set of ciphertexts which are all encrypted under the same key. Using Matsui's Algorithm 2 for DES S-box transformations as an example, we seek to determine a suitable linear approximation of the quasigroup block cipher, the number of plaintext–ciphertext pairs to test, and the amount of time and space required to mount a known-plaintext attack on the quasigroup block cipher. Our research showed that no key material could be recovered, and therefore, we conclude that the quasigroup cipher is resistant to linear cryptanalysis. Since the quasigroup does not use a Feistel network with S-box transformations as the basis of encryption, the focus of the linear cryptanalysis was on the keyed transformation during table lookups of the quasigroup, in order to 1) determine how the key bits used during encryption impact the ciphertext, and from this 2) find a linear approximation that is non-negligible.

Keywords: Quasigroup encryption, Low-energy encryption, Resource constrained algorithm.

6.1 Introduction

The recent proliferation of low-powered devices ranging from handheld smart phones to the growing Internet of Things has brought out unique challenges for data confidentiality and security. In some ways, given the hardware and power constraints of the devices, the clock has turned backward. The commonly used cryptographic algorithms such as AES or 3DES consume too much power and therefore may not be practical for these newer devices. Although there has been a lot of research done on developing low-powered encryption systems, none of the cryptosystems has stood out as a winner. As a result, most companies use proprietary encryption systems that have not been openly vetted by the community.

The abundance of network-connected cyber systems built on embedded computing platforms requires the development of inexpensive encryption algorithms. The large amounts of data that is being received, transmitted, and gathered by these devices lead to an enormous drain on battery power. Further, since embedded systems are often mass-produced, a small increase in the cost of one component is multiplied many times over the millions of devices that are being built. Examples of such embedded systems are found in power grids, water purification systems, and oil and gas delivery systems, as well as trains and transportation systems and other Supervisory Control and Data Acquisition (SCADA) systems. As a result, manufacturers often entirely leave out encryption from SCADA systems in consideration of extending battery life or use deprecated resource-inexpensive deprecated algorithms with small key sizes.

This chapter details our development efforts for a new encryption system based on quasigroups that holds promise for these low-cost and low-power embedded systems applications. We detail the software implementation of the proposed algorithm as well as single-chip quasigroup device based on Field Programmable Gate Arrays (FPGAs). The proposed quasigroup algorithm requires minimal complexity and is based on table-lookup operations along with bit-shifting operations. Tests show that the algorithm destroys existing structure of the data, making the output look like that from a true random number generator.

6.2 Background—Low Energy Cryptosystems

NIST has standardized three major cryptosystems starting with DES [48], then 3DES, and the more modern AES cryptosystem. AES commonly uses key sizes of 128 or 256 bits, although a 192-bit key size is also supported. However, AES requires much more computational power than its predecessors do. This minor increase in horsepower needed by AES translates into millions of dollars in hardware cost across the world. As a result, manufacturers and Web site developers use some form of stream cipher, most notable RC4, with low computational requirements and high throughput. However, recent research has shown several vulnerabilities in RC4. At the same time, due to increasing popularity of portable devices, RFID tags, and sensor networks, several new lightweight cryptosystems have been developed in recent years. An organized effort to develop new stream ciphers for widespread adoption was started by the EU ECRYPT network under the eSTREAM project [1]. A number of very good algorithms have resulted from this effort. However, they did not look at block ciphers. Other researchers have modified several of the known lightweight cryptosystems such as DES [2] and ECC [3, 4] and stripped them down in order to increase their throughput and decrease computational requirements. For example, [2] proposes a lightweight version of DES for RFID chips and uses a single S-box repeated eight times in order to improve on storage.

The PRINT cipher was proposed for integrated circuit printing [5]; however, its cryptanalysis has revealed some weaknesses [6, 7]. An ultra-lightweight block cipher, called Piccolo [8], for RFID tags and sensor nodes was proposed. However, its fault analysis [9, 10] has shown that the key candidates can be reduced significantly based on a few correct and faulty cipher texts.

TWINE was proposed [11] as a lightweight cipher for multiple platforms and low-end microcontrollers. Some cryptanalytic attacks have been proposed on it with slight improvements than brute force attacks [12]. PRESENT is another general-purpose block cipher developed for low-powered consumption and high chip efficiency [13, 14]. International Organization has also adopted it for Standards, and International Electrotechnical Commission as a standard algorithm. Some weaknesses have been discovered in LED as well that enable attacks on it [15–17].

These are only a few efforts related to development of lightweight cryptosystems. Their main applications have remained in the RFID and sensor network and not specifically toward SCADA systems or cyber-physical systems (CPS). In SCADA and CPS, additional hardening against reverse engineering, cloning, and side-channel attacks may be required.

Quasigroup structures have been used for error correction and in the construction of message authentication systems [17]. They have also been applied to building encryption systems in the past [18–20]. For example, stream ciphers built using quasigroups, and public key implementations were investigated by Gligoroski and others [21–24]. Satti and Kak [25] looked at multi-level stream cipher implementation with indices and nonces in order to improve the strength of the resulting encryption. An all-or-nothing system was implemented by Marnas et al. [26]. But they used quasigroup encryption only to replace the exclusive-OR operation and the actual encryption was done using other cryptosystems. The quasigroup encryption system is primarily a substitution and permutation system—a type of cryptosystem used frequently in the encryption of speech [27–30].

6.3 Overview of Quasigroup Encryption

Quasigroups can be represented in the form of a matrix and are similar to Latin Squares. A quasigroup matrix contains elements from 0 to $n - 1$ such that no element repeats in any row or column. In general, one can use any elements instead of numbers. An inverse quasigroup matrix can be defined that basically reverses the mapping of elements. Quasigroups have been used in cryptography for a long period [20], and a simple exclusive-OR is an example of a commonly used quasigroup. The XOR operation is used as a basis for one-time pads, which is the only cryptosystem providing perfect secrecy. The XOR table is based on a restrictive structure $v_{ij} = r_i + c_j$ mod 2; indices r_i and c_j start at zero. However, this structure can be generalized to any value of modulus. Further, the structure allows for commutative operations, i.e., $a \oplus b = b \oplus a$. But in general, quasigroup operations are not necessarily commutative and this provides greater security for shorter key lengths, which is desirable for embedded systems. Higher order quasigroups that main the commutative property are a generalization of XOR matrix and can be viewed as $v_{ij} = (r_i + c_j) \bmod n$.

Conventionally, quasigroups have only been used to build stream ciphers and the entire quasigroup is kept secret. A single-byte seed is used to initiate the encryption process, and the cipher system processes a single byte at a time. Although this method has high throughput, it is vulnerable to a known-plaintext attack.

6.4 The Preliminary Block Cipher Design

The proposed quasigroup block cipher is designed based upon the ideas of confusion and diffusion. In order to conform the current standardized practices, we used a 256-bit key with a 128-bit block size. This design was proposed in Battey [18, 19].

We divide the 256-bit encryption key into 32 one-byte seeds and use a 256×256 quasigroup. Each of the rounds of encryption uses one seed as follows:

1. Generate a random 256-bit encryption key and divide it into 32 one-byte (8 bit) seeds. Each seed will be used once per round of encryption.
2. Divide the source data into 128 bit (16-byte) blocks.
3. For each block, do the following:
    a. For each 8-bit seed byte in the key, do the following:
        i. Using the current block as a stream of 16- and 8-bit integers, apply the current 8-bit key as the quasigroup cipher seed and encrypt the block.
        ii. Rotate the currently encrypted 128-bit block left by 1, 3, 5, or 7 bits depending on the index of the current seed byte modulo 4.

We divide the plaintext into 128-bit long blocks and in each round, every block is subdivided into 16 one-byte sub-blocks. After every round of encryption, a left rotation is applied and all the bits in the sub-blocks are taken together.

6.5 Overview of Software Implementation

Our algorithm was first implemented in C# and compared against a popular implementation of AES. For generating the quasigroup table, first a $N \times N$ array was constructed such that row zero contained elements 0 to n in order, row one contained elements left shifted by one, so it started with 1 and ended in 0, and so on. The rows and columns were then shuffled using Knuth/Fisher-Yates shuffling to randomly rearrange the rows of the quasigroup.

NIST provides the Statistical Test Suite (STS) [31] and we use this suite to evaluate our quasigroup algorithm. Each individual test in the NIST-STS suite provides a P-value that represents the probability that a perfect random number generator would produce a sequence less random than the result at hand. While a plaintext will fail any test in the suite, the quasigroup scheme must pass all the tests in the suite. The tests built into the suite use varying

Table 6.1 Successes per 1,000 encryption tests

	AES				QGBC			
Test	0 × 00	E	0 × FF	Beowulf	0 × 00	E	0 × FF	Beowulf
AE	988	989	986	985	986	995	988	992
BF	992	990	994	991	991	991	986	991
CSF	990	993	990	994	988	992	996	992
CSR	994	989	991	994	986	994	994	994
FFT	990	988	989	986	984	981	990	980
FREQ	992	992	989	994	991	992	996	992
LR	991	987	991	989	990	988	987	991
Rank	989	989	996	989	994	995	982	995
Runs	994	988	993	991	987	993	989	993
Serial 1	990	992	995	995	991	990	989	994
Serial 2	986	993	990	987	984	993	991	988

block lengths. The data that was passed through the encryption algorithm included clear text made of all zeros, all ones, text containing all letter Es and text from project Gutenberg imprint of Beowulf. One thousand encryptions under different keys were run. The STS documentation suggests a confidence interval of 0.9805607, based on our data and parameters, translating to requirement to pass 980 out of 1,000 tests. Table 6.1 shows the results.

6.6 Overview of Three FPGA Implementations

Next, we wanted to take the quasigroup scheme and implement it in low-cost hardware for the world of embedded systems. Field Programmable Gate Arrays (FPGAs) are components that can be programmed for a specific purpose by the circuit designer, as opposed to collection of Common Logic Blocks (CLBs) connected by typically a grid of wires [32, 33]. Although ICs are faster than FPGAs, they are also more expensive, and for added security an FPGA can be fingerprinted so that it is specific to a certain customer [34]. Further from security perspective, FPGAs can be designed to make reverse engineering difficult. Wollinger provides a good overview [35] of the reverse engineering process and newer information is available in Huffmire [36]. As a result, FPGAs make a good candidate for encryption.

To investigate whether our quasigroup-based encryption will save costs in an embedded environment, we implemented three different algorithms in FPGAs. First was a translation of the C# software into the Verilog hardware description language. Next we tested an open-source Verilog realization of AES for comparison but found that the open-source AES implementation requires significant hardware components due to the large number of I/O pins required. Their design is very fast, but it requires the complete clear text to be available, in parallel, simultaneously; thus, a component with a very large "pin count" is required. To aid in comparison then, a third implementation is the quasigroup front-end to load the data with a minimal pin count, but with parallel AES algorithm inside the FPGA. Each of these designs is detailed below.

6.6.1 The Quasigroup Implementation

The first design is a FPGA implementation of the basic quasigroup algorithm with goal of keeping the cost of the design down at the expense of speed. The cost impact for FPGAs is primarily twofold: interconnection count and logic complexity. Here the interconnection or "pin" count dominates. As a result, efforts were made to minimize interconnections to the components. Externally, the design has a 16-bit address bus, two 8-bit data busses, and a 3-bit chip select logic, and clock and reset connections. Because the encryption algorithm requires a 256 $\times$ 256 size quasigroup table, an external 64K byte-addressable read-only memory was assumed; this accounts for the 16-bit address bus in the design. Upon reset, it initially reads in the initialization vector (IV) for CBC, the encryption key and then successive 16-byte blocks of data. While the encrypted data is read out on one 8-bit bus, the next clear text is written to the components on the other 8-bit path. Since the quasigroup algorithm is serial in nature, encryption takes 560 clock cycles. This is directly in correspondence with the translation of the C# implementation.

6.6.2 Comparison Design—Parallel AES

The second design we implemented was an open-source AES implementation. The code was obtained from "opencores.org". This design is a fully parallel AES implementation, reading 128-bit cleartext in as 128-bit ciphertext is coming out. Additionally, the key is given to the chip as well, resulting in three very wide I/O busses into and out of the device; we anticipate that this design will be fast but costly. Because the AES is preformed as one large computational block, the cleartext is encrypted in just 21 clock cycles. The design is pipelined so that successive blocks of clear text can be presented to the design as previous results exit the chip; after the first block is entered, each successive block can be sent into the design on each clock cycle. But one drawback of this design is that the potential for CBC is limited since the results of the previous encryption block cannot be exclusive-OR'd with the next block without incurring the penalty of waiting 21 cycles.

6.6.3 Hybrid Front-End/AES Design

These two designs, the quasigroup design with low pin count and the fully parallel AES implementation, are at opposite extremes. To make a fair comparison to AES, we designed a third FPGA with features from each of the two designs. We wish to use the AES encryption logic from the open-source design with an external interface similar to the first design. This will allow us to compare an AES design while eliminating the very costly element of the AES, the pin count. In this hybrid FPGA, the AES key, initialization vector for CBC, and the clear text are presented to the chip serially, one byte at a time, using a combination of addresses and chip select lines from the component. From the external standpoint, the interface to the encryption FPGA is thus the same, with one minor exception: Since the AES key is 16 bytes instead of 32, the initialization is slightly faster. But this cost is one-time and thus is not important for determining the throughput of the device. Once the entire clear text is into the chip, the design waits the requisite clock cycles, and as in the quasigroup design clocks out the cipher text while clocking in the next clear text.

6.7 Experimental Results

We desired to determine the power requirements as well as the approximate costs associated with the three designs, in order to determine whether our quasigroup design warrants additional research. Initially our plan was to actually implement the design using evaluation boards—reference designs—from FPGA vendors. However, the software simulations of the FPGAs have such high fidelity that there was no need to test with actual hardware.

We utilized the Altera software package Quartus-II as well as the Modelsim-Altera simulator to perform register transfer-level and then gate-level simulations of the designs. The package is sufficiently sophisticated that we can obtain exact timing requirements and determine maximum clock speeds as well as power consumption. The Quartus-II software will also select the specific component necessary to hold the design, based on the required pin counts as well as the functionality necessary for the implementation. This turns out to be a great feature as it directly allows us to compare actual costs of the FPGAs. Although the quasigroup encryption is sufficiently small that it could presumably sit in a very minimal FPGA, we decided to conduct an apples-to-apples comparison and stick with same families of Altera FPGAs, with the precise component selected by the software.

Specifically, we tested all three designs with an automatically selected component from Altera's Cyclone II product line, and then again with

Cyclone III product line. In all cases, we placed no restrictions on the automatic selection of the component, allowing the software to pick the mapping to I/O pins, for example. In the case of the quasigroup design implementation, we assumed that the external ROM necessary for the encryption table was not a factor in the speed calculations, since it will be internal ROM eventually.

Table 6.2 describes the result—but with the external ROM omitted for the quasigroup design. Specifically, the number of clocks required per 16-byte block and the throughput in bytes per clock are presented first. Our thinking is that this allows the reader to estimate what a faster frequency component would gain in throughput, since the number of clocks for the encryption is constant. Next, we have the actual clock period measured as per second based upon this clock period.

Clearly based on the preceding table, a quasigroup encryption does not look attractive from the perspective of throughput. However, SCADA systems typically do not require high throughput anyway. Our driving factors in areas such as the "smart grid," recall, are cost and power. In Table 6.3, the three designs from these perspectives are compared. To estimate the costs, we simply took the corresponding Altera component numbers and looked up the

Table 6.2 Performance comparison for designs

	Clocks/ 128-bit Block	Clock Period (ns)	Frequency (MHz)	Bytes/Sec
Quasigroup in Cyclone II	560	26	38.5	1098901
Quasigroup in Cyclone III	560	22	45.5	1298701
AES in Cyclone II	21	20	50	38095238
AES in Cyclone III	21	18	55.6	42328042
Hybrid in Cyclone II	37	34	29.4	12718601
Hybrid in Cyclone III	37	26	38.5	16632017

Table 6.3 Cost comparison for designs

	Chip Selected	Cost	Power (mW)
Quasigroup in Cyclone II	EP2C5T144C6	$19.20	23
Quasigroup in Cyclone III	EP3C5F256C6	$26.80	59
AES in Cyclone II	EP2C70F896C6	$384.00	239
AES in Cyclone III	EP3C40F780C6	$166.50	141
Hybrid in Cyclone II	EP2C70F672C6	$352.00	196
Hybrid in Cyclone III	EP3C40F324C6	$122.00	107

price on "digikey.com", an electronic component sales site. Digikey is a fast turnaround, any quantity vendor and as such, it is expensive. However, the reader should examine the figures relative to each other.

Clearly implementing the fully parallel AES scheme has a significant impact on the cost; the hybrid approach minimizes the pin count for the device, but does not save enough to justify AES. It appears that the quasigroup design will utilize less power than the AES implementation. It also appears that if the data throughout is not a primary concern that the quasigroup encryption algorithm has a lower cost.

6.8 Toward a Single-Chip Implementation

To this point, we have developed a block cipher based on quasigroups which was implemented in a simulated Field Programmable Gate Array (FPGA) and reported in [18]. However, the preliminary design used a "two-chip" solution where the FPGA ran the encryption algorithm and an ancillary read-only memory contained the quasigroup table. This is a direct result of the encryption scheme operating on eight bits at a time and requiring a 256×256 size quasigroup—thus 64 Kbytes of memory. Another drawback is that creating large random quasigroups is extremely difficult.

This section details our initial efforts to place the encryption tables in the FPGA along with the encryption algorithm itself thus eliminating the additional chip. While one can find an FPGA containing a large amount of internal memory, these chips suffer from two drawbacks: First, the FPGAs with large memory frequently have large number of input–output pins and increased internal capacity, thereby increasing their cost, and second, the memory is not contiguous. Therefore, if we wanted to place a quasigroup table along with the encryption algorithm within the internal memory of a low-cost FPGA, we aim to reduce the size of the quasigroup tables to the point where they can be implemented with combinational logic within the FPGA.

This opens up many questions: What is the effect of smaller quasigroup tables on the quality of encryption? Will, for example, more tables that are $2^4 \times 2^4$, working on the block four bits at a time, yield encryption that is on par with the larger $2^8 \times 2^8$ method?

6.9 Algorithm Results for B = 2 to 8

In our previous work, we used a key size of 256 bits and a block size of 128 bits. Following suit with previous work, we executed similar tests: We created three input files, each 50,000 bytes long. One was composed of all

0×00, one of all $0 \times$ ff, and a third file created by copying the first 50,000 bytes from /usr/shar/dict/words, the list of dictionary words on a typical Linux distribution. These "zeros," "ones," and "English" files were then encrypted 1000 times with key and initialization vectors created from /dev/urandom. We repeated this for slice of sizes 2 through 8, creating 21,000 files. The NIST–STS suite was used to access the randomness of resulting cipher. If 980 files of 1000 pass an individual test, then it considered a success by NIST. Our results are shown in Table 6.4.

We see 33 tests for each slice size, made up of 11 NIST tests for each of the three input files, for a total of 231 tests. We note that of these tests there are three that are not considered a "pass" according to the NIST specifications. These are all "Approximate Entropy" tests and we are investigating a possible cause for these results. Note that two of the three involve bit slices that do not evenly divide the key and block size.

6.10 Generating Quasigroups Fast

The proposed quasigroup encryption algorithm requires generating random quasigroups. While starting out with a Vigenere square and then shuffling it may be straightforward, creating a truly random quasigroup such that is chosen from all possible quasigroups at random is not trivial. Since our emphasis is on reducing the size of the tables, why dwell on constructing larger tables? Because even for small values of $B = 6$, for instance, the table will be 64×64 entries of [0, 63].

Battey [18, 19] proposed two quasigroup generating algorithms. However, both methods were focused on improving the speed of generation and minimizing the storage requirement. As a practical idea, it only utilizes a small number of possible quasigroups and hence is not ideal. Therefore, we looked at various other methods to generate a random quasigroup that have no restrictions. Of these, Jacobson is considered the best algorithm to generate random quasigroups [37] and a Java implementation is readily available [38]. Based upon these thoughts, we examined the methods for constructing improved quasigroup tables.

Consider a $N \times N$ square matrix containing digits from $[1, N]$ and that $N = 9$ as in a Sudoku puzzle. In Figure 6.1, the unmarked area "A" represents any 9×9 square that contains only [1, 9], in any order, and with any number of each element. There are $\mathrm{N}^{\mathrm{N}^2}$ such possible squares. This set is of little interest to us. More attractive is the area "B," where the square contains only the proper quantity, nine, of each digit [1, 9]. For example, the square can contain a total of nine zeros, nine ones, nine twos, and so on. This is

Table 6.4 Pass/Fail results from the NIST–STS tests. B is word size in bits. P/F is Pass/Fail

NIST-STS		B=2		B=3		B=4		B=5		B=6		B=7		B=8	
		Success		Success		Success		Success		Success		Success		Success	
Test	File	Rate	P/F	Rate	P/F	Rate	P/F	Rate	P/F	Rate	P/F	Rate	P/F	Rate	P/F
Approximate	Zeros	981	P	986	P	987	P	985	P	976	*F*	976	*F*	981	P
Entropy	Ones	977	*F*	987	P	983	P	986	P	983	P	983	P	981	P
	English	984	P	986	P	985	P	984	P	982	P	986	P	983	P
Block	Zeros	986	P	994	P	992	P	988	P	995	P	992	P	994	P
Frequency	Ones	992	P	990	P	993	P	995	P	990	P	991	P	993	P
	English	990	P	992	P	993	P	991	P	985	P	988	P	984	P
Cumulative	Zeros	990	P	993	P	988	P	987	P	991	P	987	P	994	P
Sums	Ones	996	P	989	P	990	P	987	P	995	P	996	P	988	P
Forward	English	992	P	991	P	983	P	990	P	988	P	991	P	991	P
Cumulative	Zeros	992	P	993	P	989	P	988	P	987	P	987	P	994	P
Sums	Ones	994	P	987	P	990	P	986	P	992	P	991	P	992	P
Reverse	English	991	P	991	P	985	P	991	P	986	P	990	P	987	P
FFT	Zeros	991	P	988	P	995	P	988	P	990	P	986	P	986	P
	Ones	986	P	993	P	984	P	985	P	978	F	985	P	986	P
	English	990	P	988	P	989	P	990	P	986	P	989	P	995	P
Frequency	Zeros	994	P	991	P	988	P	986	P	989	P	985	P	995	P
	Ones	995	P	990	P	987	P	989	P	991	P	990	P	987	P
	English	989	P	992	P	983	P	988	P	988	P	991	P	988	P
Longest	Zeros	995	P	997	P	991	P	990	P	983	P	985	P	990	P
Run	Ones	989	P	991	P	991	P	990	P	987	P	993	P	988	P
	English	987	P	996	P	984	P	996	P	986	P	988	P	994	P
Rank	Zeros	991	P	988	P	987	P	994	P	993	P	991	P	989	P
	Ones	991	P	995	P	990	P	996	P	990	F	991	P	991	P
	English	993	P	994	P	987	P	994	P	988	P	995	P	991	P
Runs	Zeros	989	P	994	P	992	P	992	P	986	P	987	P	985	P
	Ones	990	P	989	P	985	P	990	P	936	P	990	P	990	P
	English	993	P	992	P	987	P	994	P	990	P	989	P	984	P
Serial 1	Zeros	992	P	988	P	989	P	987	P	990	P	986	P	990	P
	Ones	994	P	989	P	983	P	994	P	990	P	982	P	983	P
	English	987	P	986	P	995	P	992	P	990	P	990	P	992	P
Serial 2	Zeros	990	P	989	P	992	P	993	P	993	P	989	P	992	P
	Ones	993	P	992	P	990	P	988	P	993	P	986	P	990	P
	English	987	P	990	P	989	P	986	P	990	P	990	P	993	P

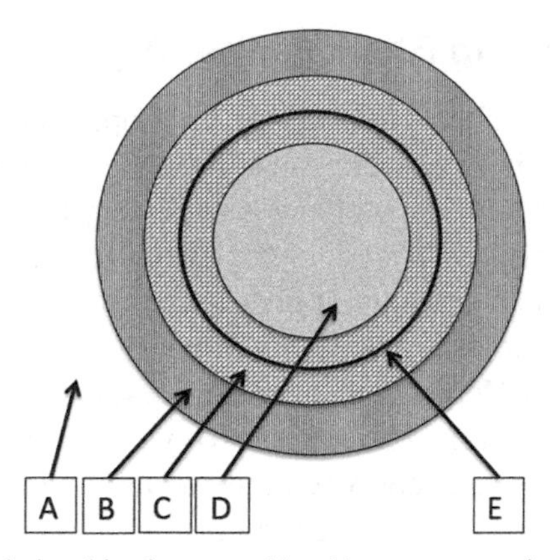

Figure 6.1 Relationships between $N \times N$ squares containing $0 \ldots N-1$.

essentially the number of ways a combination of different elements can be chosen given by $\frac{N^2!}{N!N!N!}$. These have the correct number of digits but are not necessarily Latin Squares. The area marked "C" represents Latin Squares; they contain the proper quantity of each digit and satisfy the necessary row/column uniqueness properties. Not all of the set "B" satisfies the requirements to be a Latin Square but area "C" does, and there are 5.52×10^{27} Latin Squares for $N = 9$ [39]. Since in this example we assumed $N = 9$, area "D" contains valid solutions to Sudoku puzzles, about 6.67×10^{21} squares [40]. Finally, we show the ring marked by "E" which splits "C" into two areas and which will be explained next.

Suppose one starts with a valid Sudoku solution and then shuffles the rows and columns. Because of this shuffling, the 3×3 block restriction on Sudoku puzzles is eliminated and the resulting square is "pushed out" from "D" to "C." However, not all cubes in area "C" can be created in this way. There are Latin Squares in "C" that cannot be created by shuffling a Sudoku puzzle and thus the ring indicated by "E" splits this set into those that can and cannot be created this way.

The crucial point is that creating quasigroups with a large value for N can be problematic; merely starting with some valid and possibly trivial table, followed by rearranging only rows and columns, does not reach the complete set of quasigroups, and thus does not reach the entire key space for encryption. For a full description on the number of Latin squares (and rectangles), see McKay [41] and also Stones [42]. For a genetic algorithm approach for the creation of quasigroups, see Snasel [43].

We next look at modifying the sizes of the quasigroup tables.

6.11 Our Quasigroup Block Cipher Algorithm

The original quasigroup encryption algorithm is operated on the 128-bit block and the 256-bit key in byte units, because it operates on bytes, $B = 8$, for this algorithm. There are a number of changes necessary to the original algorithm in order to support arbitrary "slices" of the data. For example, if $B = 3$, giving quasigroup tables that are $2^3 \times 2^3$, it is necessary to have a block size of 129 so that the clear text block is evenly divisible by B. Similarly, for $B = 5$ the blocks must be 130 bits. Here we have two choices: We can treat the plaintext as bits, encrypting 129 at a time, for example, or we can input blocks of 128, add random bits to the excess, and encrypt. Since we assume in the SCADA domain that the data being delivered to the encryption process will be handed over a byte at a time, the second option makes more sense regardless of the fact that we are encrypting in blocks and not a stream. So this method has the curious feature that the encrypted data is slightly larger than the unencrypted data in cases where B does not evenly divide 128—the space overhead is a result of the extra padding. When the encrypted data is unencrypted, the excess bits are discarded and the original data is recovered. Since the receiver (decryptor) knows the block sizes being used and the size of quasigroup tables being used, there is no need to include any flags indicating the start and end of padding.

The actual algorithm we have used for quasigroup encryption with a given setting for N—the number of bits per slice—is shown in Figure 6.2. Here the key, data block size, and Cipher Block Chaining (CBC) size are all rounded up according to B. We assume that there are a number of smaller quasigroup

```
set the key size to ⌈(256+(N-1))/B⌉ * B
set the 256-bit key
set the CBC size and block size to ⌈(128+(B-1))/B⌉ * B
set the CBC to an initial value
set rot[ ] to 1, 3, 5, 7
for each block in the cleartext do
  xor clear text with CBC
  for i = 0 to key size / B do
    seed = slice number i from the key
    for x = 0 to block size / B
      use quasigroup table x for this iteration
      data = slice x from current block
      seed = table[ seed ][ data ]
      current block slice x = seed
  rotate current block left by rot[ i mod 4 ]
CBC = the resulting block
```

Figure 6.2 Quasigroup block encryption.

tables, enough so that there are tables of size $2^B \times 2^B$ for each slice of the data in the plaintext block.

Note that the block is considered as a collection of (block size/*B*) slices, for the encryption utilizing the quasigroup tables, but that the rotation step rotates the entire block.

The time requirement for this algorithm when implemented in software is heavily dependent on the number of bits per slice, because of the nested "for" loops that are based on (1) the key size over B and (2) the block size over *B*. With a small value for *B*, two for example, the encryption of a block will take much longer than the original algorithm where $B = 8$; encryption with this algorithm and iterating through the slices will operate more slowly in the case where *B* is small.

When the authors first implemented the quasigroup encryption algorithm, there was only one table, for $N = 8$, and the same table is used for each byte in the plaintext. This necessitates that the hardware implementation, with the external 256×256 quasigroup table, accesses the table iteratively. Since the table was large, it was to be stored externally and this results in a number of clock cycles spent encrypting each block of the data. However, a motivation for exploring the encryption quality with smaller values of N is to—if the encryption does not suffer—implement the algorithm in hardware more efficiently. A ramification of smaller tables is to avoid the off-chip-timing penalty as well as the additional cost of the ROM, but another benefit is that the iteration can be avoided entirely to an extent. Specifically, the loop in the algorithm is shown in Figure 6.2,

```
for x = 0 to block size/N
   use quasigroup table x for this iteration
   data = slice x from current block
   seed = table[seed][data]
   current block slice x = seed
```

can be unrolled in the hardware implementation to be:

$seed_0$ = slice number i from the key
$data_0$ = slice 0 from current block
$seed_1 = table_0[seed_0][data_0]$
current block slice 0 = $seed_1$
$data_1$ = slice 1 from current block
$seed_2 = table_1[seed_1][data_1]$
current block slice 1 = $seed_2$
. . .

which will eliminate the inner “for” loop and can be implemented as combinational logic with lookup tables.

In our future work, we planned to explore this tradeoff. For small values of N, two for example, the unrolling will cause $data_{63}$ to be dependent on $data_{62}$, which in turn is back to $data_0$. What is the point at which the more significant loop unrolling causes unacceptable delays in the encryption of the block? Is a middle value, say $N = 4$, a good tradeoff between the propagation delay versus the table size? These questions are dependent on the mapping of the hardware description language versus the particular component selected for the implementation as well, since different FPGAs will have different numbers of blocks, with different amounts of contiguous ROM.

Future work also involves linear and differential cryptanalysis on the newly developed algorithm with smaller quasigroup sizes.

6.12 Cryptanalysis and Improvements in the Block Cipher

A preliminary cryptanalysis of the quasigroup block cipher [44] was previously proposed. We identified the odd-bit and identical-word problems with the cipher and recommend configurations of QGBC to counter these. Following this analysis, we proposed an improved variant of the QGBC, which doubles the block size and quarters the total number of operations by per block.

Our goal was to develop a version of QGBC cryptosystem that not only maintains the statistical and analytical profile of previous work, but also reduces the number of clock cycles in implementations. Since a FPGA allows for highly parallel calculations, we target our design toward them. We demonstrate an algorithm that is highly efficient in FPGA implementations and that can perform the QGBC encryption with CBC in only 19 clock cycles (as compared to 21 clock cycles of AES that we discussed before) plus the cycles to exchange each block with the data bus. Results of this cryptanalysis can be found in Battey [44] along with the improved algorithm.

6.13 Overview of a General Linear Cryptanalytical Attack

The linear cryptanalytical attack is a known-plaintext attack where the cryptanalyst knows how a set of random plaintexts relates to a corresponding set of ciphertexts [45, 47]. The design of a linear cryptanalytical attack uses a substitution permutation network (SPN) where the substitution box (S-box) is a simple one-dimensional array structure. The input to the S-box is mapped to the index of the array. Substitution values are found by taking a predetermined index value and looking up the element stored at this index value. The size of the S-box array indicates the total number of elements that must be considered when conducting a statistical analysis of the probability that any one bit is active when entering and an intermediate round S-boxes [45]. Heys' use of a simple cipher with an SPN structure is based on the DES cipher [45]. All values are converted from their hexadecimal value to binary in order to determine which bits are active during round S-box transformations. This attack requires three main events which include creating a linear approximation table (LAT), determining the linear probability bias of the entire cipher encryption, and then determining which key bits are active during the second to last round of the entire cipher.

In order to conduct an attack based on linear cryptanalysis, one has to determine the linear expressions with the highest probabilities, or occurrences [45]. A linear expression is defined as one in which the input and output bits of an S-box exhibit equality. Heys uses the XOR operator to determine which S-box input and output values are equal. The general form of a linear expression is as follows:

$$X_{\mathrm{i}1} \oplus X_{\mathrm{i}2} \oplus \cdots \oplus X_{\mathrm{i}u} \oplus Y_{\mathrm{j}1} \oplus Y_{\mathrm{j}2} \oplus \ldots Y_{\mathrm{j}v} = 0 \tag{6.1}$$

where X_{i} is the i-th bit position of the input to the S-box and Y_{j} is the j-th bit position of the output from the S-box, and Equation (6.1) represents a sum of u input bits along with v output bits [45].

Determining the linear expressions of each S-box allows for the construction of the linear probability biases of each round. A linear probability bias is the likelihood of a linear expression being true [45]. To extract key bits, one has to find the linear approximation of the entire round based on the the linear probability of each subround of the cipher's S-boxes [45]. Let P_i be a plaintext, U_{ij} be the input to an intermediate j-th round S-box, and V_{ij} be the output of an intermediate j-th round S-box. In addition, let K_{ij} be the key bits of an intermediate round i. The linear probability of the cipher for all rounds is defined as follows:

$$U_{\mathrm{ij}} = P_{\mathrm{j}} \oplus K_{\mathrm{ij}} \tag{6.2}$$

$$V_{\mathrm{ij}} = U_{\mathrm{ij}} \tag{6.3}$$

Matsui's [46] pilingup lemma is typically used to find the number of plaintext—ciphertext pairs needed for the attack. The final stage of the attack then looks for active key bits based on the linear approximation of the entire cipher from a previous step and counts how many times this is true.

One of the primary goals of this research model is to determine whether linear cryptanalysis models based on one-dimensional SPN ciphers can accommodate the quasigroup structure in order to determine key bits during round transformations. In this research model, each round was chained to the previous using the CBC mode method. The model of the linear cryptanalytical attack also omitted the left-shift when considering the probability bias between rounds. This was due to the fact that a shift to a bit position that was greater than the order of N caused the resulting value to be beyond the scope of *N*, and thus any value in the linear approximation table (LAT). For example, if the order of the quasigroup was $N = 16$, a permutation shift that resulted in the cipher output of 17 would exceed all the possibilities calculated in the LAT over the sixteen possible values that the cipher substitution would allow.

6.14 The LAT Design

In keeping with Heys' [45] definition of a linear approximation of an S-box commonly found in substitution—permutation networks (SPN), the overall structure of the quasigroup table build had to be considered. The relationship between the inputs and outputs of the quasigroup substitution transformation saw that every possible value in *N*, the order of the quasigroup, had to be enumerated for each of the row and column values used to determine the output of the quasigroup substitution. Several intermediate truth tables were used to determine which bits in the binary representation of each row, column, and table substitution values combined with every possible value in N allowed for equality of bits as shown in Equation (6.1). The construction of an adequate LAT is the foundation of the entire attack. The LAT is the primary means by which the linear approximation of each quasigroup table row can be enumerated and further analyzed.

Let N be the order of the quasigroup. Let a_i represent the binary row value bias, where *i* is the *i*-th bit position of the given row value. Each binary row value bias was determined by taking the logical AND operator and combining a_i with every value in *N*. Let b_j, where *j* is the *j*-th bit position of the given column value, represents the binary column value bias. Each binary column value bias was determined by taking the logical AND operator and combining b_j with every value in *N*. For the purpose of this chapter, one can liken this AND operation for both a_i and b_j over *N* as a truth table "compression." This compression is a fairly quick way to take all the active, true bits in a binary representation of a (row, column) value which is later used in calculating the magnitude of each linear bias in *N*. Note that in Figures 6.3, 6.4, and 6.6, the compression occurs with aX_i, bY_j, and cZ_k respectively.

a	a_1	a_2	a_3	a_4	a_5	a_6	X_1	X_2	X_3	X_4	X_5	X_6	aX_1	aX_2	aX_3	aX_4	aX_5	aX_6	a_iXOR
1	0	0	0	0	0	1	0	0	0	0	0	1	0	0	0	0	0	1	1
1	0	0	0	0	0	1	0	0	0	0	1	0	0	0	0	0	0	0	0
1	0	0	0	0	0	1	0	0	0	0	1	1	0	0	0	0	0	1	1
1	0	0	0	0	0	1	0	0	0	1	0	0	0	0	0	0	0	0	0
1	0	0	0	0	0	1	0	0	0	1	0	1	0	0	0	0	0	1	1
1	0	0	0	0	0	1	0	0	0	1	1	0	0	0	0	0	0	0	0

Figure 6.3 Compressing $a = 1$ over $N = 6$.

b	b_1	b_2	b_3	b_4	b_5	b_6	Y_1	Y_2	Y_3	Y_4	Y_5	Y_6	bY_1	bY_2	bY_3	bY_4	bY_5	bY_6	b_jXOR
1	0	0	0	0	0	1	0	0	0	0	0	1	0	0	0	0	0	1	1
1	0	0	0	0	0	1	0	0	0	0	1	0	0	0	0	0	0	0	0
1	0	0	0	0	0	1	0	0	0	0	1	1	0	0	0	0	0	1	1
1	0	0	0	0	0	1	0	0	0	1	0	0	0	0	0	0	0	0	0
1	0	0	0	0	0	1	0	0	0	1	0	1	0	0	0	0	0	1	1
1	0	0	0	0	0	1	0	0	0	1	1	0	0	0	0	0	0	0	0

Figure 6.4 Compressing $b = 1$ over $N = 6$.

a_iXOR	b_jXOR		a XOR b = X_iY_j
1	1		0
0	0		0
1	1		0
0	0		0
1	1		0
0	0		0

Figure 6.5 XOR A and B for $N = 6$.

Once X_iY_j is found, then the row–column lookup used to determine the quasigroup "s-box" transformation, C_k, can be found. Each possible value in N was taken against the current i-th row in the quasigroup table. It was believed that in doing so, the construction of the quasigroup LAT would mirror the LAT creation process as in Heys [45]. The quasigroup row column values are known as Z and are different between the values of N depending on how the quasigroup was constructed.

From this, a_i and b_j over N are compared for equality using the logical XOR operator, and called X_iY_j (see Figures 6.3–6.5).

After all values of C have been compressed into c_kXOR (see Figure 6.5) with the possible quasigroup table values for the given i-th row, the XOR of

c	c1	c2	c3	c4	c5	c6	Z1	Z2	Z3	Z4	Z5	Z6	cZ1	cZ2	cZ3	cZ4	cZ5	cZ6	ckXOR
1	0	0	0	0	0	1	0	0	0	0	0	1	0	0	0	0	0	1	1
1	0	0	0	0	0	1	0	0	0	0	1	1	0	0	0	0	0	1	1
1	0	0	0	0	0	1	0	0	0	0	1	0	0	0	0	0	0	0	0
1	0	0	0	0	0	1	0	0	0	1	1	0	0	0	0	0	0	0	0
1	0	0	0	0	0	1	0	0	0	1	0	0	0	0	0	0	0	0	0
1	0	0	0	0	0	1	0	0	0	1	0	1	0	0	0	0	0	1	1

Figure 6.6 Compressing $c = 1$ over $N = 6$.

aX_i, bY_j, and cZ_k, known as abcXOR, can be determined. Note that the bias and likelihood values are computed by counting the total number of times abcXOR equals zero. This sum is then divided by N to yield the bias for that particular row–col–value quasigroup combination. The likelihood value shown in Figure 6.7 is the bias $- {}^1/_2$.

6.15 Pilingup Attempts for N = 16, 32, and 64

When computing the pilingup lemma on the 8 subrounds found within a two-byte block, it was noted that the value of the pileup was exceedingly large (greater than 1). Rather than treating each sub-block as its own separate bias in the pileup, it was decided to instead focus on the subround key bit, U1 input bias for the very first round and the V8 output bias for the very last round. Since U1 is the key for that round, one should only care about the key value bias and the final output bias for the purpose of this research. The key schedule included eight bits from 0×02 through 0×09 and was the same for every round. Three rounds were used in this research model for each of the $N = 16$, 32, and 64 ordered quasigroups. Figure 6.8 includes sample data from the $N = 16$ ordered quasigroup linear probability analysis for rounds 1 and 2.

To begin the pileup, let the input to the i-th round be equivalent to the vi-1 round output as in Heys' definition of the linear cryptanalytical attack.

aXOR	bXOR		a XOR b	cXOR		abcXOR	Bias	Likelihood
1	1		0	1		1	1/2	0
0	0		0	1		1		
1	1		0	0		0		
0	0		0	0		0		
1	1		0	0		0		
0	0		0	1		1		

Figure 6.7 Compression of $a = 1$, $b = 1$, $c = 1$ for $N = 6$.

Round 1 – Block 0

Byte	Ki	m1	c16	Bias
1	0x02	0x0a	0x02	5/8
2	0x03	0x0d	0x01	1/2
3	0x04	0x01	0x07	1/2
4	0x05	0x0e	0x08	5/8
5	0x06	0x0b	0x0f	3/8
6	0x07	0x02	0x06	3/8
7	0x08	0x02	0x05	5/8
8	0x09	0x06	0x0d	1/2

Round 2 – Block 1

Byte	Ki	m1	c16	Bias
1	0x02	0x06	0x07	5/8
2	0x03	0x0c	0x0b	5/8
3	0x04	0x02	0x0e	5/8
4	0x05	0x0c	0x0f	3/8
5	0x06	0x0d	0x03	1/2
6	0x07	0x06	0x04	3/8
7	0x08	0x09	0x05	5/8
8	0x09	0x01	0x01	5/8

Figure 6.8 Results of the pilingup attempts for $N = 16$ using Equation (6.4).

For each round, let the following represent the round components included during the pileup:

1. K_i represents the subround key value.
2. $m_1 = U_1$ represents the target input value as the first plaintext message value.
3. $C_8 = V_8$ represents the output from the final subround.
4. Using Matsui's [46] pilingup lemma, with p_i being the probability of a given S-box input value:

$$1/2 + 2^{n-1} \left(\prod_{i-1}^{n} \right) (p_i - 1/2) \tag{6.4}$$

6.16 Analysis of the Attack on the Quasigroup

The use of a linear cryptanalytical attack on the quasigroup structure as defined in this research model is not applicable; in order to determine the active key bits associated with round S-boxes, the use of Matsui's pilingup lemma is required. Without the results of the pilingup lemma, one cannot proceed to determine the total number of plaintext and ciphertext pairs necessary to begin the attack.

The pilingup lemma is a useful tool that saves time for the cryptanalyst. Without this tool, the work required increases substantially as the cryptanalyst is left with insufficient statistical information of where in the cipher the active key bits will occur.

Upon creating a LAT for each quasigroup order $N = 16$, $N = 32$, and $N = 64$, it was found that the final pileup results for all three orders was $^1/_2$.

6.17 The Issue of a Total Linear Bias of 1/2

Heys mentions that the attack is based on the premise that the rounds are independent of each other. If the bias is $^1/_2$, then the rounds may be dependent on each other [45] and that would suggest that the quasigroup in this model provides substantially random substitutions for even smaller sized quasiqroups. Matsui [46] indicates that in order for a suitable probability bias to be found in order to mount the attack, it must be the furthest distance from $^1/_2$. When considering the pileup from this research model, the linear probability bias stays at $^1/_2$, and the magnitude of the effectiveness of the probability bias is zero. Heys also states that correct partial subkey is derived from a linear approximation with a significantly different bias from $^1/_2$. If one were to proceed with trying to extract key bits using this model, incorrect key bits will be guessed. This is because the input bits to the S-box of the second to last round will be close to random if the piledup bias is close to $^1/_2$ [45].

6.18 Attack Complexity

The goal of a pragmatic cryptanalytical attack is to do less work for more information. The basis of linear cryptanalysis is to do just that: Find the linear probability bias $|p - {}^1/_2|$with the highest magnitude that can be used to represent the cipher, which in turn allows for fewer plaintexts required for the attack [45]. Heys [45] describes the total attack complexity to include the number of plaintext and ciphertext pairs required for the attack. Therefore, fewer plaintexts can reduce the space required to mount the attack [45]. While Heys [45] and Matsui [46] provide examples that demonstrate a successful attack on an SPN using linear cryptanalysis in less work than exhausting the key space, using an adapted linear cryptanalytical attack on a nonstandard substitution network does not yield any key bits since the total probability bias is $^1/_2$.

This research has found that the probability bias derived from the attack model used against a lower-ordered quasigroup is statistically insignificant and therefore trivial. No key bits can be extracted using the general tenants of linear cryptanalysis on an SPN-like structure. Therefore, the attack complexity of this model can be considered on par with making random guesses as to which key bits are active within each round of the quasigroup table-lookup transformation.

6.19 Possible Changes that Could Be Made in the Design of This Attack Model

This attack requires the use of a linear approximation table (LAT) that in effect represents all the possible outcomes of a quasigroup substitution given the order N. Much like a gambler counting cards at a blackjack table, the LAT becomes the primary source for choosing how to proceed with the attack. This research model assumes that in order for the linear cryptanalysis to work properly, the S-box input values must be compressed against all possible values of N. This model also sees that the LAT is constructed where each of the possible row and column values in the order of the quasigroup N is compressed using the t logical AND operator against each of the values in N. Instead, one could simply compress the binary representation of each row and column first, and then compress this value against each of the possible values in N.

Another point to consider when evaluating the efficacy of this research model includes the use of a single, generic plaintext. It could be argued that more can be done to test a slightly larger pool of plaintexts against an N ordered quasigroup. Doing this may point to other plaintext transformations that could generate nonzero linear probability biases once compared to the LAT used in this research model. Ideally, this would not be required as the LAT technically already defines all the possibilities of all values within N, the order of the quasigroup.

6.20 Which Quasigroup Order Is Best?

Since only one probability of $^1/_2$ needs to appear in the transformation within a quasigroup subround on a block, it appears that for storage and computational complexity, an $N = 16$ order quasigroup may be sufficient for applications that require low power and memory. The limiting factor is still the total cost of mounting the brute force attack on a quasigroup cipher key space. Figure 6.8 shows that the second round has a probability of $^1/_2$ over the 5th byte. As long as the round prior to the last round has a total bias of $^1/_2$ (see Figure 6.9), it can be assumed that the entire bias for the linear cryptanalytical attack is $^1/_2$, which is close to making a random guess at the likely key bit that will be in the final round of the quasigroup substitution.

Bias	Count	- 1/2	Bias - 1/2	Bias - 1/2 * Count
1/2	4	1/2	0	0
3/8	4	1/2	- 1/8	- 1/2
5/8	8	1/2	1/8	1
			Total Bias	1/2

Figure 6.9 Final pileup on $N = 16$.

An additional implication to explore in further quasigroup block cipher research should evaluate the possibility of keeping the quasigroup table secret. From a practical standpoint, this may not be possible if the source code that constructs the quasigroup table is made public. An attacker could theoretically use such source code to reconstruct all possible quasi tables. But an attacker will probably look for other low-hanging fruit when mounting an attack on a cipher of this type. From the results of a Matsui-like linear cryptanalytical attack on a quasigroup structure, smaller-ordered quasigroups may provide enough randomness in the actual round transformations that when the cryptanalyst knows the quasigroup table structure, the attack could not be mounted with a suitable pool of plaintext and ciphertext pairs.

6.21 Conclusions

Linear cryptanalysis is an effective method for enumerating all the possible values in an S-box that are most likely to occur given a standard SPN design. The complexity of mounting a linear cryptanalytical attack on DES-like SPN networks can be less than a brute force attack on the keyspace itself. When this attack method is adapted to suit a cipher that includes a nonstandard substitution–permutation component, linear cryptanalysis may fail to yield key bits. If no key bits can be deduced from the statistical analysis of the attack, then it can be assumed that the cipher transformations are sufficiently random. More research will need to be conducted to address other methods for constructing the linear approximation table (LAT), which is the fundamental component of a linear cryptanalytical attack. In the event that smaller quasigroups where the order of N is constrained significantly by memory and power usage, research should be conducted that can explore different LAT constucts. By examining how random the smaller quasigroups may be should assist further cryptanalysis and cipher development. A key area of cryptanalytical research should seek to find an ordered quasigroup that is small in size and that has a sufficiently complex key space.

References

[1] Klein, A. (2013). Stream ciphers, The estream project. pp. 229–239, Springer.

[2] Poschmann, A., Leander, G., Schramm, K., and Paar, C. (2007). New light-weight crypto algorithms for RFID. *IEEE International Symposium on Circuits and Systems, 2007. ISCAS 2007.*, pp. 1843, 1846, 27–30 May. 2007.

[3] Cheol-Joong, K., Sung-Yeol, Y., and Seok-Cheon, P. (2010). A lightweight ecc algorithm for mobile rfid service. *2010 Proceedings of the 5th International Conference on Ubiquitous Information Technologies and Applications (CUTE),* pp. 1, 6, 16–18 Dec. 2010.

[4] Sojka, A., Piotrowski, K., and Langendoerfer, P. (2011). Symbiosis of a lightweight ECC security and distributed shared memory middleware in wireless sensor networks. *2011 30th IEEE Symposium on Reliable Distributed Systems Workshops (SRDSW)*, pp. 36, 41, 4–7 Oct. 2011.

[5] Knudsen, L., Leander, G., Poschmann, A., and Robshaw, M. (2010). PRINTcipher: A block cipher for ic-printing. cryptographic hardware and embedded systems, CHES 2010, *Lecture Notes in Computer Science*, Vol. 6225, pp. 16–32.

[6] Leander, G. et al. (2011). A cryptanalysis of PRINTcipher: The invariant subspace attack, advances in cryptology, CRYPTO 2011, LNCS 6841, pp. 206–221.

[7] Bagheri, N., Ebrahimpour, R., Ghaedi, N. (2013). Differential fault analysis on PRINTcipher. *Networks, IET*, 2 (1), 30, 36, March.

[8] Shibutani, K. (2011)."Piccolo: an ultra-lightweight blockcipher. Cryptographic Hardware and Embedded Systems, CHES 2011, LNCS 6917, pp. 342–357.

[9] Sheng, L., Dawu, G., Zhouqian, M., and Zhiqiang L. (2012). Fault analysis of the piccolo block cipher. *2012 Eighth International Conference on Computational Intelligence and Security (CIS),* pp. 482, 486, 17–18 Nov. 2012.

[10] Jeong, K. "Differential Fault Analysis on Block Cipher Piccolo", Cryptology ePring Archive, Report 2012/399.

[11] Suzaki, T., Minematsu, K., Morioka, S. and Kobayashi, E. (2013). TWINE: A lightweight block cipher for multiple platforms. Selected Areas in Cryptography, LNCS 7707, pp. 339–354, 2013.

[12] Coban, M., Karakoc, F. and Boztas, O. Biclique cryptanalysis of TWINE. Cryptology and Network Security, LNCS 7712, pp. 43–45, 2012.

[13] Bogdanov, A. et al. (2007). PRESENT: An ultra-lightweight block cipher. Cryptographic Hardware and Embedded Systems, CHES 2007, LNCS 4727, pp. 450–466.

[14] Ultra-lightweight encryption method becomes international standard: http://www.kuleuven.be/english/news/ultra-lightweight-encryptionmethod-becomes-international-standard. Retrieved on Jan 13, 2014.

[15] Guo, J., Peyrin, T., Poschmann, A. and Robshaw, M. (2011). The LED block cipher. Cryptographic Hardware and Embedded Systems, CHES 2011, LNCS 6917, pp. 326–341.

[16] Jeong, K. (2013). Weakness of lightweight block ciphers mCrypton and LED against biclique Cryptanalysis. Peer-to-Peer Networking and Applications, May 2013.

[17] Bakhtiari, S., Safavi-Naini, R., and Pieprzyk, J. (1997). A message authentication code based on latin squares. In Vijay, V., Josef P., and Yi M. (Eds.). *Proceedings of the second australasian conference on information security and privacy* (ACISP '97), Springer-Verlag, London, UK, pp. 194–203.

[18] Battey, M., and Parakh, A. (2012). An efficient quasigroup block cipher, Wireless Personal Communications, pp. 1–14, Springer.

[19] Battey, M., and Parakh, A. (2012). Efficient quasigroup block cipher for sensor networks. In *21st IEEE International Conference on Computer Communications and Networks (ICCCN)*, pp. 1–5, Munich, Germany, July 30–Aug. 2, 2012.

[20] Dvorsky, J., Ochodkova, E., and Snasel, V. (2010). Quasigroups with good statistical properties. In *2010 International Conference on Computer Information Systems and Industrial Management Applications (CISIM)*, pp. 244–249, 8–10 Oct.

[21] Gligoroski, D. (2004). Stream cipher based on quasigroup string transformations in Zp. *Contributions Sec. Math. Tech. Sci.*

[22] Gligoroski, D. (2005). Candidate one-way functions and one-way permutations based on quasigroup string transformations, *Cryptology ePrint Archive*, Report 2005/352.

[23] Gligoroski, D., Markovski, S., and Knapskog, S. J. (2008). Public key block cipher based on multivariate quadratic quasigroups, Updated and extended version of the paper presented at MATH'08—Cambridge, MA, USA, March 24–26, 2008. Last revised August 2, 2008.

[24] Gligoroski, D., Markovski, S., and Kocarev, L. (2007). Error-correcting codes based on quasigroups. In *Proceedings of 16th international conference on computer communications and networks*, pp. 165–172.

[25] Satti, M., and Kak, S. (2009). Multilevel indexed quasigroup encryption for data and speech. *IEEE Trans. Broadcast.*, 55 (2), 270–281, June.

[26] Marnas, S. I., Angelis, L., and Bleris, G. L. (2007). An application of quasigroups in all-or-nothing transform, *Cryptologia*, 31 (2), pp. 133–142.

[27] Borujeni, S. E. (2000). Speech encryption based on fast Fourier transform permutation. In *The 7th IEEE International Conference on Electronics, Circuits and Systems (ICECS 2000)*, vol. 1, pp. 290–293.

[28] Mosa, E., Messiha, N. W., and Zahran, O. (2009). Chaotic encryption of speech signals in transform domains. *International Conference on Computer Engineering & Systems* (*ICCES 2009*), pp. 300–305.

[29] Bernstein. D. J. (2013). Failures of secret-key cryptography, *20th International Workshop on Fast Software Encryption*, 10–13 March, Singapore.

[30] Rosenhouse, J., and Taalman, L. (2011). Taking sudoku seriously: The math behind the world's most popular pencil puzzle, USA: Oxford University Press.

[31] Rukhin, A., Soto, J., Nechvatal, J., Barker, E., Leigh, S., Levenson, M., et al. (2001). A statistical test suite for random and pseudorandom number generators for cryptographic applications, NIST, Special Publication 800–22, Revision 1a.

[32] Hauck, S., Andre DeHon (2007). Reconfigurable computing: The theory and practice of FPGA-based computation (systems on silicon). Morgan Kaufmann.

[33] Kilts, S. (2007). Advanced FPGA design: architecture, implementation, and optimization. Wiley-IEEE Press.

[34] Mangione-Smith, W. H., and Potkonjak, M. (1998). FPGA fingerprinting techniques for protecting intellectual property. In *Proceedings of the Custom Integrated Circuits Conference*, IEEE.

[35] Wollinger, T., and Paar, C. (2003). How secure are FPGAs in cryptographic applications. *Field Programmable Logic and Application.* Lecture Notes in Computer Science, vol. 2778, pp. 91–100.

[36] Huffmire, T., et al. (2008). Managing security in FPGA-based embedded systems. *IEEE Design Test Comput.*, 25 (6), 590–598.

[37] Jacobson, M. T., and Matthews, P. (1996). Generating uniformly distributed random latin squares. *J. Combinatorial Designs*, 4 (6), pp. 405–437.

[38] Sagastume, I. G. (2014). Generation of random latin squares step by step and graphically. *XX Congreso Argentino de Ciencias de la Computación (Buenos Aires, 2014)*, October 2014.

[39] OEIS, The On-line Encyclopedia of Integer Sequences, sequence number A002860, Number of Latin squares of order n; or labeled quasigroups.

[40] Delahaye, J.-P. (2006). The science behind sudoku. *Scientific American*, June, pp. 81–87.

[41] McKay, B., Wanelss, I. M. (2005). On the number of latin squares. *Annals of Combinatorics*, 9 (3), 335–344, October.

[42] Stones, D. S. (2010). The many formulae for the number of latin rectangles. *Electr. J. Combinator.*, 17, January.

[43] Snasel, V., Abraham, A., Dvorsky, J., Kromer, P., and Platos, J. (2010). Evolving quasigroups by genetic algorithms. *Proceedings of the Dateso 2010 Workshop, Databases*, Information Retrieval, Algebraic Specification and Object Oriented Programming, pp. 108–117.

[44] Battey, M., Parakh, A., and Mahoney, W. (2015). Cryptanalysis and improvements of the quasigroup block cipher journal of information assurance and security. ISSN 1554–1010 Vol. 10, pp. 031–039.

[45] Heys, H. (2001). A tutorial on linear and differential cryptanalysis. Waterloo, Ont.: Faculty of Mathematics, University of Waterloo.

[46] Matsui, M. (1993). Linear cryptanalysis method for DES cipher. Advances in Cryptology – EUROCRYPT '93, 386–397.

[47] Swenson, C. (2012). Modern cryptanalysis: Techniques for advanced code breaking. Indianapolis, IN: Wiley Pub.

[48] Data Encryption Standard (DES) Fips Pub 46–3. (1999, October 25). Retrieved December 10, 2014, from http://csrc.nist.gov/publications/fips/fips46–3/fips46–3.pdf.

7

Measuring Interpretation and Evaluation of Client-side Encryption Tools in Cloud Computing

Md. Alam Hossain[1], Ahsan-Ul-Ambia[2], Md. Al-Amin[1] and Rahamatullah Khondoker[3]

[1]Department of Computer Science & Engineering,
Jessore University of Science & Technology, Jessore, Bangladesh
[2]Department of Computer Science & Engineering,
Islamic University, Kushtia, Bangladesh
[3]Fraunhofer Institute for Secure Information Technology,
Darmstadt, Germany

Abstract

Everything of our life is migrating into virtualization, and we are depending on information technology. We are using information and program extensively day by day. To access data globally and timely, we are storing data into cloud storage space. Cloud service providers maintain authentication and authorization mechanism, but there are no mechanisms for data security if the users' credentials are compromised. For ensuring data security, several client-side encryption tools have been developed and widely used with their own features and security schemes. Client-side encryption tools are becoming popular in the field of cloud computing every day. It is very important for users to know which tool is the best for encrypting and storing data in the cloud. "Which one provides more security among those tools? Which one consumes less time? In a single word, which one is the most effective?" These are the open questions to the users, in present days. To find out the best tool, we have analyzed the performance of the selected tools namely AxCrypt, nCrypted Cloud, SafeBox, SpiderOak, and Viivo. For performance measuring, we have calculated the upload time including both encryption and synchronization time for different sizes of data files for each tool. After analyzing and comparing the performances, we have found SafeBox as the best client-side encryption tool.

Keywords: Cloud computing, Cloud service providers, Client-side encryption tools, Encryption, Decryption, DropBox, OneDrive, Google Drive, CloudMe.

7.1 Introduction

In the past, people would store information and run applications or programs from software on a physical local computer or server; cloud computing allows people to do the same kinds of activities through Internet. Cloud computing increases efficiency, helps improve cash flow, and offers many more benefits such as flexibility, disaster recovery, automatic software updates, capital–expenditure free increased collaboration, work from anywhere, document control, security, competitiveness, and environmentally friendly. Cloud is a set of hardware, networks, applications, storages, and interfaces. Cloud service providers are a kind of third-party organization such as DropBox, OneDrive, Google Drive, and CloudMe, which provides online storage services [1–15]. Now people are using cloud computing mainly for storing information and application services where application services are not available sufficiently. Cloud service providers use username and password for a user authentication and authorization. Many cloud service providers also provide a two-way authentication option as optional for validating a legitimate user which ensures more security. But CSPs do not provide any kind of service for data security. If a user's username and password as well as two-way authentication token are compromised by the intruders such as hackers, then the information may be stolen and could be used maliciously. As all the communications are occurred in a real-time communication which means through Internet, users' credentials could be compromised easily by the hackers [16–32].

For avoiding these kinds of security threats and for ensuring data security, several new kind of tools are developed which are called client-side encryption tools (CSETs) and widely used with their own features and security schemes [33–44]. Client-side encryption tool is application software which encrypts the files and data to be uploaded in the CSPs storage through the process called encryption where the original data converted to a special form called ciphertext. This ciphertext is not understandable to anyone without decrypting through decryption. Decryption makes the opposite operation of encryption that means converts the ciphertext into original data. So, the client-side encryption tools are mainly used for encrypting user files and uploading the encrypted files in the cloud and also downloading the files from the cloud and decrypting the encrypted files to the original files.

Now-a-days, various CSETs are available in market such as AxCrypt, nCrypted Cloud, SafeBox, SpiderOak, Viivo, Boxcryptor, Ensafer, Cloud-fogger, SharedSafe, and SafeMonk. For user convenience, it is important to know which tool has the best efficiency. Which tool provides the best security services? In a word, which tool should they use? In this chapter for finding the best tool, we select five popular client-side encryption tools namely AxCrypt, nCrypted Cloud, SafeBox, SpiderOak, and Viivo to analyze and compare their performances based on uploading time including both encryption and synchronization time for different sizes of data files. After analyzing and comparing the performances, we have found SafeBox as the best client-side encryption tool.

The remaining part of this chapter is structured as following. Section 7.2 discusses about cloud service providers. Section 7.3 presents the deployment model of cloud service provider. Section 7.4 describes the methodology for measuring performance of the tools. Section 7.5 derives the attributes of existing tools. Section 7.6 compares the features of studied tools. Section 7.7 studies the characteristics of the tools. Section 7.8 discusses the strength and security of the encryption algorithm and key generation mechanisms of the studied tools. Section 7.9 measures and analyzes the performances of the tools by calculating uploading time including both encryption and synchronization time. A summary of the results is reported in Section 7.10. And finally in Section 7.11, the decision of this research work is made and future works are mentioned.

7.2 Cloud Service Providers (CSPs)

Cloud service provider is a third-party organization which offers some components of cloud computing such as storage space or software services by using private or public cloud network. Customers can access the storage space and software that is available through the Internet. So, all these service providers are known as cloud service providers.

Services in a cloud are mainly designed to provide easy, scalable access to applications, resources, and services that are fully managed by a cloud service provider. The required hardware and software for any kind of operation will be provided by the service provider. There are various types of cloud services such as database processing, online data storage and backup solutions, hosted office suites, and document collaboration services.

CSPs provides some features of cloud computing, and they are based on services that include infrastructure as a service (IaaS) which includes virtual servers, virtual storage, and virtual desktops, or computers; software as a service (SaaS) which indicates delivery of simple to complex software through the Internet; or platform as a service (PaaS) that is a combination of IaaS and SaaS and delivered as a unified service [45–50].

Cloud provider is also known as a utility computing provider. IT infrastructure that is commercially distributed and sourced across several subscribers is provided by cloud providers. In accordance with demand, cloud providers deliver cloud solutions. Users access cloud resources with the help of Internet. There are many types of cloud provider which includes public cloud provider, private cloud provider, hybrid cloud provider, or community cloud provider.

7.3 Deployment Model of Cloud Service Provider

Cloud service providers [51–54] are mainly classified based on the services that are provided by the service providers to the users. In the following section, we describe briefly the classification of cloud service providers with examples:

Public Cloud Provider: Public cloud provider provides the service resources over a network and is open for public use so that users can easily use the resources [47]. Normally, public cloud service providers such as Amazon AWS, Microsoft, and Google operate the infrastructure and users access them through the Internet.

Private Cloud Provider: Private cloud provider is a type of cloud provider which involves a distinct and secure cloud-based environment. In a private cloud, only the authorized user can operate and it is managed by a third party and hosted internally or externally. As self-run data are generally capital intensive, they have much impact physical footprint, requiring allocations of space, hardware, and environmental controls. These systems need to refresh periodically for better services, and for this, additional capital expenditures are required. High control system and privacy are imposed by a single organization for accessing the cloud. Private cloud services are provided from Microsoft, VMware, OpenStack, CloudStack, Platform9, etc. [55]. Their services are compared in terms of management, compatibility, complexity, and security.

Community Cloud Provider: Community cloud is a cloud service model that provides a cloud computing solution to a limited number of individuals that are managed by a third party and hosted internally or externally [47]. It shares infrastructure between several organizations from a specific community with common concerns such as security and compliance that may be related to performance requirements such as hosting applications that require a quick response time. So, community clouds are a hybrid form of private clouds and they operate specifically for a targeted group.

Hybrid Cloud Provider: Hybrid cloud is a composition of two or more clouds such as private, community, or public that remains unique entities but is bound together and offers the benefits of multiple deployment models [47]. It has the ability to connect the managed and dedicated services with cloud resources.

DropBox, OneDrive, and Google Drive are the most commonly used cloud service providers.

DropBox: DropBox also known as the online backup service, and this is a personal cloud storage and file hosting service. It is frequently used for file sharing and collaboration. Windows, Macintosh, and Linux are supported operating system for it [56]. There are also applications for iPhone, iPad, Android, and BlackBerry devices. Here, user data are protected with secure sockets layer (SSL) and advanced encryption system (AES) 256-bit encryption.

OneDrive: OneDrive is a personal cloud storage and file hosting service from Microsoft Corporation. It permits users to upload data and synchronize them to cloud storage and then access them from a web browser or their local devices [57]. It allows users to keep the data private, share them with contacts as user's wish, or make the data public. If user wants to share the data publicly, he never requires a Microsoft account to access. It is officially known as Microsoft OneDrive previously SkyDrive, Windows Live SkyDrive, and Windows Live Folders. Windows and MAC operating systems and mobile devices such as smartphones and tablets, including Windows phone 7 and 8 devices and Apple iOS powered iPhones and iPads, are supported by it.

Google Drive: Google Drive is personal and premium cloud storage service from Google that helps users to store and synchronize digital content and information. Computers, laptops, and some mobile devices such as Android, Apple iOS-powered iPhones, and iPads use this service to store information in the cloud [58]. The operating system Windows, MAC, Android, and iOS are supported by it.

7.4 Methodology

The problem to be solved can be stated in short as follows: Now-a-days, we have different client-side encryption tools with different features (encryption technique, cross platform, file sharing, etc.). It is necessary to find out the best client-side encryption tool among the existing tools in terms of user requirements. The best client-side encryption tool could be found in different ways as their encryption speed, uploading speed, key length, and file sharing technique differ from one another. In this chapter, we measured the efficiency of different tools for encrypting and uploading the data files in the cloud to find out the best client-side encryption tool.

Here, we compared the performances (uploading time = encryption time + synchronization time) of six popular client-side encryption tools which are available in the market namely AxCrypt, Wuala, nCrypted Cloud, SafeBox, SpiderOak, and Viivo. We select these six tools based on their popularity and user-friendly functionalities. Here, encryption time is the time taken by a tool for encrypting a file completely, where file is an object that stores information and data in a computer system. After putting the enCrypted Cloud file in the cloud service provider's folder in our system, it takes some time for synchronizing with the cloud server. This time is known as synchronization time. We took various sizes of file (256 KB, 512 KB, 1 MB, 3 MB, and 5 MB), encrypted them with these tools, and stored them in a cloud service provider like DropBox. The total uploading time including both encryption and synchronization time for an individual tool was measured by a program written in C language. Encryption starting and synchronization ending time are updated from the system time. And finally, total uploading time is calculated by the difference of synchronization ending and encryption starting time. In Figure 7.1, the role of client-side encryption tools has been shown.

7.5 Deriving the Attributes of Existing Tools

7.5.1 AxCrypt

In modern days, file encryption, decryption, compression, storing, and sharing are necessary. Encryption is used to protect data from being seen by others. But it cannot protect data from being lost. This can be overcome by keeping the backup of the necessary and required files.

For windows-based operating system, one of the top file encryption softwares is AxCrypt [59–61]. It is integrated to Windows Explorer. AxCrypt works as a complement to DropBox, Google Drive, OneDrive, Live Mesh,

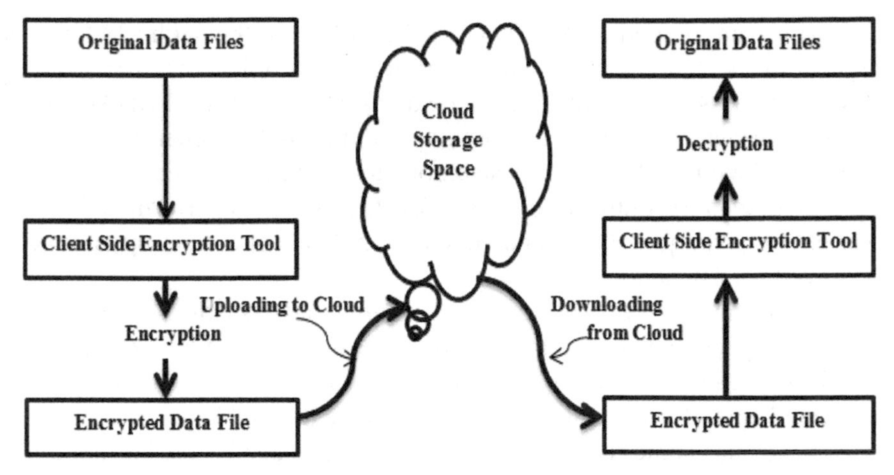

Figure 7.1 Role of client-side encryption tool in cloud computing.

and Box.net which is free for using and works in standard desktop mode. It uses strong open-source file encryption for different windows-based OS which is easy.

On November 19, 2001, the first version of AxCrypt was released for the general public which was a pre-released beta. Now, there are different setup packages for 32-bit and 64-bit environment where the cryptographic standards are AES-128 and SHA-1.

For encryption, AxCrypt easily and safely sends file to others using e-mail or any other means. It operates with an in-memory cache of used passphrases. The cache is cleared every time when the computer is restarted or new user is logged on which it is more comfortable to use and saves us from re-entering the passphrase every time. If we do not want to use passphrase every time, we can use stronger one which is more secured. The crypto logical primitives are AES with 128-bit keys for encryption and SHA-1 for hashes.

AxCrypt algorithms are deemed secure by both the US Government and the Internet. In AES Key Wrap, key wrapping is done by using the NIST specification. The key-encrypting key is the key derived from the passphrase for SHA-1. We use NIST AES Key Wrap Algorithm to wrap out the key. Key wrapping is done with at least 10,000 iterations which increase the work effort with approximately 13 bits. The actual iteration count is determined dynamically, adding 16–18 bits of effective key length. The key file is linked together with the provided passphrase. Before using key file as an encryption key, it also hashed with the passphrase.

During data encryption, AxCrypt uses the AES algorithm with 128-bit keys in cipher block chaining mode with a "random" IV. For integrity verification, it uses hash message authentication code using SHA-1 (HMAC-SHA-1–128) with 128-bit output and key. The pseudorandom number generator (PRNG) is integrated with SHA-1 as the hash algorithm. The PRNG seed is a constant accumulating value. Entropy collection for PRNG seed is performed through a variety of techniques. One hundred and eight bytes of the entropy pool are also saved persistently in the registry.

The data in the file may consist of many header sections. The header sections may contain information about the file name, file size, and file modification times as well as version information, integrity checksum, etc. These are kept enCrypted Cloud under a separate derived key.

AxCrypt has some features for recovery of damaged files, but there are some limitations. We must also be careful and start by marking the original (damaged) file as read only and then make a copy of it—and always only work on the copy!

There are some limitations and factors to be aware of also.

The way AxCrypt encrypts means that in the best of case, a one-bit error will cause 32 bytes (256 bits) of damage.

AxCrypt includes optional automatic compression, which is invoked if it is deemed worthwhile by the program, i.e., if the savings are enough to outweigh the extra time. If the data were compressed before encryption and the file gets damaged, we will probably lose at least 64 K of data. In earlier versions, we may effectively not be able to recover any data because the decompression will never get back in sync after an error. Later versions output sync blocks every 64 K, at a slight expense decreased compression ratio.

Even if the attempt to decrypt the damaged file appears successful, there may not be any recoverable data left, and it may also be the case that the original program that created the file is incapable of working with damaged files, and you thus effectively lose all anyway.

7.5.2 nCrypted Cloud

To protect all of our cloud-based files, nCrypted Cloud is used. It is a key management and sharing system, designed for security and privacy in cloud [62–64]. For securing our data, nCrypted Cloud uses Zip format and additional software is used for doing this. User account and password are used for authentication and authorization during the operation in that software. nCrypted Cloud uses AES-256 to protect data in Zip container. The individual password is used to encrypt the files. The steps of nCrypted Cloud key recovery are as follows:

1. This file password is enCrypted Cloud with user recovery key (URK).
2. ZIP file is enCrypted Cloud with unique password using AES-256 bit algorithm.
3. A unique password is derived for each file from a key value and additional entropy using PBKDF2.

nCrypted Cloud uses different methods to share files. It uses unique symmetric key for secure sharing of a folder. If we keep a file in the folder for sharing, a password is required with the key to encrypt the data. The key is stored in encrypted cloud server as well as local key store. If the folder is removed, then the key is also removed from the local key store. nCrypted Cloud supports world-class privacy to store data on cloud storage provider such as DropBox, Google Drive, Sky Drive, Box accounts, and One Drive.

Restoring data: nCrypted Cloud never stores users' private keys, but there is a smart failsafe mechanism that uses another set of generated keys. Besides the user personal key, the system will also create a public/private key set dubbed user recovery key. This key is stored locally on the client's computer, but user can store an encrypted variant of it on nCrypted Cloud servers as well. When user encrypt a file, its unique password gets enCrypted Cloud as well by using the user recovery key. This value is then stored in the comment file of the resulting zip archive. This allows users to recover an enCrypted Cloud file if they still have the recovery key.

7.5.3 SafeBox

SafeBox is a tool that works for encrypting files [65]. It synchronizes data files with reserving and sharing services of cloud service provider such as DropBox, OneDrive, SugarSync, ZumoDrive, Carbonite, and F-Secure. SafeBox is designed with less complexity and easy to use with cloud service provider. Before user data go anywhere, SafeBox encrypts and secures it. SafeBox provides end-to-end security so user data are completely secured. It ensures the privacy of user data by encrypting and masking the names of data files and folders.

SafeBox uses AES-256 (advanced encryption standard with a 256-bit key) to secure user data [28], since the user is the only one who knows the key to decrypt encrypted file and no one else can access the content. This includes hosting staff and government agencies.

SafeBox uses the following algorithms for the operation:

1. PBKDF2 for key strengthening. Iteration count 10,900 for AES engine and 10,500 for HMAC signing.
2. AES-256 with CBC mode and PCKCS7-RFC5652 (CMS) padding.
3. HMAC-SHA-256 for file authentication.
4. Random unit depends on operating system for generating IV.

SafeBox works with Mac OS X, Windows XP, Windows Vista, and Windows 7. It is compatible with both 32-bit and 64-bit operating system.

7.5.4 SpiderOak

SpiderOak simultaneously works as a client-side encryption tool and cloud storage provider [66, 67]. Therefore, if users use SpiderOak as encryption tool, they need not require additional cloud storage service to store data. Besides storage, SpiderOak also provides sync among all connected devices. It provides encryption backup, sync, storage space, and share option. Backup and sync occurred automatically.

It can be used for local backup with 2-GB free storage, and additional storage space can be rented monthly or annually. Auto-synchronization of the backup file from any folder even from any external drive is possible. File accessing from browser is possible, but uploading file from browser is not possible. There is no file type and file size limitation. This tool allows folder sharing publicly and privately. It supports the OS of Windows, Mac, Linux, iPhone, iPad, Android, and Windows mobile. This tool is complex in use.

True Privacy with SpiderOak

To protect sensitive user data, SpiderOak uses correspondingly 256-bit AES and 2048-bit RSA encryption algorithm for making the file and password secured. SpiderOak encrypts data in local device before uploading it to the server. In this system, the password is never transmitted to server in plain text format, so the password is only accessible by the user alone. Analyzing user password SpiderOak creates a key using derivation/strengthening algorithm PBKDF2 (using sha256) which has a minimum of 16384 rounds and 32 bytes of random data ("salt"). By this method, all brute force and pre-computation or database attacks against the key can be defeated. Afterwards, the password key is used to encrypt/decrypt a set of robust encryption keys which are used to encrypt/decrypt main data. Using PBKDF2 and the salt, the user create the outer level encryption key from the known password, then deciphering the outer level keys, user machine decrypt the main data.

Significant Advantages:

- **"Zero-Knowledge" Privacy Policy**: During encryption process, SpiderOak applies a "zero-knowledge" privacy policy which means only the user will have the accessibility to the stored files by knowing the password. On the other hand, SpiderOak server cannot reset any password, because they have no information about password. So the user has to be quiet responsible for saving his password.
- **Auto-synchronization**: Easy and secure auto-synchronization of files through multiple computers and devices on multiple platforms (Windows, Mac, Linux, and Android) occurs in SpiderOak.
- **Easy Sharing**: With the sharing option, SpiderOak allows sharing files with family and friends by creating ShareRooms which facilitates auto-updating during changes occur.
- **Access File from Anywhere**: Users can easily access and download their files from server though SpiderOak's excellent Web interface and "View" tab of SpiderOak software by simply logging in with the username and password.
- **Utilization of Storage**: For saving storage space and reducing storage costs, SpiderOak uses rigorous and sophisticated compression method and data redundancy prevention system.
- **Infinite Version Recovery**: SpiderOak always saves every editing version history. So user can restore files at any point in time.
- **File Recovery**: SpiderOak follows Recycle Bin system which assists user to recover his deleted or accidentally damaged files.
- **Single Account can Access Multiple Drives**: SpiderOak application on various computers/external drives can be used by a single SpiderOak account.
- **Multi-directory Support**: With SpiderOak, user can synchronize many different local directories as user wants.

Limitations:

- Scheduled backup is not still confirmed.
- There is no easy drag and drop restores option via Explorer drive icons.
- There is no system for making default configuration of important files.

7.5.5 Viivo

Viivo is a client-side encryption tool used in Windows, Mac, iOS, and Android operating system [68–70]. It is most widely used by accountants, attorneys, and Govt. & Health Cares. Viivo uses a multi-level hybrid cryptography approach when securing files.

Some key features of Viivo:

- **Security for the Cloud**: Files are secured at the source with strong encryption and authentication, protecting data from hackers, breaches, and user error.
- **Visibility into Cloud Usage**: The Viivo for business administrative console provides cloud usage reports, audits, and monitoring for cloud connections to business systems.
- **Sync and Share in Major Public Clouds**: Viivo for business automatically syncs and shares with public cloud services such as DropBox, Box, OneDrive, and Drive.
- **Reduce Storage Space and Costs**: Data compression of up to 95% happens before files are shared in the cloud, resulting in efficient storage and cost savings.
- **Direct Customer Support**: Viivo for business comes with full-time online and phone support.
- **Access from any Operating System**: Secure access to documents on Mac, Windows, iOS, and Android.

Viivo for business was created by PKWARE, the inventor of the ZIP file standard and innovator of data security and performance software products for nearly three decades. It is the commercial-grade version of Viivo that is already in use for securely storing and sharing files in the public cloud.

Viivo security uses industry standards such as RSA-2048 and AES-256 to lock down data regardless of hackers, data snoopers, or any other unauthorized user [32]. At the base level, Viivo creates a 2048 RSA key pair to safely exchange keys between collaborators and devices. User-given password is considered as RSA private key and strengthened using PBKDF2 HMAC SHA-256. All files in client side to be uploaded are enCrypted Cloud using symmetric AES-256 before they leave the physical device.

The PBKDF2 key derivation function has five input parameters:

$$DK = PBKDF2\ (PRF, Password, Salt, c, dkLen)$$

where

- PRF is a pseudorandom function of two parameters with output length hLen (e.g., a keyed HMAC).

- Password is the master password from which a derived key is generated.
- Salt is a cryptographic salt.
- c is the number of iterations desired.
- dkLen is the desired length of the derived key.
- DK is the generated derived key.

7.6 Comparison of the Studied Tools

After analyzing the studied tools, we have found some similarities (encryption, decryption, sharing, etc.) and dissimilarities which are represented in Table 7.1. We can see that almost all the tools use AES and RSA algorithms for encryption and key generation. But Viivo uses in addition PKWARE and ZIP file standards. All the tools except SafeBox have authentication feature. Without this authentication, no encryption or decryption is possible in that tool. Almost all the tools support DropBox as CSP, and some of them support other CSPs such as Google Drive and OneDrive. All the tools support Windows OS, and some of them support MAC, iOS, Android, etc. In case of sharing, AxCrypt and Ensafer support all kinds of sharing (both file and folder). SafeBox supports folder sharing only. Ensafer and SafeMonk support offline encryption and decryption and once it is authenticated for the first time in a device. SafeBox performs offline encryption to a linked safe, and SafeBox can be accessed over Internet by using safe key from the recipient end. Among these tools, SafeBox only can create safe key in offline. AxCrypt supports encryption and decryption even though the device is currently offline after a successful logged in and until the system shutdown.

7.7 Characteristics of the Studied Tools

The important characteristics of the studied client-side encryption tools are highlighted in the following:

Advantages of AxCrypt:

1. AxCrypt integrates seamlessly with windows to compress, encrypt, decrypt, store, send, and work with individual files.
2. AxCrypt is free and open-source software; user can redistribute it and modify it under the terms of the GNU General Public License as published by the Free Software Foundation.

Table 7.1 Comparison of the characteristics of studied tools

Parameter	Viivo	AxCrypt	SpiderOak	Safebox	nCrypted Cloud
Encryption Algorithm	AES-256 encryption using PBKDF2 HMAC SHA256 [69, 70].	AES-128, HMAC-SHA1–128, FIPS 186-2 [61].	AES256 in CFB mode and HMAC-SHA256, PBKDF2 (using sha256), RSA-2048 [67].	AES-256, PBKDF2 [65].	AES-256, SSL [62].
Supported Platforms	Microsoft Windows XP/Vista/7/8/10, Mac OS X 10.4 and greater, iOS 6.0 and later, Android 4.0 and later [68].	Microsoft Windows Vista/7/8/10 (both 32- and 64-bit compatible) [59, 60].	Windows, OS X, Debian, Fedora, Slackware, Android, iOS [66].	Mac OS X Lion 10.7 or higher and Windows XP SP3 or higher(both 32-bit and 64-bit) [65].	Microsoft Windows, Mac OS X, Android, iOS [63, 64].
Sharing	File Sharing	File Sharing	File and Folder Sharing, but not Encrypted file	File Sharing	File Sharing
Authentication required	Yes	Yes	Yes	Yes	Yes
Offline encryption	Yes	Yes	No	No	Yes

Disadvantages of AxCrypt:

1. During the encryption, a single-bit error will cause 32 bytes (256 bits) of damage.
2. AxCrypt includes optional automatic compression which is invoked if it is deemed worthwhile by the program.

Advantages of nCrypted Cloud:

1. It provides enterprise strength security capabilities to enable businesses to securely share files and collaborate over the Internet using managed or unmanaged devices.
2. Maintaining enterprise security, privacy, and compliances, i.e., auditable and regulatory compliant infrastructure; ease of deployment, administration, and maintenance.
3. Empowers consumers while maintaining corporate governance, oversight, and compliance. Minimal-to-no footprint to existing IT infrastructure.
4. Eliminates concerns and risks of personal and enterprise data.

Disadvantages of nCrypted Cloud:

1. It has never been easier to store and share data on the Internet with the help of this tool.
2. It does not give enough security about information.
3. It has no application on mobile yet.

Advantages of SafeBox:

1. SafeBox provides a quality service to customers for the growing demand of safety boxes for the protection of business and personal information.
2. SafeBox adds military-grade security by encrypting user data before it leaves user computer so only user can decrypt it.

Disadvantages of SafeBox:

1. SafeBox provides only desktop service.
2. It does not provide offline data file encryption.

Advantages of SpiderOak:

1. During encryption process, SpiderOak applies a "zero-knowledge" privacy policy which means only the user will have the accessibility to the stored files by knowing the password.
2. Besides its really strong "zero-knowledge" security features, SpiderOak is very versatile: It can sync not only our desktop or mobile device but your external drive or network volume as well.

3. SpiderOak offers a lot of information backups, uploads, and syncs.
4. SpiderOak has a well-publicized "no password storage" policy which ensures that users data even filenames are inaccessible to the company.
5. Protect files with encryption from hackers, competitors, and cloud service provider.
6. Easy and secure auto-synchronization of files through multiple computers and devices on multiple platforms (Windows, Mac, Linux, and Android) occurs in SpiderOak.
7. For saving storage space and reducing storage costs, SpiderOak uses rigorous and sophisticated compression method and data redundancy prevention system.
8. SpiderOak always saves every editing version history. So user can restore files at any point and at any time.
9. SpiderOak application on various computers/external drives can be used by a single SpiderOak account.

Disadvantages of SpiderOak:

1. Control over the data files that user shares offline or online is limited.
2. User activity history is not available.
3. EnCrypted Cloud file sharing is not available, only zero-knowledge file sends.
4. SpiderOak server cannot reset any password, because they have no information about password. So the user has to be quiet responsible for saving his password.
5. Scheduled backup is not still confirmed.
6. File-level digital rights management is not available. Policy-based access limitations are there also, any folder Sync system is limited.
7. BlackBerry and Windows Phone system is not available.

Advantages of Viivo:

1. Viivo secures documents before they are synchronized with cloud service provider.
2. Viivo security uses industry standards such as RSA-2048 and AES-256.
3. It uses a multi-level hybrid crypto approach when securing all files; whether they are personal or shared, it does not have to be re-enCrypted Cloud.
4. Viivo creates a 2048 RSA key pair to safely exchange keys between collaborators and devices.
5. It has very clear and easy user interface.

6. Viivo supports passphrase recovery through a secure process that uses data on the server with data on Viivo-enabled device.
7. Viivo has no size limit. It will encrypt any size of file. DropBox (CSPs) may not be able to upload it, but it will be enCrypted Cloud and placed into the local DropBox (CSPs) folder.

Disadvantages of Viivo:

1. It is not completely free.
2. It cannot be used in BlackBerry or Linux-type OS.

7.8 Security of Encryption and Key Generation Mechanisms of the Studied Tools

Viivo, SpiderOak, and SafeBox use AES-256 algorithm for file encryption with PBKDF2 [71–74] as key derivation function. PBKDF2 stands for password-based key derivation function 2 and replaces an earlier standard, PBKDF1. This approach prevents brute force and pre-computation or database attacks against the key. In client-side encryption tool, key generation is very important. Security of the encrypted file depends on the key. Key should be generated randomly. If a client uses long key that is generated with less randomness, then they cannot ensure security of the data properly. PBKDF2 applies a pseudorandom function, such as a cryptographic hash, cipher, or HMAC, to the input password or passphrase along with a salt value and repeats the process many times to produce a derived key, which can then be used as a cryptographic key in subsequent operations. The added computational work makes password cracking much more difficult and is known as key stretching. When the standard was written in 2,000, the recommended minimum number of iterations was 1,000, but the parameter is intended to be increased over time as CPU speeds increase. As of 2005, a Kerberos standard recommended 4096 iterations, Apple iOS 3 used 2,000, iOS 4 used 10,000, while in 2011, LastPass used 5,000 iterations for JavaScript clients and 1,00,000 iterations for server-side hashing. Having a salt added to the password reduces the ability to use pre-computed hashes (rainbow tables) for attacks and means that multiple passwords have to be tested individually, not all at once. The standard recommends a salt length of at least 64 bits. One weakness of PBKDF2 is that while its number of iterations can be adjusted to make it take an arbitrarily large amount of computing time, it can be implemented with a small circuit and very little RAM, which makes brute force attacks using ASICs or GPUs relatively cheap.

AxCrypt uses the advanced encryption standard with 128-bit keys in cipher block chaining mode with a "random" IV for the data encryption. For integrity verification, AxCrypt uses HMAC-SHA-1–128, i.e., hash message authentication code using SHA-1 with 128-bit output and key. Key wrapping of the passphrase is done using the NIST specification for AES Key Wrap. The key derived from the passphrase with SHA-1 is only used as a key-encrypting key. As a brute force counter measure, key wrapping is done with at least 10,000 iterations, increasing the work effort with approximately 13 bits. The actual iteration count is determined dynamically, a typical value is 1,00,000 to 2,00,000, adding 16–18 bits of effective key length. When a key file is used, this is concatenated with the provided passphrase, and hashed together with it, before using it as a key-encrypting key as above. The pseudorandom number generator (PRNG) is described in FIPS 186-2, with SHA-1 as the hash algorithm. The Federal Information Processing Standard (FIPS) [75–76] is an U.S. government computer security standard used to accredit cryptographic modules. FIPS defines four levels of security, simply named "Level 1" to "Level 4." It does not specify in detail what level of security is required by any particular application.

Researchers have found a weakness in the AES algorithm [77–83]. They managed to come up with a clever new attack that can recover the secret key four times easier than anticipated by experts. The attack is a result of a long-term cryptanalysis project carried out by Andrey Bogdanov (KU Leuven, visiting Microsoft Research at the time of obtaining the results), Dmitry Khovratovich (Microsoft Research), and Christian Rechberger (ENS Paris, visiting Microsoft Research). In the last decade, many researchers have tested the security of the AES algorithm, but no flaws were found so far. In 2009, some weaknesses were identified when AES was used to encrypt data under four keys that are related in a way controlled by an attacker; while this attack was interesting from a mathematical point of view, the attack is not relevant in any application scenario. The new attack applies to all versions of AES even if it used with a single key. The attack shows that finding the key of AES is four times easier than previously believed; in other words, AES-128 is more like AES-126. Even with the new attack, the effort to recover a key is still huge: The number of steps to find the key for AES-128 is an 8 followed by 37 zeroes. To put this into perspective: On a trillion machines, that each could test a billion keys per second, it would take more than two billion years to recover an AES-128 key. Note that large corporations are believed to have millions of machines, and current machines can only test 10 million keys per second.

Because of these huge complexities, the attack has no practical implications on the security of user data; however, it is the first significant flaw that has been found in the widely used AES algorithm and was confirmed by the designers. The attack has been confirmed by the creators of AES, Dr Joan Daemen, and Professor Dr Vincent Rijmen, who also applauded it.

7.9 Performance Measurement and Analysis

In the previous section, the selected client-side encryption tools namely AxCrypt, nCrypted Cloud, SafeBox, SpiderOak, and Viivo and their characteristics are explained. In this section, we will measure the performances of these tools.

7.9.1 System Setup

In this section, the application tools which are necessary for finding the best client-side encryption tool will be examined. Further, the testing environment is described.

7.9.1.1 Application tools

In this section, the application tools used for the analysis are listed below:

1. AxCrypt,
2. nCrypted Cloud,
3. SafeBox,
4. SpiderOak, and
5. Viivo.

These are the client-side encryption tools which are installed in the testing environment.

7.9.1.2 Cloud service provider

DropBox is used as cloud service provider for storing the encrypted data.

7.9.1.3 Testing environment

The specifications of the computer used are as follows:

1. Operating System—Windows 7.
2. Processor—Intel(R) Core(TM) i7-2630QM CPU @ 3.10 GHz.
3. RAM—6 GB (DDR-3).
4. Clock—Core Speed 3093.0 MHz, Bus Speed 99.8 MHz

5. Main Board—Intel (Model—DH61WW).
6. Bandwidth—1 Mbps.

7.9.2 Analysis

In this section, the performance of each selected tool is described with corresponding table and performance graph. At initial stage, we installed every tool and by using these tools, different sizes of data were enCrypted Cloud and uploaded. The total uploading time including both encryption and synchronization time for an individual tool was measured by a program written in C language. Encryption starting and synchronization ending time are updated from the system time. And finally, total uploading time is calculated by the difference of synchronization ending and encryption starting time.

We performed the operation for 100 times for each file size for every tool. Then, we calculated the average uploading time for a specific file size. Finally, we calculated the average upload time from average time for different sizes of data files. Measured values are shown in the table for every client-side encryption tool and the performance graph was obtained by plotting the value of data size (KB) in the Y axis and the upload time (sec) in the X axis.

A. We took 256 KB, 512 KB, 1 MB, 3 MB, and 5 MB sizes of data for encrypting and uploading them in DropBox (cloud service provider) with the help of AxCrypt, and the measured values are shown in Table 7.2. The average upload time of AxCrypt is 33.06 sec (from Table 7.2).

A graphical representation of the performance of AxCrypt is given in Figure 7.2.

B. We took 256 KB, 512 KB, 1 MB, 3 MB, and 5 MB sizes of data for encrypting and uploading them in DropBox (cloud service provider)

Table 7.2 The average upload time of AxCrypt

Tool	Data Size	Average Upload Time (100 Times)	Average Data Size	Average Upload Time
Axcrypt	256 KB	05.47 sec	1996.8 KB	33.06 sec
	512 KB	15.40 sec		
	1 MB	30.38 sec		
	3 MB	44.05 sec		
	5 MB	70.03 sec		

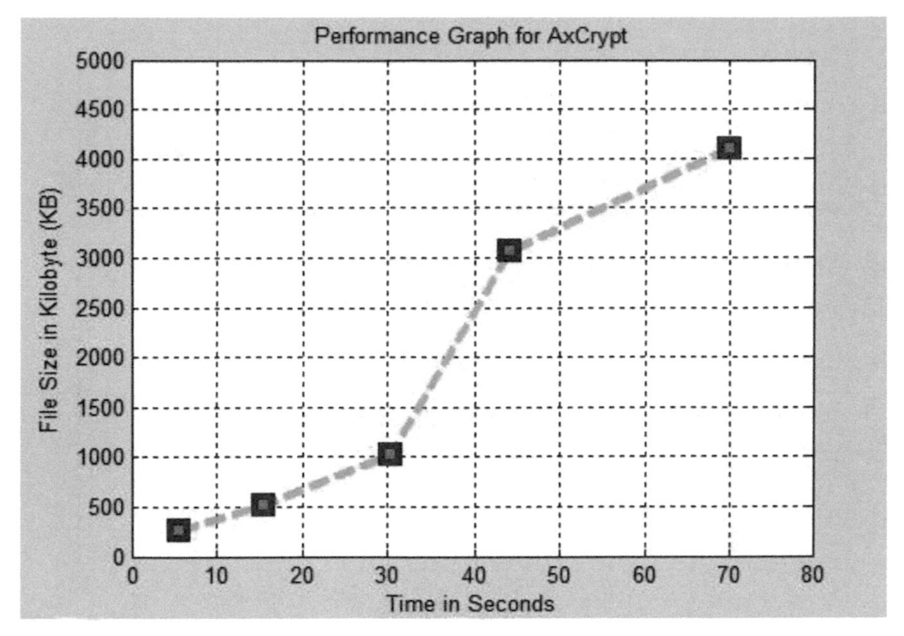

Figure 7.2 Graphical representation of the performance of AxCrypt.

Table 7.3 The average upload time of nCrypted Cloud

Tool	Data Size	Average Upload Time (100 Times)	Average Data Size	Average Upload Time
nCrypted Cloud	256 KB	05.57 sec	1996.8 KB	34.06 sec
	512 KB	15.41 sec		
	1 MB	30.23 sec		
	3 MB	44.02 sec		
	5 MB	70.10 sec		

with the help of nCrypted Cloud, and the measured values are shown in Table 7.3. The average upload time of nCrypted Cloud is 33.16 sec (from Table 7.3).

A graphical representation of the performance of nCrypted Cloud is in Figure 7.3.

C. Here, 256 KB, 512 KB, 1 MB, 3 MB, and 5 MB sizes of data are taken by the SafeBox for encrypting and uploading data in DropBox and

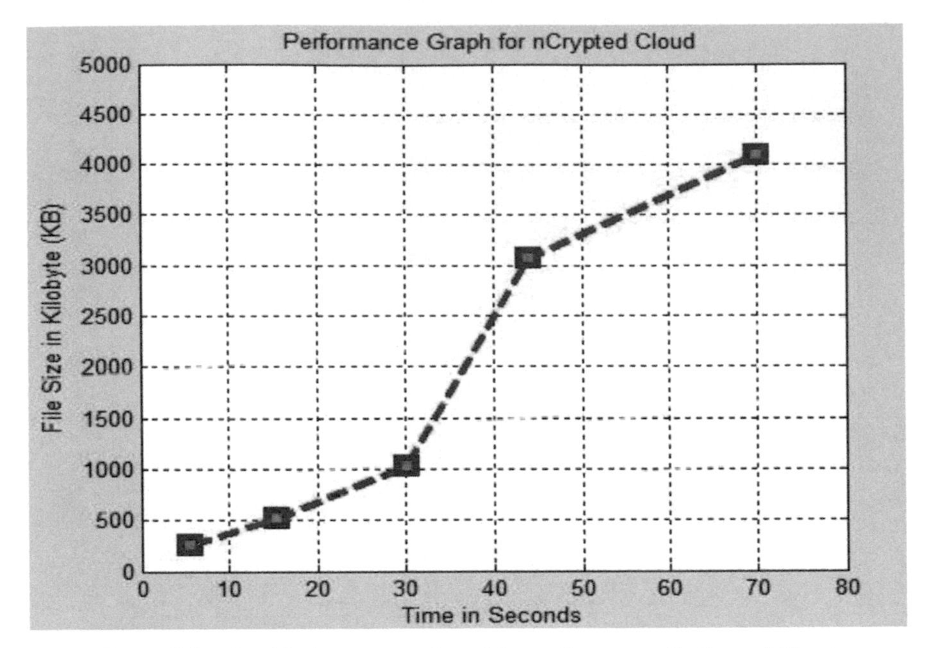

Figure 7.3 Graphical representation of the performance of nCrypted Cloud.

Table 7.4 The average upload time of SafeBox

Tool	Data Size	Average Upload Time (100 Times)	Average Data Size	Average Upload Time
SafeBox	256 KB	06.14 sec	1996.8 KB	32.53 sec
	512 KB	15.30 sec		
	1 MB	30.67 sec		
	3 MB	43.36 sec		
	5 MB	67.19 sec		

measuring the average upload time. The average upload time of SafeBox is 32.53 sec (from Table 7.4).

A graphical representation of the performance of SafeBox is in Figure 7.4.

D. Various sizes of data such as 256 KB, 512 KB, 1 MB, 3 MB, and 5 MB are taken by the SpiderOak for measuring the average upload time after

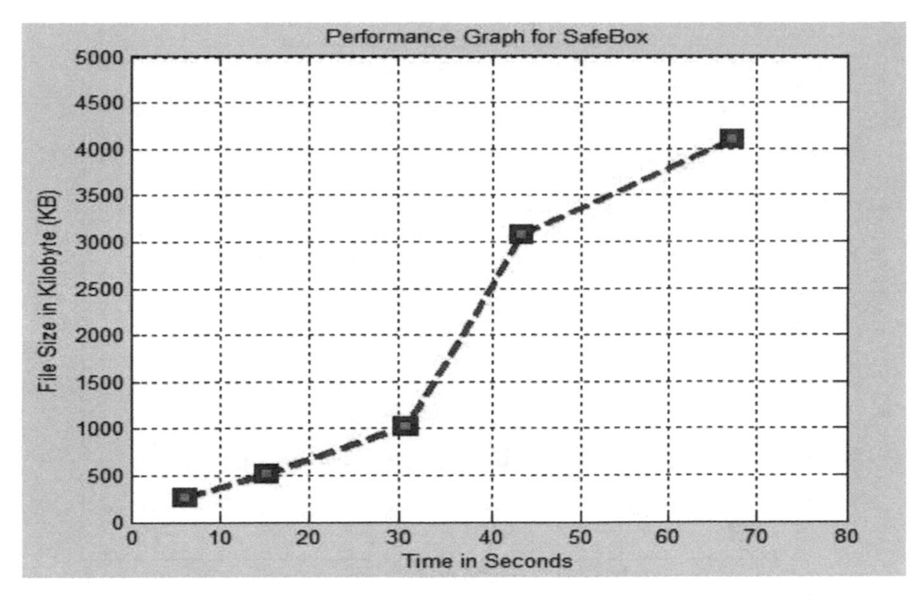

Figure 7.4 Graphical representation of the performance of SafeBox.

Table 7.5 The average upload time of SpiderOak

Tool	Data Size	Average Upload Time (100 Times)	Average Data Size	Average Upload Time
SpiderOak	256 KB	06.52 sec	1996.8 KB	37.00 sec
	512 KB	18.31 sec		
	1 MB	34.05 sec		
	3 MB	48.67 sec		
	5 MB	77.45 sec		

encrypting and uploading data in DropBox. The average upload time of SpiderOak is 37.0 sec (from Table 7.5).

A graphical representation of the performance of SpiderOak is in Figure 7.5.

E. Various sizes of data such as 256 KB, 512 KB, 1 MB, 3 MB, and 5 MB are taken by the Viivo for measuring the average upload time after encrypting and uploading data in DropBox. The average upload time of Viivo is 33.2 sec (from Table 7.6).

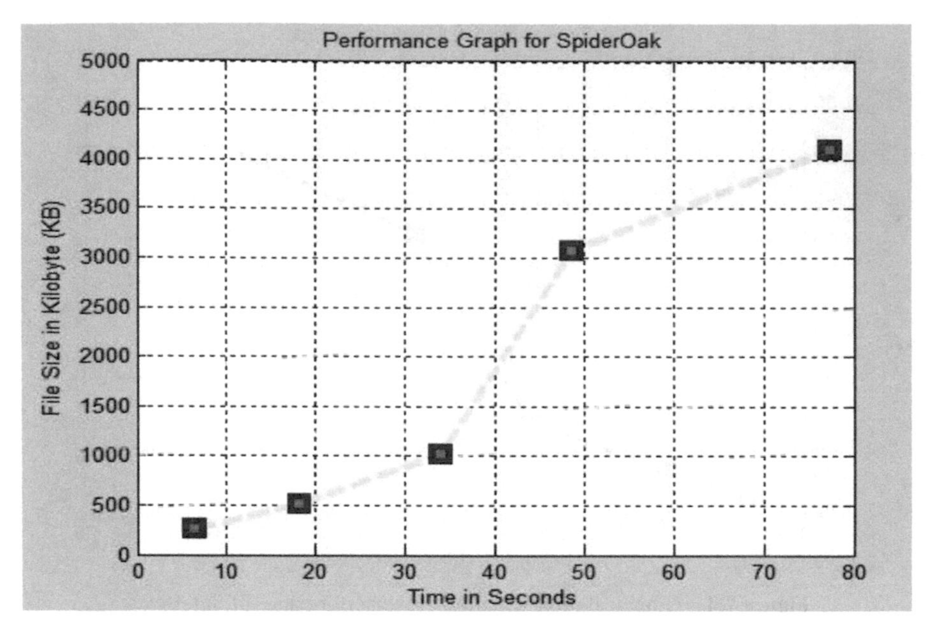

Figure 7.5 Graphical representation of the performance of SpiderOak.

Table 7.6 The average upload time of Viivo

Tool	Data Size	Average Upload Time (100 Times)	Average Data Size	Average Upload Time
Viivo	256 KB	04.39 sec	1996.8 KB	33.198 sec
	512 KB	15.67 sec		
	1 MB	30.27 sec		
	3 MB	44.13 sec		
	5 MB	71.53 sec		

A graphical representation of the performance of Viivo is in Figure 7.6.

7.10 Results and Discussion

From Tables 7.2–7.6, we get the average upload times of five different client-side encryption tools. Using these reading, we can detect the best tool among these tools by comparing the values (average upload time). We get the graphical representation of the average uploads time of different tools for comparison which is given in Figure 7.7 and Table 7.7.

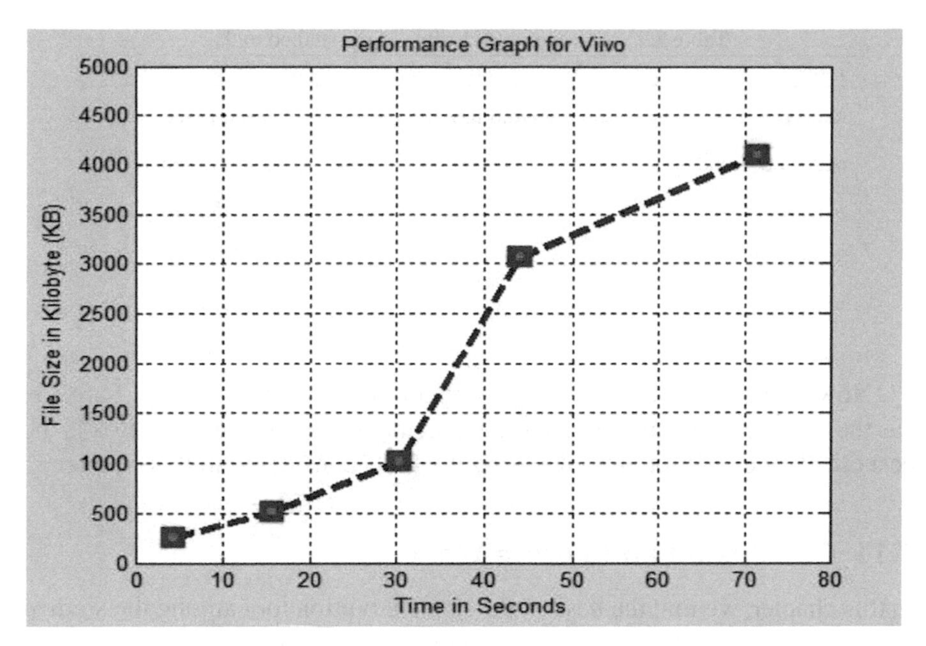

Figure 7.6 Graphical representation of the performance of Viivo.

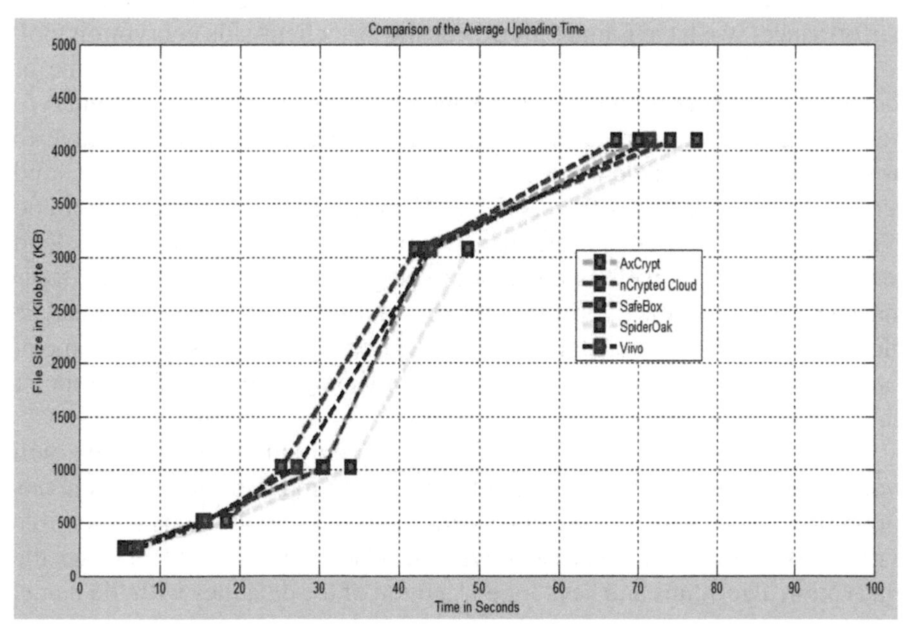

Figure 7.7 Graphical representation of average upload times of different tools.

Table 7.7 Average uploads time of the studied tools

Tools	Average Data Size	Average Upload Time
Axcrypt	1996.6 KB	33.06 sec
nCrypted Cloud		34.06 sec
SafeBox		32.53 sec
SpiderOak		37.00 sec
Viivo		33.19 sec

Now, if we compare the average upload times, we can say that SafeBox has the minimum average upload time than the other tools. So, SafeBox is the best client-side encryption tool among the tools considered in this chapter.

7.11 Conclusion and Future Work

In this chapter, we find the best client-side encryption tool among the studied tools. We measure the performance based on the uploading time including encryption and synchronization time. The performance of each tool was calculated and compared with each other. After analyzing and comparing the performances, we have found SafeBox as the best client-side encryption tool.

During this work, we have got some important issues such as some of the client-side encryption tools simultaneously work as a client-side encryption tool and cloud storage provider. In this scenario, user do not have to use additional client-side encryption tools or cloud storage provider. Cloud service provider provides the services of storage as well as encryption and synchronization. But other tools are not compatible with these types of cloud service providers as well as client-side encryption tools. Most of the tools manage the encryption algorithms and keys during the encryption of the data files. User cannot select or change the encryption algorithms or keys. Some of the tools work only in windows platform, and some work only for computer not for mobile devices.

To overcome the problems that founded in our study, in future, we will work to design architecture of a client-side encryption tool for heterogeneous cloud providers where a single client-side encryption tool will be compatible with most of the cloud service providers. Also, a user would manage the encryption algorithms and keys for encryption of the data files with file name.

References

[1] Dillon, T., Wu, C., and Chang, E. (2010). Cloud computing: issues and challenges. In *24th IEEE international conference on advanced information networking and applications*, 550-445X/10.

[2] Anitha. Security issues in cloud computing–a review. *Int. J. Thesis Projects Dissertations (IJTPD)*. 1 (1), 1–6, October–December 2013.

[3] Pring et al. (2009). Forecast: Sizing the cloud; understanding the opportunities in cloud services. Gartner Inc., Tech. Rep.G00166525, March 2009.

[4] Zeng, W., Zhao, Y., Ou, K., and Song, W. (2009). Research on cloud storage architecture and key technologies. In *Proceedings of the 2nd International Conference on Interaction Sciences: Information Technology, Culture and Human*, Seoul, Korea, pp. 1044–1048.

[5] Weinman, J. (2011). The future of cloud computing. In *IEEE Technology Time Machine Symposium on Technologies Beyond 2020 (TTM)*, pp. 1–2, June, 2011.

[6] Dikaiakos, M. D., Katsaros, D., Mehra, P., Pallis, G., and Vakali, A. (2009). Cloud computing: distributed internet computing for IT and scientific research. *IEEE Internet Comput. J.*, 13 (5), 10–13, September.

[7] Armbrust, M., Fox, A., Grith, R., Joseph, A. D., Katz, R., Konwinski, A., Lee, G., Patterson, D., Rabkin, A., Stoica, I., and Zaharia, M. (2010). A view of cloud computing. *Commun. ACM*, 53 (4), 50–58.

[8] Harmer, T., Wright, P., Cunningham, C., and Perrott, R. (2009). Provider independent use of the cloud. In *The 15th International European Conference on Parallel and Distributed Computing*, p. 465.

[9] Nurmi, D., Wolski, R., Grzegorczyk, C., Obertelli, G., Soman, S., Youseff, L., and Zagorodnov, D. (2008). The eucalyptus open-source cloud computing system. Presented at the *Proceedings of Cloud Computing and Its Applications*.

[10] Armbrust, M., Fox, A., Griffith, R., Joseph, A. D., Katz, R., Konwinski, A., Lee, G., Petterson, D., Rabkin, A., Stoica, I., and Zaharica, M. (2010). A view of cloud computing. *Commun. ACM*, 53 (4), April.

[11] Mell, P., and Grance, T. (2011). The NIST definition of cloud computing, Computer Security Division, IT Laboratory, National Institute of Standards and Technology, Gaithersburg.

[12] Jensen, M., Schwenk, J., Gruschka, N., and Iacono, L. (2009). On technical security issues in cloud computing. *IEEE Int. Conf. Cloud Comput. Bangalore*, 21–25 Sep. 2009, pp. 109–116.

[13] Lenk, A., Klems, M., Nimis, J., Tai, S., and Sandholm, T. (2009). What's inside the cloud? An architectural map of the cloud landscape. *Proceedings of the 2009 ICSE Workshop on Software Engineering Challenges of Cloud Computing*. Washington DC, 23 May 2009, pp. 23–31.

[14] Shamsi, J., Khojaye, M., and Qasmi, M. (2013). Data-intensive cloud computing: requirements, expectations, challenges, and solutions. *J. Grid Comput.,* 11 (2), 281–310.

[15] Maggiani, R. (2009). Communication consultant, Solari communication. Cloud computing is changing how we communicate. *2009 IEEE International Professional Conference, IPCC*, pp. 1–4, Waikiki, HI, USA, July 19–22, 2009.

[16] Sabahi, F. (2011). Cloud computing security threats and responses. In *3rd International Conference on IEEE*, May 2011, 245–249.

[17] Ibrahim, A. S., Hamlyn-Harris, J., and Grundy, J. (2011). Emerging security challenges of cloud virtual infrastructure. *APSEC 2010 Cloud Workshop*, Sydney, Australia, 30th Nov 2011.

[18] Sabahi, F. (2011). Virtualization-level security in cloud computing. In *3rd International Conference on IEEE*, May 2011, 250–254.

[19] Mathisen, E. (2011). Security challenges and solutions in cloud computing. *Proceedings of the 5th IEEE International Conference on Digital Ecosystems and Technologies (DEST)*, pp. 208–212, June, 2011.

[20] Jensen, M., Schwenk, J., Gruschka, N., and Iacon, L. L. (2009). On technical Security Issues in Cloud Computing. In *Proceedings of IEEE International Conference on Cloud Computing (CLOUD-II, 2009)*, pp. 109–116, India.

[21] Kamara, S., and Lauter, K. (2010). *Cryptographic cloud storage*. Lecture Notes in Computer Science, Financial Cryptography and Data Security, pp. 136–149, vol. 6054.

[22] Kaufman, L. M. (2009). Data security in the world of cloud computing. *IEEE Secur. Privacy J.*, 7 (4), 61–64, July–Aug 2009.

[23] Gaur, T., and Kharb, N. (2015). Security of data storage in cloud computing. *Int. J. Comput. Appl. (0975–8887)*. 110 (10), January 2015.

[24] Padhy, R. P., Patra, M. R., and Satapathy, S. C. (2011). Cloud computing: security issues and research challenges. *Int. J. Comput. Sci. Inform. Technol. Security (IJCSITS),* 1 (2), 136–146, ISSN: 2249–9555.

[25] Kalaichelvi, C. R., Shanmuga Priya, S., and Arockiam. L. (2012). Research challenges and security issues in cloud computing. *Int. J. Comput. Intelligence Inf. Security* 3.3, 42–48.

[26] Kant, C., and Sharma. Y. (2013). Enhanced security architecture for cloud data security. *Int. J. Adv. Res. Comput. Sci. Soft. Eng.* 3.5, 571–575.

[27] Subashini, S., and Kavitha, V. (2011). A survey on security issues in service delivery models of cloud computing. *J. Network Comput. Appl.* 34 (1), 1–11.

[28] Md Tanzim Khorshed, A. B. M., Shawkat Ali, Wasimi, S. A. (2011). Trust issues that create threats for cyber attacks in cloud computing. In *IEEE 17th International Conference on Parallel and Distributed Systems*, pp. 900–905.

[29] Grobauer, B., Walloschek, T., and Stocker, E. (2011). Understanding cloud computing vulnerabilities. *IEEE Security Privacy*, 9 (2), 50–57.

[30] Almorsy, M., Grundy, J., and Müller, I. (2010). An analysis of the cloud computing security problem. In *Proceedings of the 2010 Asia Pacific cloud workshop*, Australia, 30 November.

[31] Popovic, K., and Hocenski, Z. (2010). Cloud computing security issues and challenges. In *Proceedings of the 33rd International Convention in MIPRO*, 2010, pp. 344–349.

[32] Jamil, D., and Zaki, H. (2011). Cloud computing security. *Int. J. Eng. Sci. Technol.*, 3 (4), 3478–3483.

[33] Subrata Kumar Das, Md. Alam Hossain, Md. Arifuzzaman Sardar, Ramen Kumar, B., and Prolath Dev Nath. (2014). Performance analysis of client side encryption tools. *Int. J. Adv. Comput. Res.,* (ISSN (print): 2249–7277 ISSN (online): 2277–7970) 4 (3), 16 September.

[34] Aleem, A., and Sprott, C. R. (2013). Let me in the cloud: Analysis of the benet and risk assessment of cloud platform. *J. Financial Crime.*, 20 (1), 6–24.

[35] Yunqi Ye, Liangliang Xiao, I-Ling Yen, and Farokh Bastani (2010). Secure, dependable, and high performance cloud storage. In *2010 29th IEEE International Symposium on Reliable.*

[36] Ramgovind, S., Elo, M., and Smith, E. (2010). The management of security in cloud computing. In *Information Security for South Africa*, Sandton, 2–4 August 2010, pp. 1–7.

[37] Soghoian, C. (2010). Caught in the cloud: privacy, encryption, and government back doors in the web 2.0 era. *J. Telecommun. High Technol. Law*, 8 (2), 359–424.

[38] Chen, D., and Zhao, H. (2012). Data security and privacy protection issues in cloud computing. *Int. Conf. Comput. Sci. Electr. Eng.*, 1, Hangzhou, 23–25 March 2012, pp. 647–651.

[39] Sotomayor, B., Montero, R., Llorente, I., and Foster, I. (2009). Virtual infrastructure management in private and hybrid clouds. *IEEE Int. Comput.*, 13, 14–22.

[40] Zissis, D., and Lekkas, D. (2012). Addressing cloud computing security issues. *Future Gen. Comput. Syst.*, 28 (3), 583–592.

[41] Wang, C., Wang, Q., Ren, K., and Lou, W. (2010). Privacy-preserving public auditing for data storage security in cloud computing. *Proceed. IEEE INFOCOM*, San Diego, 14–19 March 2010, pp. 1–9.

[42] Pearson, S. (2009). Taking account of privacy when designing cloud computing services. In *CLOUD '09 Proceedings of ICSE Workshop on Software Engineering Challenges of Cloud Computing*, pp. 44–52, IEEE Computer Society Washington, DC, USA, May 2009.

[43] Lin, H. C., Babu, S., Chase, J. S., and Parekh, S. S. (2009). Automated control in cloud computing: opportunities and challenges. In *Proceedings of the 1st workshop on automated control for data centres and clouds*, New York, NY, USA, pp. 13–18, 2009.

[44] Chen, S., Nepal, S., Liu, R. (2011). Secure connectivity for intra-cloud and inter-cloud communication. In *2011 40th international conference on parallel processing workshops (ICPPW)*, pp. 154–159, 13–16 Sept. 2011.

[45] Bhardwaj, S., Jain, L., and Jain, S. (2010). Cloud computing: A study of infrastructure as a service (IAAS). *Int. J. Eng. Inf. Technol.,* 2 (1), 60–63.

[46] Zhang, S., Zhang, S., Chen, X., and Huo, X. (2010). Cloud computing research and development trend. *International Conference on Future Networks*, pp. 93–97, China, 2010.

[47] Rimal, B. P., Eunmi, C., and Lumb, I. (2009). A taxonomy and survey of cloud computing systems. *International Joint Conference on INC, IMS and IDC*, pp. 44–51, Seoul, Aug, 2009.

[48] Zhang, Q., Cheng, L., and Boutaba, R. (2010). Cloud computing: state of the art and research challenges. *J. Int. Ser. Appl.*, 1 (1), pp. 7–18, Feb.

[49] Wang, L., Laszewski, G., Kunze, M., and Tao, J. (2008). Cloud computing: a perspective study. *New Generation Computing-Advances of Distributed Information Processing*, pp. 137–146, vol. 28, no. 2.

[50] Kandukuri, B. R., Paturi, V. R., and Rakshit, A. (2009). Cloud security issues. In *Proceedings of the 2009 IEEE International Conference on Services Computing,* Washington DC, 21–25 September 2009, pp. 517–520.

[51] Cloud Provider, http://www.webopedia.com/TERM/C/cloud_provider.html, Accessed: 12.07.2015.
[52] Cloud Provider, http://searchcloudprovider.techtarget.com/definition/cloud-provider, Accessed: 12.07.2015.
[53] Cloud Provider, http://www.techopedia.com/definition/133/cloud-provi der, Accessed: 18.08.2015.
[54] Cloud Services, http://www.webopedia.com/TERM/C/ cloud_services.html, Accessed: 19.08.2015.
[55] Private Cloud Providers, http://www.tomsitpro.com/articles/private-cloud-providers-comparison,2-899.html, Accessed: 02.09.2015.
[56] Dropbox, http://searchconsumerization.techtarget.com/definition/Drop box, Accessed: 25.09.2015.
[57] OneDrive, http://www.webopedia.com/TERM/S/OneDrive.html, Accessed: 10.10.2015.
[58] Google Drive, http://www.webopedia.com/TERM/G/google_drive.html, Accessed: 13.11.2015.
[59] AxCrypt, http://www.axantum.com/axcrypt/, Accessed: 13.11.2015.
[60] AxCrypt Features, http://www.axcrypt.net/documentation/features/, Accessed: 27.11.2015.
[61] AxCrypt Security, http://www.axantum.com/AxCrypt/Security.html#Algorithms, Accessed: 27.11.2015.
[62] Impact of nCrypted Cloud, https://www.encryptedcloud.com/wpcontent/uploads/GWG46JVE_451_research_ncrypted_cloud_impact_report_12_mar_2013.pdf, Accessed: 10.12.2015.
[63] Secure Cloud Collaboration, http://edtechtimes.com/2014/02/26/ncryp ted-cloud-google-drive-microsoft-onedrive/, Accessed: 10.12.2015.
[64] nCrypted Cloud, https://www.encryptedcloud.com/apps, Accessed: 15.12.2015.
[65] Safebox, http://safeboxapp.com/index.html, Accessed: 15.12.2015.
[66] Spideroak, https://spideroak.com/features/zero-knowledge, Accessed: 18.12.2015.
[67] Spideroak, https://spideroak.com/features/private-bydesign#engineer ing-matters-10-ribbon, Accessed: 20.12.2015.
[68] Viivo supported Device and Platforms, https://forums.viivo.com/show thread.php?3-Supported-device-and-cloud-platforms, Accessed: 20.12.2015.
[69] Viivo, http://www.nextofwindows.com/viivo-to-easily-encrypt-and-secure-your-cloud-data-in-dropbox, Accessed: 20.12.2015.
[70] Viivo, https://viivo.com/how-our-security-works, Accessed: 20.12.2015.

[71] Password Iterations (PBKDF2), https://helpdesk.lastpass.com/account-settings/general/password-iterations-pbkdf2/, Accesed: 12.03.2016.
[72] Pseudo-Random Number Generator, http://www.lavarnd.org/faq/prng.html, Accesed: 12.03.2016.
[73] Random Number Generator Attack, http://www.wow.com/wiki/Random_number_generator_attack, Accesed: 12.03.2016.
[74] Statistical Analysis of Random Number, https://www.random.org/analysis/, Accesed: 12.03.2016.
[75] Federal Information Processing Standards, http://www.nist.gov/itl/, Accesed: 12.03.2016.
[76] Federal Information Processing Standards Publications, http://www.nist.gov/itl/fips.cfm, Accesed: 12.03.2016.
[77] AES Cracking, http://www.theinquirer.net/inquirer/news/2102435/aes-encryption-cracked, Accesed: 12.03.2016.
[78] First Security Flaws of AES, http://www.kuleuven.be/english/newsletter/newsflash/encryption_standard.html, Accesed: 12.03.2016.
[79] Security of AES, http://www.eetimes.com/document.asp?doc_id=1279619, Accesed: 12.03.2016.
[80] New Attack on AES, http://www.cryptosystem.net/aes/, Accesed: 12.03.2016.
[81] AES New Attack, https://www.schneier.com/blog/archives/2009/07/another_new_aes.html, Accesed: 12.03.2016.
[82] AES Hacking, https://www.kotfu.net/2011/08/what-does-it-take-to-hack-aes/, Accesed: 12.03.2016.
[83] Advanced Encryption Standard (AES), http://csrc.nist.gov/publications/fips/fips197/fips-197.pdf, Accesed: 12.03.2016.

8

Kolmogorov–Smirnov Test-based Side-channel Distinguishers: Constructions, Analysis, and Implementations

Yongbin Zhou[1,2], Chao Zheng[1] and Huan Wang[2]

[1]State Key Laboratory of Information Security, Institute of Information Engineering, Chinese Academy of Sciences, Minzhuang Rd. 89-A, Haidian District, Beijing, 100093, People's Republic of China
[2]University of Chinese Academy of Sciences, Yuquan Rd. 19-A, Shijingshan District, Beijing, 100049, People's Republic of China

Keywords: Side-channel analysis, Generic distinguisher, MPC-KSA, Kolmogorov–Smirnov test.

8.1 Introduction

An implementation of a cryptographic algorithm on embedded devices often produces physical leakages, such as power consumption [1], electromagnetic radiation [2], or execution time [3]. Secret information can be derived by analyzing these leakages using statistic tools (distinguishers). This type of analysis is called side-channel analysis (SCA). Differential power analysis (DPA) introduced by Kocher [1] is a typical representative of SCA. The core idea is to compare key-dependent predictions of physical leakages with the practical measurements, in order to identify the most likely hypothesis of (part of) the cryptographic key within a set of candidates. In practice, it requires to model the leakages based on sufficient knowledge about the particular dependency between the processed data and the corresponding leakages in order to make the predictions of physical leakages accurately. Meanwhile, it requires to have a good distinguisher to efficiently reveal the keys. For example, DPA employs difference of means [1], while correlation power analysis (CPA) employs Pearson's correlation coefficient [4]. However, these typical SCA may be inefficient without specific knowledge of, or assumptions about, cryptographic device.

In order to deal with such cases, mutual information analysis (MIA), a generic side-channel distinguisher, was proposed at CHES 2008 [5]. Requiring no specific knowledge of the target device or the particular dependencies between leakages and processed data, MIA is still sufficient. Even though MIA is generic, previous works such as [6] and [7] showed that the performance of MIA depends on the accuracy of the estimation methods for the probability density function (PDF). In fact, the PDF estimation is wildly accepted to be a difficult problem. Considering this, Veyrat-Charvillon gave an alternative to MIA, Kolmogorov–Smirnov analysis (KSA) [6], while Liu gave partial Kolmogorov–Smirnov analysis (PKS) [8]. KSA and PKS are similar to MIA but do not require accurate density estimation, and they both use empirical cumulative density function (CDF) instead to avoid PDF estimation. In [9], Zhao et al. systematically construct nine new variants of Kolmogorov–Smirnov (KS) test-based distinguishers by combining different construction strategies in KSA and PKS. In the nine distinguishers, they found that the multiple p-value and cumulative partition method-based KSA (MPC-KSA) is the best one. Also, they submitted an attack based on MPC-KSA to DPA Contest V2 [10], which is a global academic and technical contest to make it possible for researchers to compare in an objective manner their different attack algorithms.

According to the results published on the website of DPA Contest V2, this attack based on MPC-KSA ran so slowly that the average time per trace revealing one byte key was 600s, while other submitted attack took only a few seconds. The slow implementation of MPC-KSA made it impractical. In order to address this problem, we analyze the naive method and find out that the amount of sorting and counting operations is the main reason for the slow implementation. We observe that there is an essential relationship between two leakage samples that one sample is always a subset of the other. By this observation, we reduce the sorting times and use the sorting result to accelerate the counting operations in the KS test for MPC-KSA. The recovery of whole key of an AES-128 implementation with 15,000 traces provided by DPA Contest V2 was significantly faster. It takes only 3.4 ms per trace by the optimized method to reveal 16 key bytes, while it takes 85.5 ms per trace by the naive method.

8.2 Preliminaries

In this section, KS test will be recalled first. Then, how KSA and PKS tell the correct key from wrong keys will be explained.

8.2.1 Kolmogorov–Smirnov Test

In statistics, the Kolmogorov–Smirnov test (KS test) is a nonparametric test whose main target is to determine whether two distributions differ significantly [11].

Assume X denotes a random variable containing n samples. Its empirical cumulative distribution function (CDF) is $F_n(x) = \frac{1}{n}\sum_{i=1}^{n} I_{x_i \leqslant x}, x_i$ donates a sample of X and if $x_i \leqslant x$, then $I_{x_i \leqslant x}$'s value is 1; otherwise, it is 0.

For one random variable X whose CDFs are $F_n(x)$ and a given CDF $F(x)$ of reference probability distribution M, Formula (8.1) is used to testing their similarity.

$$D_{KS}(X, M) = sup_x|F_n(x) - F(x)| \tag{8.1}$$

The sup_x denotes the supermom of the set of distances. Specifically, the largest distance between two distributions represents the similarity between them. On the other hand, p-value can also be used to measure the similarity of two distributions. The smaller the p-value is, the less similar they are.

For two random variables X and Y whose CDFs are $F_{X,n}(x) - F_{Y,m}(x)$, Formula (8.2) is used to testing their similarity.

$$D_{KS}(X, Y) = sup_x|F_{X,n}(x) - F_{Y,m}(x)| \tag{8.2}$$

Here, an example of one-sample *KS* test is given. Assume $\{1, 2, 3, 4, 5\}$ is a set of sample selected from variable A randomly. Figure 8.1 shows the result of KS test between A and standard normal distribution $N(0,1)$. Red and blue lines each correspond to A's empirical distribution function and standard normal cumulative distribution function, and the green arrow is the one-sample KS statistic.

$D_{KS}(A,N(0,1)) = 0.8447$. Obviously, A is difference from standard normal distribution $N(0,1)$ significantly.

Following is an example is blow to show how two-sample KS test works. Assume that $\{1, 2, 3, 4, 5\}$, $\{1, 3, 5, 7, 9\}$, and $\{1, 2, 3, 3, 4, 4, 5, 5\}$ are sets of sample selected from variable A, variable B, and variable C, respectively,

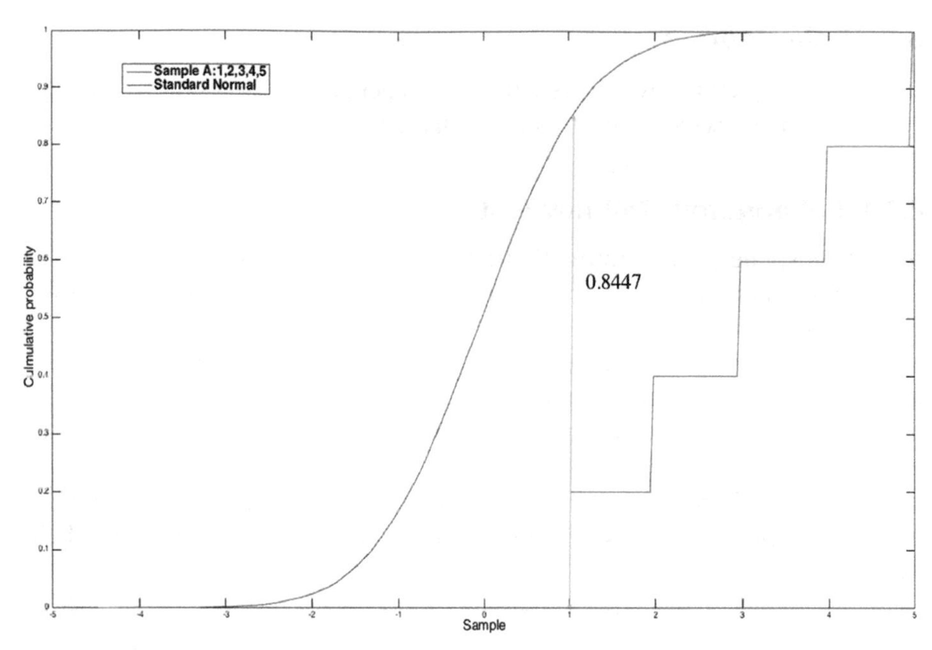

Figure 8.1 Illustration of the one-sample Kolmogorov–Smirnov statistic.

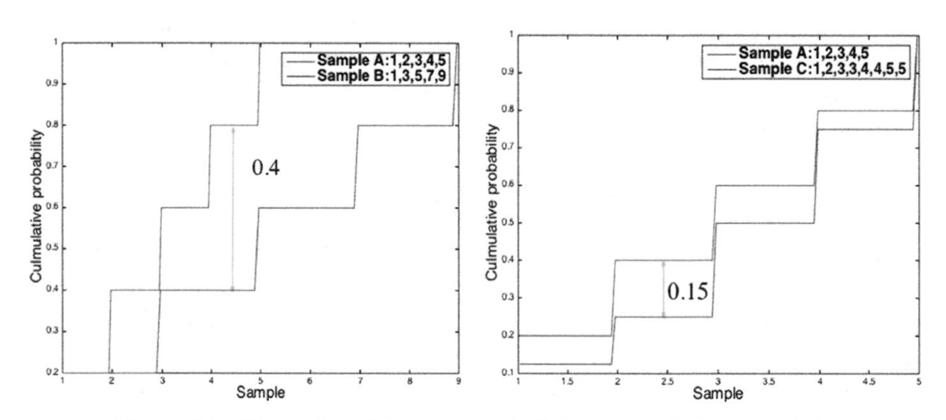

Figure 8.2 Illustration of the two-sample Kolmogorov–Smirnov statistic.

and all samples are randomly selected. The left part of Figure 8.2 shows the result of KS test between A and B, while the right part shows that between A and C.

$D_{KS}(A,B) = 0.4$, $D_{KS}(A,C) = 0.15$. So, C has a more similar distribution with A than B.

8.2.2 KSA Distinguisher

KSA distinguisher is based on two-sample KS test. KSA's central idea is to measure the maximum distance between the global leakage L's distribution and the conditional leakage $L|H$'s distribution, and then average the distances over the prediction space, where H denotes hypothetical leakage model. Assume that l denotes a leakage measurement value and h denotes a hypothetical leakage value. KSA is shown in Formula (8.3).

$$E_{h \in H}(D_{KS}(L|H = h, L)) \tag{8.3}$$

From Figure 8.3, it is obvious that KSA will produce a large average difference when the key hypothesis is correct.

8.2.3 PKS Distinguisher

PKS distinguisher [8] is based on single-sample KS test. Its central idea is to measure the p-value produced by comparing normal distribution and part of conditional trace distribution $L|H$. For convenience, leakages L and the hypothetical power consumptions H are usually processed by Z-score transformation in PKS. p is an empirical parameter in PKS from zero to one. $N(0,1)$ represents standard normal distribution. PKS, a two-partial KS test distinguisher, is shown in Formula (8.6) [9].

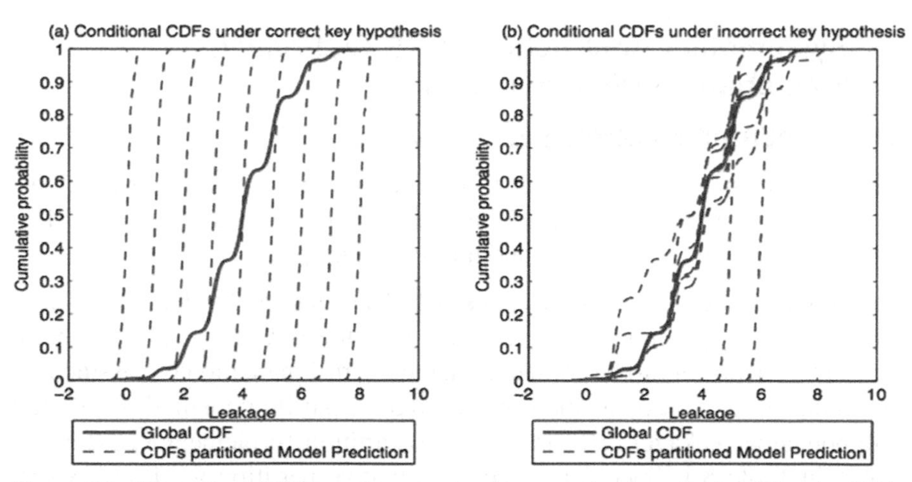

Figure 8.3 Illustration of CDFs under correct and incorrect key hypothesis while using KSA.

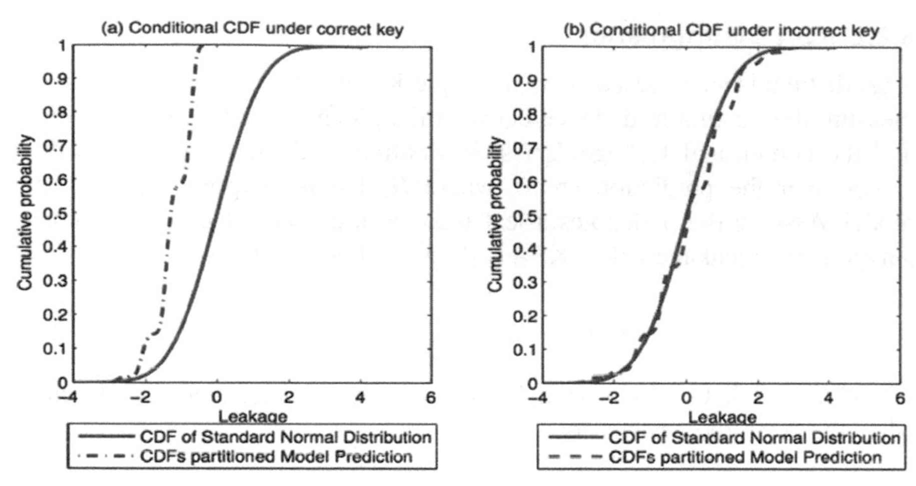

Figure 8.4 Illustration of CDFs under correct and incorrect key hypothesis while using PKS.

$$D_{ksl} = P_{value}(D_{KS}(L|H \leq p, N(0,1))) \tag{8.4}$$

$$D_{ksr} = P_{value}(D_{KS}(L|H > p, N(0,1))) \tag{8.5}$$

$$D_{PKS} = D_{ksl} \times D_{ksr} \tag{8.6}$$

PKS will return the smallest product of p-values when the key hypothesis is correct.

8.3 Systematic Construction of KS Test-based Side-channel Distinguishers

8.3.1 Construction Strategies of KSA and PKS

In this subsection, the construction differences between KSA and PKS will be compared in four aspects: partition method, similarity measure used by KS test, assumption about leakages, and normalization [9].

Partition Method: In a partition attack [12], leakages are divided into several sets $p_k^1, p_k^2, \ldots, p_k^n$, according to each key hypothesis k. These sets are built according to a power model H. In this chapter, partition method is classified as non-cumulative partition method and cumulative partition method. Examples of hypothetical leakages that can be used to partition 16-element leakages are shown in Table 8.1. Specifically, non-cumulative partition used by KSA is shown in the left part of Table 8.1, while cumulative partition used by PKS is shown in the right part of Table 8.1.

Table 8.1 Examples of non-cumulative partition (left) and cumulative partition (right)

Partition	Leakages
p_k^1	l_5
p_k^2	$l_2\ l_7\ l_9\ l_{16}$
p_k^3	$l_1\ l_4\ l_8\ l_{10}\ l_{11}\ l_{15}$
p_k^4	$l_3\ l_6\ l_{12}\ l_{13}$
p_k^5	l_{14}

Partition	Leakages
p_k^1	l_5
p_k^2	$l_5\ l_2\ l_7\ l_9\ l_{16}$
p_k^3	$l_5\ l_2\ l_7\ l_9\ l_{16}\ l_1\ l_4\ l_8\ l_{10}\ l_{11}\ l_{15}$
p_k^4	$l_5\ l_2\ l_7\ l_9\ l_{16}\ l_1\ l_4\ l_8\ l_{10}\ l_{11}\ l_{15}\ l_3\ l_6\ l_{12}\ l_{13}$
p_k^5	$l_5\ l_2\ l_7\ l_9\ l_{16}\ l_1\ l_4\ l_8\ l_{10}\ l_{11}\ l_{15}\ l_3\ l_6\ l_{12}\ l_{13}\ l_{14}$

Similarity measure used by KS test: Distance is used by KSA to measure the similarity of two distributions. In contrast, p-value is adopted in PKS to indicate whether or not partial leakages follow a normal distribution.

Assumption about leakages: PKS distinguisher considers that leakages follow a normal distribution, while KSA makes no assumption about leakages.

Normalization: [6] suggests that normalization can improve the performance of KSA. So when it comes to constructing KS test-based distinguisher, the normalization of KS test should also be to taken into consideration.

8.3.2 Nine Variants of KS Test-based Distinguishers

For convenience, each strategy that was used by KSA and PKS is labeled. Denote A0 the non-cumulative partition, and A1 the cumulative partition. Denote B0 the expectation of distance as the similarity measure of KS test, and B1 the product of p-values as the similarity measure of KS test. Denote C0 the distinguisher that makes no assumption about leakage distribution, and C1 the distinguisher that assumes the leakage follows a normal distribution. Denote D0 that normalization is performed on a distinguisher, and D1 that it is not.

By combining these strategies systematically, one can, in total, construct sixteen ($16 = 2^4$) KS test-based distinguishers. Among these sixteen distinguishers, three are existing and they are KSA (A0, B0, C0, D1), PKS (A1, B1, C1, D1), and norm-KSA (A0, B0, C0, D0). On the other hand, note that B1 and D0 conflict with each other; therefore, four combinations (A1, C1, B1, D0; A1, C0, B1, D0; A0, C1, B1, D0; and A0, C0, B1, D0) do not make any sense. Additionally, three combinations, which are (A0, B0, C1, D1), (A0, B0, C1, D0), and (A0, B1, C1, D1), fail to work in the key recovery attacks. The limitations of Z-score on hypothetical power consumptions of D-PKS (A1, B0, C1, D1), norm-D-PKS (A1, B0, C1, D0), and PKS (A1, B1, C1, D1) to form C-PKS (A1, B0, C1, D1), norm-C-PKS (A1, B0, C1, D0), and MPC-PKS (A1, B1, C1, D1) are freed. Finally, the remaining combinations are MP-KSA (A0, B1, C0, D1), C-KSA (A1, B0, C0, D1), norm-C-KSA (A1, B0, C0, D0), and MPC-KSA (A1, B1, C0, D1). Therefore, only nine ($9 = 2^4 - 4 - 3 - 3 + 3$) variants are newly constructed. These nine new distinguishers are summarized in Table 8.2. In these nine new variants of KS test-based distinguishers, the distinguisher which contains B0 strategy will return the largest expected distance under the correct key hypothesis, while the distinguisher which contains B1 strategy will return the smallest product of p-values. Additionally, in order to avoid arithmetic underflow, one typically applies the logarithm to the distinguisher, which contains B1 strategy.

8.4 An Experiment Analysis of All Twelve KS Test-based Side-channel Distinguishers

In article [9], Zhao et al. compared all the twelve KS test-based side-channel distinguishers using the power trace sets provided by DPA Contest V2 [10]. The input of the S-box of the last round of AES operation is chosen as the target. Their experiment result is in Figure 8.5.

In Group A, Figure 8.5(a) shows that both PKS and MIA can reveal the correct key, while KSA and norm-KSA fail to do that. The empirical parameter in PKS can largely improve the performance of PKS. Therefore, PKS (p = 0.01) is selected as the benchmark for finding the most promising variants in this case. In Group B, Figure 8.5(b) shows that MPC-KSA and MPC-PKS outperform PKS (p = 0.01) in terms of achieving a partial success rate of 80%. In Group C, other KS test-based distinguishers are less efficient than the benchmark, so we do not discuss them in more

Table 8.2 Nine new variants of KS test-based distinguishers

MP-KSA	$log_2(\prod_{m \in M} P_{value}(D_{KS}(L\|H = h,L)))$
C-KSA	$E_{m \in M}(D_{KS}(L\|H \leq h,L))$
norm-C-KSA	$E_{m \in M}\left(\frac{1}{\|L\|M=m\|} D_{KS}(L\|H \leq h,L)\right)$
MPC-KSA	$log_2(\prod_{m \in M} P_{value}(D_{KS}(L\|H \leq h,L)))$
D-PKS	$E(D_{KS}(L\|H \leq p, N(0,1)))$
norm-D-PKS	$E\left(\frac{1}{\|L\|H \leq p\|} D_{KS}(L\|H \leq p, N(0,1))\right)$
C-PKS	$E_{m \in M}(D_{KS}(L\|M \leq m, N(0,1)))$
norm-C-PKS	$E_{m \in M}\left(\frac{1}{\|L\|H \leq h\|} D_{KS}(L\|H \leq h], N(0,1))\right)$
MPC-PKS	$log_2\left(\prod_{m \in M} P_{value}\left(\frac{1}{\|L\|H \leq h\|} D_{KS}(L\|H < h, N(0,1))\right)\right)$

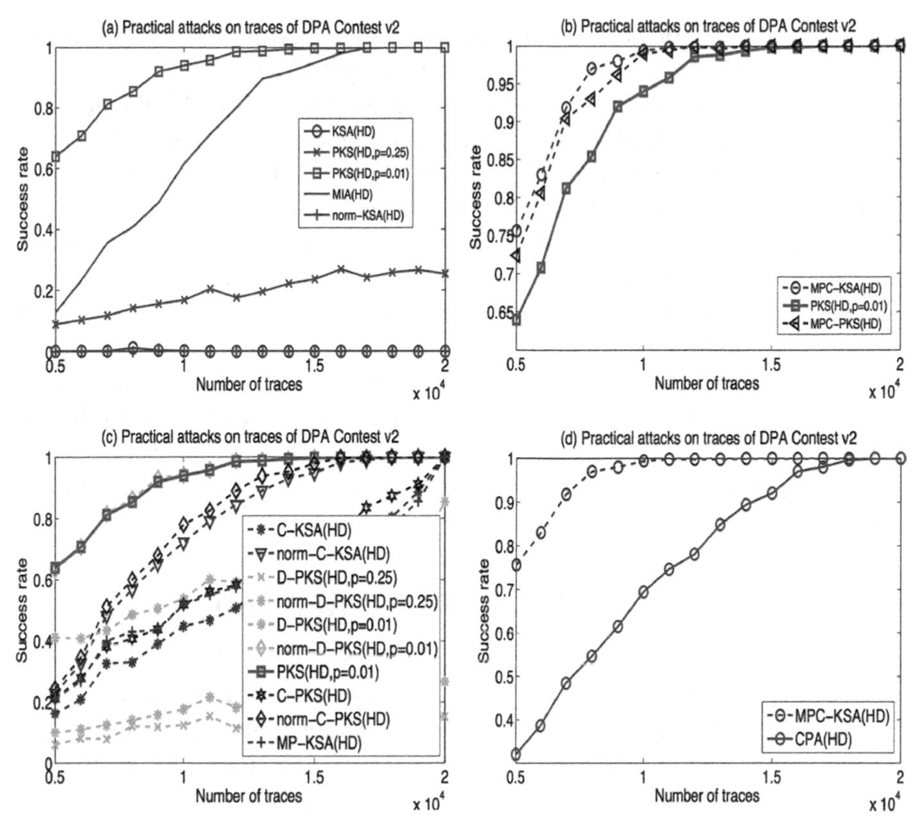

Figure 8.5 SRs for twelve KS test-based distinguishers, MIA, and CPA with HD model in attacks against the first AES S-box of the last round.

details. In Group D, Figure 8.5(d) shows that MPC-KSA is even better than CPA.

In order to enhance the understanding of whether or not MPC-KSA is a reasonable alternative for CPA, they perform attacks on all sixteen bytes of AES encryption. Table 8.3 shows the number of traces required to achieve

Table 8.3 Number of traces required to achieve partial SR of 80% on individual byte

Distinguisher	byte 1	byte 2	byte 3	byte 4	byte 5	byte 6	byte 7	byte 8
MPC-KSA	5,300	6,100	5,700	9,800	9,600	5,500	4,800	6,800
CPA	12,500	10,000	6,900	7,000	12,700	6,000	5,900	7,400

Distinguisher	byte 9	byte 10	byte 11	byte 12	byte 13	byte 14	byte 15	byte 16
MPC-KSA	4,500	5,200	9,200	3,500	4,100	14,500	6,000	5,500
CPA	6,800	3,600	10,000	3,000	6,600	16,900	15,000	5,100

partial SR of 80% of attacks on individual bytes. Although CPA is more efficient than MPC-KSA on four bytes (byte 4, byte 10, byte 12, and byte 16), it is less efficient on other twelve bytes. For example, for byte 15, the number of required traces for MPC-KSA to achieve partial SR of 80% is 6,000, while that of CPA is 15,000. However, for byte 4, the number of required traces for MPC-KSA to achieve partial SR of 80% is 9,800, while that of CPA is 7,000. Although MPC-KSA does not perform consistently better than CPA, it performs better than CPA on 75% of sixteen bytes. As a whole, MPC-KSA is more efficient than CPA in terms of the required number of traces to achieve the global SR of 80%. In summary, MPC-KSA is the best choice in this case. This experimental result indicates that when the leakages of a cryptographic device could not been accurately characterized, MPC-KSA exhibits better performance than CPA in terms of SR, as the former is capable of measuring the total dependency between hypothetical power consumptions and physical leakages.

8.5 Implementation Methods of MPC-KSA [13]

According to the definition of MPC-KSA in Section 8.4, a naive attack implementation based on the MPC-KSA is given in Algorithm 8.1 to reveal one key byte. The inputs are a set of messages *Msg* (often plaintext or ciphertext) and the corresponding leakage *L* in traces. The traces are collected from a certain device that uses the same key to encrypt each message in *Msg*. (*HL (k, x)* is a hypothetical leakage function that can return a hypothetical leakage value, ranging from 0 to $g - 1$, using key hypothesis *k* and message *x*).

As shown in Algorithm 8.1, from line 2 to line 6 for each candidate key *k* and each message msg_i, we compute the hypothetical leakage value. And then from line 8 to line 18 for each possible hypothetical leakage value *h*, we need to select the conditional leakage (line 11 to 13) to do KS test with the global leakage (line 14). While $h = g - 1$, the condition leakage is equal to the global leakage. So we only do KS test while h is less than $g - 1$.

We assume that the length of the leakage L and that of $\hat{L}$ are *n* and *m*, respectively. Two-sample KS test D_{KS} with the leakage L and $\hat{L}$ is listed in Algorithm 8.2.

In the counting procedure (line 4 to 15 of Algorithm 8.2), there should be one more interval $[sl_{n+m}, +\infty]$ to make sure every item in L and $\hat{L}$ can fall in one interval. In this case, the CDF value of the largest leakage value will be calculated. However, it is always one. So we do not counting in this interval or even calculate the CDF value of the largest leakage value.

Algorithm 8.1 Naive implementation method of MPC-KSA

Input: a set of messages and corresponding leakage $Msg[[msg_1, msg_2, \ldots, msg_n]]$, $L[[l_1, l_2, \ldots, l_n]]$ and the function *HL(k, x)* calculating hypothetical leakage

Output: the most possible key byte k_c

```
1.  Score[[s_1, s_2, ..., s_256]], s_i = 0, i ∈ [1, 256]
2.  for k ∈ [0, 255] do
3.          for j ∈ [1, n] do
4.                  h_kj = HL(k, msg_j)
5.          end for
6.  end for
7.  kc = 0, max = 0
8.  for k ∈ [0, 255] do
9.          L̂ = [[ ]]
10.         for h ∈ [0, g − 2] do
11.                 for j ∈ [l, n] do
12.                         if h_k j = h then L̂ = L̂ ∪ [l_j]
13.         end for
14.         t = log(P_value(D_KS(L, L̂)))
15.         s_{k+1} = s_{k+1} + abs(t)
16.         end for
17.         if max < s_{k+1} then kc = k, max = s_{k+1}
18.  end for
19.  return kc
```

8.5.1 Analysis of the Naive Method

As shown in Algorithm 8.2, the general KS test implementation contains three main procedures: sorting, counting, and maximizing. Firstly, leakages are sorted to acquire ordered leakages list, and every two adjacent items in the list is used as the counting interval. Then, the program counts how many leakages fall in each interval. Finally, the CDF is calculated based on the counting result, and then, the maximum distance between the global leakage distribution and the conditional leakage distribution is returned.

As the conditional leakage is selected from the global leakage, the length of the conditional leakage m is no larger than the length of the global leakage n. So the average time complexity of the sorting operation in line 1 is $O(n \log n)$. And the average time complexity of the counting operation in line 4 to 15 is $O(n^2)$, since the average time complexity of counting in one interval is $O(n)$. So we conclude that the time complexity of the general two-sample KS test implementation in Algorithm 8.2 is $O(n^2)$.

Algorithm 8.2 Two-sample KS test $D_{KS}(L, \hat{L})$

Input: two leakages $L[[l_1, l_2, \ldots, l_n]]$, $\hat{L}[[\hat{l}_1, \hat{l}_2, \ldots, \hat{l}_m]]$
Output: the largest distance *dis* between two distributions of leakages

1. $C[[c_1, c_2, \ldots, c_{n+m}]]$, $c_i = 0,\ i \in [1, n+m]$
2. $\hat{C}[[\hat{c}_1, \hat{c}_2, \ldots, \hat{c}_{n+m}]]$, $\hat{c}_i = 0,\ i \in [1, n+m]$
3. $SL[[sl_1, sl_2, \ldots, sl_{n+m}]] = mergeAndSort\ (L, \hat{L})$
4. for $i \in [1, n+m-1]$ do
5.   for $idx \in [1, n]$ do
6.   if $sl_i \leqslant l_{idx} < sl_{i+1}$ then
7. $c_i = c_i + 1$
8.   end if
9.   end for
10.   for $idx \in [1, m]$ do
11.   if $sl_i \leqslant \hat{l}_{idx} < sl_{i+1}$ then
12. $\hat{c}_i = \hat{c}_i + 1$
13.   end if
14.   end for
15. end for
16. $cdf = c_1/n$, $c\hat{d}f = \hat{c}_1/m$
17. $dis = |cdf - c\hat{d}f|$
18. for $i \in [2, n+m]$
19. $cdf = cdf + c_i/n, c\hat{d}f = c\hat{d}f + \hat{c}_i/m$
20. $t = |cdf - c\hat{d}f|$
21.   if $dis < t$ then $dis = t$
22. end for
23. return *dis*

In the *naive method,* the number of candidate key is 256, and the number of possible hypothetical leakage values is a constant while the hypothetical leakage function is given. Then the whole procedure to do KS test in Algorithm 8.1 (line 8 to 18) in time $O(n^2)$. The time complexity of computing hypothetical leakage values is $O(n)$. So the time complexity of Algorithm 8.1 is $O(n^2)$.

On the other hand, based on experiments, we find out that more than 90% execution time of the naive MPC-KSA implementation method is spent on KS test. So we mainly focus on the KS test to optimize the implementation. The KS test uses counting intervals to calculate two samples' distributions. The counting intervals are gotten from sorting combining leakages. It is widely accepted that sorting operation takes longer time than other operations. Meanwhile, the counting operations take quite a long time, for the time complexity is $O(n^2)$ where *n* denotes the length of the leakage. As a large number of KS tests needed to be done in MPC-KSA, the *naive method* is obviously slow.

8.5.2 Optimized Method I

The analysis in last subsection points out that there are too many sorting and counting operations in the whole MPC-KSA process, which is the main reason for the low efficiency. If we can find some kind of strategy that we only can sort the leakage once and use the result to accelerate the counting operation for doing KS test each time, then the efficiency of MPC-KSA will be optimized.

In two-sample KS test, sorting combining leakage is used to get the same counting interval for both samples. Procedures in line 4 to 15 of Algorithm 8.2 are to count how many leakages fall in each left-closed and right-open interval. However, the interval is gotten from sorting leakages, and two endpoints of the interval are two adjacent items of the ordered leakages. So values bigger than the left endpoint and smaller than the right endpoint do not belong to the leakages. Then, only the leakage value equal to the left endpoint falls in every left-closed and right-open interval, while the two endpoints of the interval are not the same. If they are the same, the interval is an empty set.

For example, if the global leakage is $L[[2, 3, 2, 4]]$ and the conditional leakage is $\hat{L}[[4, 2]]$, then the sorted list $SL = <2, 2, 2, 3, 4, 4>$. The corresponding interval is as follow:

$$I_1 = [2, 2) \qquad I_2 = [2, 2) \qquad I_3 = [2, 3)$$
$$I_4 = [3, 4) \qquad I_5 = [4, 4) \qquad I_6 = [4, +\infty).$$

Table 8.4 shows that intermediate values in procedure of calculating the CDF of L and $\hat{L}$.

Counting value of the global leakage L is $c_1 = 0$, $c_2 = 0$, $c_3 = 2$, $c_4 = 1$, $c_5 = 0$, $c_6 = 1$, while counting value of the conditional leakage $\hat{L}$ is $\hat{c}_1 = 0$, $\hat{c}_2 = 0$, $\hat{c}_3 = 1$, $\hat{c}_4 = 0$, $\hat{c}_5 = 0$, $\hat{c}_6 = 1$. The intervals I_1, I_2, I_5 are empty sets, so there are not any items falling in them, and also, the results in these intervals do not affect the CDF computing. And there is only the item equaling to their left endpoint falling in I_3, I_4, I_6.

Table 8.4 Intermediate values in procedure of calculating the CDF of L and $\hat{L}$

Intervals	c	CDF	$\hat{c}$	$\hat{\text{CDF}}$
I_1	$c_1 = 0$	0	$\hat{c}_1 = 0$	0
I_2	$c_2 = 0$	0	$\hat{c}_2 = 0$	0
I_3	$c_3 = 2$	0.5	$\hat{c}_3 = 1$	0.5
I_4	$c_4 = 1$	0.75	$\hat{c}_4 = 0$	0.5
I_5	$c_5 = 0$	0.75	$\hat{c}_5 = 0$	0.5
I_6	$c_6 = 1$	1	$\hat{c}_6 = 1$	1

Actually, the counting operation is to count how many unique numbers appear. Then, the counting operation is to count how many times every unique value appears. So every time we perform KS test, we can sort leakages and then remove the duplicate ones in them and take each item as a counting bin.

More importantly, one sample is selected from the other in MPC-KSA. In other word, the conditional leakage $\hat{L}$ is a subset of the global leakage L. We only need to sort the global leakage rather than the combining leakage, then remove the duplicate in them as counting bins denoted L_bin, while it is functionally equal to the one getting from the combing leakages. Using the data in the last example, we illustrate their equivalence. As we sort L and remove duplicate from it, we get $L\_bin = <2, 3, 4>$. The corresponding bin is 2, 3, and 4. Table 8.5 shows that intermediate values in procedure of calculating the CDF of L and $\hat{L}$ by bins.

For the global leakage L, the counting value is $c_1 = 2$, $c_2 = 0$, $c_3 = 1$, while the result of the conditional leakage $\hat{L}$ is $\hat{c}_1 = 1$, $\hat{c}_2 = 0$, $\hat{c}_3 = 1$. These results are the same as that on the intervals $\hat{I}_3, \hat{I}_4, \hat{I}_6$.

As shown in Algorithm 8.1, one of two samples in the KS test is always the global leakage L while the other, the conditional leakage $\hat{L}$, and changes with different hypothetical leakage h value. Although the conditional leakage $\hat{L}$ changes with different hypothetical leakage h value, the relationship that the conditional leakage $\hat{L}$ is a subset of the global leakage L always hold. So we only need to sort the global leakage once for the whole MPC-KSA process, and then the unique sorting result can be considered as the counting bin. Meanwhile, we can use a dictionary to have every unique value in the leakage map to its index of the counting bin, whose time complexity is nearly $O(1)$. We can build the dictionary when we compute CDF of the global leakage. And later the CDF of the global leakage and the dictionary can be used in the following KS test. Algorithm 8.3 shows the whole process of the optimized method.

In Algorithm 8.4, we present the procedures building the dictionary when we calculate the CDF of the global leakage. An empty dictionary *dic* is used

Table 8.5 Intermediate values in procedure of calculating the CDF of L and $\hat{L}$

Bins	c	CDF	$\hat{c}$	$\hat{\text{CDF}}$
B1=2	$c_1 = 0$	0.5	$\hat{c}_1 = 1$	0.5
B2=3	$c_2 = 1$	0.75	$\hat{c}_2 = 0$	0.5
B3=4	$c_3 = 1$	1	$\hat{c}_3 = 1$	1

Algorithm 8.3 Optimized implementation method I of MPC-KSA

Input: a set of messages Msg and corresponding leakage L $Msg[[msg_1, msg_2, \ldots, msg_n]]$, $L[[l_1, l_2, \ldots, l_n]]$ and the function *HL(k, x)* calculating hypothetical leakage and an empty dictionary *dic*

Output: the most possible key byte k_c

1. $Score[[s_1, s_2, \ldots, s_{256}]], s_i = 0, i \in [1, 256]$
2. for $k \in [0, 255]$ do
3.   for $j \in [1, n]$ do
4. $h_{kj} = HL(k, msg_j)$
5.   end for
6. end for
7. $CDF = Cal_{cdf}(L, dic)$, $Kc = 0$, $max = 0$
8. for $k \in [0, 255]$ do
9. $\hat{L} = [[\,]]$
10.   for $h \in [0, g-2]$ do
11.   for $j \in [l, n]$ do
12.   if $h_k j = h$ then $\hat{L} = \hat{L} \bigcup [l_j]$
13.   end for
14. $t = log(P_{value}(DI_{KS}(CDF, L, \hat{L}, dic)))$
15. $s_{k+1} = s_{k+1} + abs(t)$
16.   end for
17.   if $max < s_{k+1}$ then $kc = k, max = s_{k+1}$
18. end for
19. return kc

Algorithm 8.4 Calculating the CDF of the global leakage and building the dictionary $Cal_{cdf}(L, dic)$

Input: the global leakage $L[[l_1, l_2, \ldots, l_n]]$ and an empty dictionary *dic*

Output: the $CDF[[cdf_1, cdf_2, \ldots, cdf_x]]$ of the global leakage

1. $SL\ [[sl_1, sl_2, \ldots, sl_n]] = sort(L)$
2. $idx = 1, Lk = sl_1, c = 1$
3. for $i \in [2, n]$ do
4.   if $LK = sl_i$ then
5. $c = c + 1$
6.   else
7. $dic.add(\ll Lk, idx \gg)$
8. $cdf_{idx} = c/n$
9. $Lk = sl_i$
10. $idx = idx + 1$
11.   end if
12. end
13. $dic.add(\ll Lk, idx \gg), cdf_{idx} = c/n$
14. return CDF

Algorithm 8.5 Improved KS test $DI_{KS}(CDF, \hat{L}, dic)$

Input: the conditional leakage $\hat{L}[[\hat{l}_1, \hat{l}_2, \ldots, \hat{l}_m]]$, the $CDF[[cdf_1, cdf_2, \ldots, cdf_x]]$ of the global leakage and the dictionary dic

Output: the largest distance *dis* between two distributions of leakages

1. $\hat{C}[[\hat{c}_1, \hat{c}_2, \ldots, \hat{c}_x]], \hat{c}_i = 0$
2. for $i \in [l, m]$ do
3. $\quad idx = dic.get(\hat{l}_i)$
4. $\quad \hat{c}_{idx} = \hat{c}_{idx} + 1$
5. end for
6. $\hat{cdf} = \hat{c}_1/m, dis = |cdf_1 - \hat{cdf}|$
7. for $i \in [2, m]$
8. $\quad \hat{cdf} = \hat{cdf} + \hat{c}_i/m$
9. $\quad t = |cdf_i - \hat{cdf}|$
10. $\quad$ if $dis < t$ then $dis = t$
11. end for
12. return dis

as reference. While Algorithm 8.4 finished, *dic* has already been built to have every unique value in the leakage map to its index of the counting bin.

By Algorithm 8.4, we get the CDF of the global leakage and the dictionary for the further calculation. As Algorithm 8.4 sorts *L* in $O(n \log n)$ time and builds the dictionary in $O(n)$ time, consequently the time complexity of Algorithm 8.4 is $O(n \log n)$.

Algorithm 8.5 below presents how KS test works with the dictionary dic in MPC-KSA. The CDF of the global leakage *L*, the condition leakage $\hat{L}$, and the dictionary dic are the inputs.

As shown in line 3 to 12 of Algorithm 8.4, the counting operation only traverses the global leakage once. Also in line 2 to 5 of Algorithm 8.5, the condition leakage is traversed once. MPC-KSA can be executed with sorting only once by Algorithm 8.4, and the time complexity of KS test for MPC-KSA reduces from $O(n^2)$ to $O(n)$ as indicated in Algorithm 8.5. Therefore, the complexity of *optimized method I* for MPC-KSA is $O(n \log n)$, while that of *naive method* is $O(n^2)$.

8.5.3 Optimized Method II

In this subsection, an optimization technique is presented to further accelerate both Algorithms 8.1 and 8.5.

In line 11 to 13 in Algorithm 8.3, it is needed to search in the conditional leakage based on the result in line 4. We can add an operation after line 4 to

Algorithm 8.6 Optimized implementation method II of MPC-KSA

Input: a set of messages $Msg[[msg_1, msg_2, \ldots, msg_n]]$ and corresponding leakage $L[[l_1, l_2, \ldots, l_n]]$, the function $HL(k, x)$ calculating hypothetical leakage and a null dictionary dic

Output: the most possible key byte kc

1. $Score[[s_1, s_2, \ldots, s_{256}]], s_i = 0, i \in [1, 256]$
2. for $k \in [0, 255]$ do
3.   for $j \in [1, n]$ do
4. $h = HL(k, msg_j)$
5. $G_{k,h} = G_{k,h} \cup \{j\}$
6.   end for
7. end for
8. $CDF = Cal_{cdf}(L, dic), Kc = 0, max = 0$
9. for $k \in [0, 255]$ do
10. $\hat{L} = [[\,]]$
11.   for $h \in [0, g-1]$ do
12. $\hat{L} = \hat{L} \cup L[G_{k,h}]$
13. $t = log(P_{value}(DI_{KS}(CDF, \hat{L}, dic)))$
14. $s_{k+1} = s_{k+1} + abs(t)$
15.   end for
16.   if $max < s_{k+1}$ then $kc = k, max = s_{k+1}$
17. end for
18. return kc

include the leakage index into a set $G_{k,h}$. $G_{k,h}$ is used to record the leakage index whose hypothetical leakage under key guess k is h. In this way, the leakage partition information also can be saved. The recorded leakage index is actually the result of searching operations in line 11 to 13 both of Algorithms 8.1 and 8.3. As the leakage partition information has been saved, doing K-S test can load the leakage from the global leakage directly with an index set. In this way, leakage partition sets are calculated in advance to save the time to search for the conditional leakage. Algorithm 8.6 gives the detailed procedures.

By Algorithm 8.6, the optimized method is further improved. The experimental result shows that the efficiency of the *optimized method I* is improved more than 60% by the *optimized method II*. The complexity of Algorithm 8.6 is the same as that of Algorithm 8.3.

8.6 Implementation Results

In this section, we perform several attacks on PC against unprotected hardware AES-128 implementation on a Xilinx Vertex-5 FPGA. The power traces are from DPA Contest V2 [10]. All attacking programs are written in C language on OS X 10.9.2, and run on a Mac laptop (Intel Core i5 with CPU 1.7 GHz and 4 GB RAM). Our experiments are performed to reveal the whole last round key. And we choose five leakage points from one trace for each key byte. In practice, we sum the $score$ on each leakage point for a candidate key byte value as the final score. For each key byte, the hypothetical leakage function $HL(k, x)$ is defined to be

$$HL(k, x) = HD(iSbox(x) \oplus k, iShiftRow(x)) \tag{8.7}$$

where x denotes one ciphertext byte, $iSbox(\cdot)$ denotes the operation of inverse Sbox, $iShiftRow(\cdot)$ denotes the operation of inverse shift row, and $HD(a, b)$ denotes the Hamming distance between a and b.

For the following experiments, we use 80 leakages on different number traces and the corresponding ciphertexts as the input to reveal the whole last round key. Three types of experiments are designed. They are named "Naive Method," "Optimized Method I," and "Optimized Method II," respectively.

- *Naive Method (Algorithm 8.1)*
- Experiment to perform the attack using Algorithm 8.1.
- *Optimized Method (Algorithm 8.3)*
- Experiment to perform the attack using Algorithm 8.3.
- *Optimized Method II (Algorithm 8.6)*
- Experiment to perform the attack using Algorithm 8.6.

In each type of experiment, we perform the attack based on MPC-KSA using different number of traces (from 1,000 to 15,000, step by 500) to calculate the execution time of each method. The execution time of the attack based on MPC-KSA with different number traces is shown in Table 8.6.

Table 8.6 shows the time consumption of the *naive method*, *optimized method I,* and *optimized method II* with different number traces. Clearly, optimized method II is the most efficient of the three. The efficiency of the

Table 8.6 The execution time of the attacks based on MPC-KSA with different number traces

	Average Execution Time (in Seconds)			Speedup Rate
Number of Traces	Naive Method	Optimized Method I	Optimized Method II	Optimized II vs. Naive
2,500	225.24	32.20	12.69	17.73
5,000	452.65	59.65	21.16	21.38
7,500	676.66	85.34	28.92	23.39
10,000	902.36	113.06	36.53	24.69
12,500	1,112.58	137.07	43.81	25.39
15,000	1,281.83	163.40	51.01	25.12

optimized method I is improved more than 60% by the optimized method II. To depict the relative speed efficiencies more clearly, we draw time consumption curves for each attack based on MPC-KSA with different number traces ranging from 1,000 to 15,000. Comparing with the *naive method*, the time complexity of MPC-KSA is reduced from $O(n^2)$ to $O(n \log n)$ after optimization.

We also perform experiments to show the relationship between the execution time and the global success rate (GSR) [10] by amounting key recovering attacks with a set of 10 independent experiments. Each experiment was performed using different number of power traces (from 50 to 15,000, step by 50) randomly chosen from a trace set containing 20,000 traces corresponding to 20,000 random plaintext encryption under the same secret key. Table 8.4 gives the relationship between the execution time and the GSR. Because it takes too long time to do evaluation by the *naive method*, we only do experiments with three trace sets.

As we can gain the number of traces at each success rate by *optimized method II*, we estimated the time of running on the rest seven data sets with that on prior the three data sets. From Tables 8.6, 8.7, and Figure 8.6, it is clear that, for the attack based on MPC-KSA with different number traces, *optimized method II* is the most efficient. It is concluded that the main solution to optimize the implementation is to use the inherent relationship between these two leakage samples.

Table 8.7 The execution time at different GSR

		Average Execution Time (in Hour)		Speedup Rate
Global Success Rate	Number of Traces	Naive Method	Optimized Method II	Optimized II vs. Naive
10%	4,050	4.10	0.20	20.38
20%	5,650	12.15	0.52	23.26
30%	5,700	20.33	0.91	22.44
40%	5,900	29.10*	1.22	23.86
50%	6,000	38.16*	1.60	23.87
60%	6,600	49.11*	1.99	24.65
70%	8,050	65.35*	2.52	25.97
80%	8,100	81.79*	3.05	26.86
90%	8,900	101.67*	3.64	27.97
100%	10,900	127.53*	4.28	29.77

*Estimate.

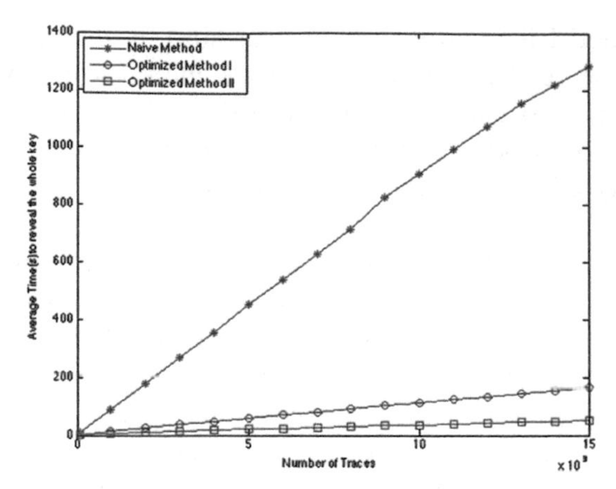

Figure 8.6 Time consumption curves for attacks based on MPC-KSA with different number traces.

8.7 Conclusions

MPC-KSA uses KS test as the major operation, and KS test is a time-consuming execution in the normal scenario. So the naive implementation of MPC-KSA is considerably slow and impractical in practice. The inherent relationship between these two leakage samples used in KS test for MPC-KSA is exploited to accelerate the naive implementation. The main work of this chapter highlights that using inherent properties and structures of processed data can optimize an algorithm to the fullest. As a result, the time complexity of KS test for MPC-KSA is reduced from $O(n^2)$ to $O(n)$, and that of MPC-KSA is reduced from $O(n^2)$ to $O(n \log n)$. By precomputation technique, the efficiency of the implementation is further improved more than 60%. It only takes less than 1 minute with 15,000 traces to recover the whole key of the AES-128 embedded in the device, which is about 25 times faster than naive method. By the optimized implementation, MPC-KSA distinguisher can be used efficiently and effectively in practice.

Acknowledgments

This work was supported in part by Strategic Priority Research Program of the Chinese Academy of Sciences (Nos. XDA06010701 and XDA06010703), National Natural Science Foundation of China (Nos. 61272478 and 61472416), and National Science and Technology Major Project (No.2014ZX01032401-001).

References

[1] Kocher, P., Jaffe, J., and Jun. B. (1999). Differential power analysis. CRYPTO 1999, LNCS 1666, pp. 388–397.

[2] Quisquater, J.-J., and Samyde, D. (2001). Electromagnetic analysis (EMA): Measures and Countermeasures for smart cards. E-smart 2001, LNCS 2140, pp. 200–210, 2001.

[3] Kocher, P. (1996). Timing attacks on implementations of Diffie-Hellman, RSA, DSS, and other systems. CRYPTO 1996, LNCS 1109, pp. 104–113.

[4] Brier, E., Clavier, C., and Olivier, F. (2004). Correlation power analysis with a leakage model. CHES 2004, LNCS 3156, pp. 16–29.

[5] Gierlichs, B., Batina, L., Tuyls, P., and Preneel, B. (2008). Mutual information analysis—a generic side-channel distinguisher. CHES 2008, LNCS 5154, pp. 426–442.

[6] Veyrat-Charvillon, N., and Standaert, F.-X. (2009). Mutual information analysis: how, when and why? CHES 2009, LNCS 5747, pp. 429–443.

[7] Prouff, E., and Rivain, M. (2009). Theoretical and practical aspects of mutual information based side channel analysis. ACNS 2009. LNCS 5536, pp. 499–518.

[8] Liu, J., Zhou, Y., Yang, S., and Feng. D. (2012). Generic side-channel distinguisher based on kolmogorov–smirnov test: Explicit construction and practical evaluation. CJE 2012, 21 (3), July 2012.

[9] Zhao, H., Zhou, Y., Standaert, F.-X., and Zhang, H. (2013). Systematic construction and comprehensive evaluation of Kolmogorov–Smirnov test based side-channel distinguishers ISPEC 2013, LNCS 7863, pp. 336–352.

[10] DPA contest v2. Available at http://www.dpacontest.org/v2, 2016.

[11] Kolmogorov–Smirnov test at https://en.wikipedia.org/wiki/Kolmogorov-Smirnov_test, 2016.

[12] Standaert, F.-X., Gierlichs, B., and Verbauwhede, I. (2009). Partition vs. Comparison side-channel distinguishers: an empirical evaluation of statistical tests for univariate side-channel attacks against two unprotected CMOS devices. In: Lee, P. J., Cheon, J. H. (eds.) ICISC 2008. LNCS, vol. 5461, pp. 253–267. Springer, Heidelberg.

[13] Zheng, C., Zhou, Y., Zheng, Y. (2015). A fast implementation of MPC-KSA side-channel distinguisher. *2015 24th International Conference on Computer Communication and Networks (ICCCN)*. IEEE, pp. 1–7.

[14] Mangard, S., Oswald, E., Standaert. F.-X. (2010). One for all–all for one: Unifying standard DPA attacks. IET 2010, 100C111. ISSN: 1751-8709; Digital Object Identifier: 10.1049/iet-ifs.2010.0096.

[15] Doget, J., Prouff, E., Rivain, M., Standaert, F.-X. (2011). Univariate side channel attacks and leakage modeling. *J. Cryptographic Eng.*, 1 (2), 123–144.

[16] Moon, Y.-I., Rajagopalan, B., Lall, U. (1995). Estimation of mutual information using kernel density estimators. *Physical Review E.*, 52 (3), 2318–2321.

[17] Walters-Williams, J., Li. Y. (2009). Estimation of mutual information: A survey. RSKT 2009. LNCS 5589, pp. 389–396.

9

Multi-antenna Transmission Technique with Constellation Shaping for Secrecy at Physical Layer

Paulo Montezuma[1,2,3] and Rui Dinis[1,2]

[1]Instituto de Telecomunicações (IT), Av. Rovisco Pais 1, 1049-001 Lisboa, Portugal
[2]DEE, Faculdade de Ciências e Tecnologia (FCT)-Universidade Nova de Lisboa (UNL), Monte de Caparica, 2829-516 Caparica, Portugal
[3]UNINOVA, Campus da FCT/UNL, Monte de Caparica, 2829-516 Caparica, Portugal

Abstract

In wireless communication systems, privacy among users' contents is a crucial security requirement such as power and spectral efficiency. Spectral and power efficiencies are attainable with power-efficient transmission schemes where constellations are decomposed into several polar components, being each one amplified and transmitted independently by an antenna. Besides that, due to constellation shaping on the desired transmission direction performed by these transmitters, some kind of physical layer security is assured as well. Under this approach, security is achieved by constellation shaping of the transmitted constellation since each user must know the transmitter configuration parameters associated to the constellation shaping, i.e., the direction in which the constellation is optimized; otherwise, the received data would be meaningless. Commonly, security is assured by encrypted algorithms implemented by higher layers, such as private and public encrypted keys. However, system security can be improved with physical layer security schemes since they can be complemented with other security schemes from higher layers. Simulation results show the effectiveness of the proposed approach.

Keywords: Information directivity, Physical security, Constellation Shaping, MIMO, Nonlinear amplifiers, Power efficiency.

9.1 Introduction

The broadcast nature of wireless communications makes it difficult to avoid interception of transmitted signals by unauthorized users, while superposition can lead to the overlapping of multiple signals at the receiver [1]. Due to the broadcast nature of the wireless channel, security is a critical issue in wireless communication systems. Commonly, security is assured by encrypted schemes from higher layers, such as key cryptography protocols [2]. However, cryptographic measures can be complemented with other techniques such as physical layer security schemes [3, 4]. Common physical layer security techniques use coding or precoding schemes and exploit channel state information to allow secrecy over a wireless medium [5]. The first one sacrifices spectral efficiency due to the use of redundant bits. In the second approach, the assumption of a highly mobile or dynamic environment with significant variations in channel characteristics is not always valid for all scenarios where wireless communications may occur, and for static channels, physical layer security schemes based on the variation of channel characteristics perform poorly [6]. For these scenarios, a channel-independent security scheme without redundant coded bits is more appropriate. Since they can operate independently of higher layers, physical layer techniques can be used in a multi-layered approach to reinforce already existing security measures.

The high data rates of communication systems can be supported by MIMO systems (multiple-input multiple-output) which can increase throughput with a reduction in the transmitted power by each antenna. One advantage of MIMO systems is the reduction in the transmitted power [7]. On the other hand, high spectral efficiency requirements of modern wireless communication systems are only attainable with the use of multilevel modulations, characterized by higher peak-to-average power ratios. This can compromise power amplification efficiency or drive up the costs of power amplifiers. With a transmission scheme, where multilevel constellations are decomposed into several biphase shift keying (BPSK), quadri phase shift keying (QPSK), or offset QPSK (OQPSK) components, being each component amplified and transmitted independently by an antenna, efficiency of power amplification is improved since it is possible to use nonlinear (NL) amplifiers in such operation [8–10] (it should be noted that we may have constant envelope components in each branch). As current MIMO transmitters, this transmitter requires a separate radio frequency (RF) chain including a power amplifier for each antenna element [11], but the key difference relies on the fact that each RF chain is associated to a BPSK component that is combined at channel level to generate the desired multilevel constellation and the elementary sub-constellations and each antenna transmits uncorrelated signals. It should be noted that despite the fact that the radiation pattern of the transmitter array remains unaffected, phase rotations between RF branches lead to an optimization of the transmitted constellation in a desired direction. Therefore, security is achieved by shaping of the transmitted constellation, since the receiver must know the constellation coefficients g_i affecting each RF chain and transmission direction θ_j, otherwise receives meaningless data. Therefore, this transmitter has an inherent security scheme if unauthorized users are unaware about the use at the transmitter.

The main motivation for this chapter is to present an extended characterization of power-efficient transmitter structure that allows at same time the physical layer security due to constellation shaping achieved by this structure. As we shall see, the constellation shaping performed by the transmitter acts as an amplitude and phase distortion of the constellation when the transmitter parameters are unknown. Thus, one might expect that distortion would affect mutual information (MI) for any user unaware about transmitter configuration. This chapter explains how it is possible to implement a scheme of physical-level security using a power-efficient multi-branch transmitter with constellation shaping. Section 9.2 characterizes the concepts inherent to the transmission technique. The rest of this chapter is organized as follows: freedom degrees for transmitter configuration, the inherent security, and the problems related with the analytical characterization of mutual information and secrecy capacity are discussed in Section 9.3. A characterization of the receiver and the impact of configuration of the transmitter array are presented in Section 9.4. Simulation results are presented in Section 9.4.1. Section 9.5 concludes this chapter.

9.2 Transmitter Structure

The basic idea is to employ a scheme with N_m antenna elements at the transmitter as depicted in Figure 9.1. Contrarily to MIMO, the N_m elements are employed to allow an efficient amplification of the signals associated to a large constellation. The data bits are mapped into a given constellation (e.g., a quadrature amplitude modulation (QAM) constellation) characterized by the ordered set $\mathfrak{S} = \{s_0, s_1, \ldots, s_{M-1}\}$ following the rule $(\beta_n^{(\mu-1)}, \beta_n^{(\mu-2)}, \ldots, \beta_n^{(1)}, \beta_n^{(1)}) \mapsto s_n \in \mathfrak{S}$, with $(\beta_n^{(\mu-1)}, \beta_n^{(\mu-2)}, \ldots, \beta_n^{(1)}, \beta_n^{(1)})$

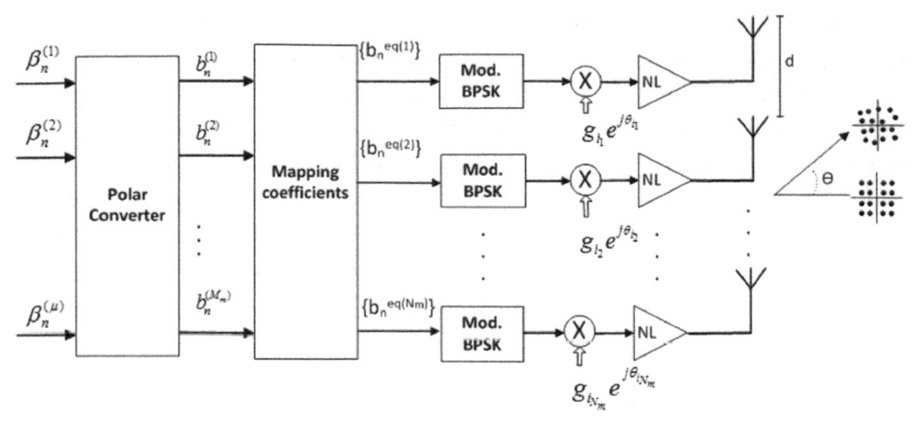

Figure 9.1 Structure of constellation directive transmitter.

denoting the binary representation of n with $\mu = \log_2(M)$ bits. Next, the constellations' symbols are decomposed in M_m polar components, i.e.,

$$
\begin{aligned}
s_n &= g_0 + g_1 b_n^{(1)} + g_2 b_n^{(2)} + g_3 b_n^{(1)} b_n^{(2)} + g_4 b_n^{(3)} + \dots \\
&= \sum_{i=0}^{M-1} g_i \prod_{m=0}^{\mu-1} (b_n^{(m)})^{\gamma_{m,i}} = \sum_{i=0}^{M-1} g_i b_n^{eq(i)},
\end{aligned} \tag{9.1}
$$

with $(\gamma_{\mu,i}\ \gamma_{\mu-1,i}\ \dots\ \gamma_{2,i}\ \gamma_{1,i})$ denoting the binary representation of i, $b_n^{eq(i)} = \prod_{m=0}^{\mu-1}(b_n^{(m)})^{\gamma_{m,i}}$ and $b_n^{(m)} = (-1)^{\beta_n^{(m)}}$ is the polar representation of the bit $\beta_n^{(m)}$. Since we have M constellation symbols in $\mathfrak{S}$ and M complex coefficients g_i, (9.1) is a system of M equations that can be used to obtain the coefficients g_i, $i = 0, 1, \dots, M-1$ [12]. If we denote N_m the number of nonzero coefficients g_i, then it is clear that a given constellation can be decomposed as the sum of $N_m \leq M$ polar components. Each one of the N_m polar components is modulated as a BPSK signal (or other polar modulation format with constant envelope). Being each one of these N_m BPSK signals a serial representation of an OQPSK signal [13], with reduced envelope fluctuations and compact spectrum (e.g., a Gaussian minimum shift keying (GMSK)), higher power efficiencies are achieved due to the use of nonlinear amplifiers. The corresponding signals are separately amplified by N_m nonlinear amplifiers before being transmitted by N_m antennas. Combination losses are also avoided, since the outputs of the N_m amplifiers are combined at channel level (in Figure 9.1, g_{i_j} denote the N_m coefficients selected to define the constellation, but for the sake of representation simplicity, we will use g_i along this chapter).

Since the BPSK components in RF branches are uncorrelated, the radiation pattern remains unchanged. However, the phase differences between N_m successive antenna elements affect the BPSK components that generate the transmitted constellation at channel level. This means that the constellation symbols are rearranged according to a desired direction θ (under these conditions, the constellation's shaping in the desired direction is assured by phase rotations of the BPSK components). Consequently, we may say that this transmitter introduces information directivity at constellation level, which can be increased by changing dynamically phase's rotations between RF branches.

Several possibilities exist for the arrangement of BPSK components along the N_m RF elements, but we restrict the BPSK components' arrangements for 16-QAM constellations to the linear arrangement of Table 9.1.

9.3 Transmitter Configuration Possibilities and Security

Using (9.1), several mapping rules can be defined for M-QAM constellations and Voronoi constellations. For instance, for a 64-QAM constellation with Gray mapping, we only need 6 nonzero g_i coefficients: $g_4 = 4$, $g_6 = 2$, $g_7 = 1$, $g_{32} = 4j$, $g_{48} = 2j$, and $g_{56} = j$). 16-QAM constellations with Gray mapping are the sum of 4 BPSK signals and can be defined by the set of nonzero complex coefficients $g_2 = 2j$, $g_3 = j$, $g_8 = 2$, and $g_{12} = 1$ (actually, this corresponds to only two QPSK constellations). On the other hand, the energy optimized Voronoi constellations need $M - 1$ non-null coefficients g_i. Other mapping rules or constellations can be easily obtained by changing the set of coefficients g_i. Clearly, there is an inherent security since a sequence of bits from the sender is converted into symbols on the constellation space using the set of coefficients g_i and an antenna array configuration only known by the sender and the intended receiver. Since the intended receiver knows

Table 9.1 16-QAM: g_i arrangements and distances between antennas for uniform and non-uniform arrays

	Antenna Order			
16-QAM	1	2	3	4
Coefficients g_i	g_2	g_{12}	g_8	g_3
Coefficients values	2j	1	2	j
Spacing between antennas: $d_{i,j}$ denotes spacing between antennas *i* and *j*				
	Uniform		Non-Uniform	
$d_{2,1}$	$\lambda/4$		$\lambda/2$	
$d_{3,2}$	$\lambda/4$		$3\lambda/4$	
$d_{4,3}$	$\lambda/4$		$\lambda/2$	

the set of coefficients g_i and the array configuration, it will be able to decode with success the original data. This means that in every transmission, both the sender and the intended receiver use a custom constellation mapping, which may act as a secret key to any interception from an eavesdropper (also known as Eve). For the particular case of M-ary constellations, there are $M!$ possible mappings, which becomes impractical for the eavesdropper to decode when transmissions are based on constellations with sizes equal or greater than 64.

Here, we do not explore the potential of mapping diversity since we restrict our analysis to the security assured by information directivity that results from antenna array's configuration (however, the secrecy level assured by constellation mapping diversity should be conveniently addressed in further work). Another factor that also increases the complexity of any non-authorized interception relies on the relation between constellation shaping and the configuration of the transmitter array. For any set of coefficients g_i, the order of the transmit antennas can be changed. Since the number of active antennas is the same of BPSK components, i.e., N_m components, for each mapping rule and uniform configuration of the antenna array, there are $N_m! - N_m$ different array configurations and consequently $N_m! - N_m$ different spacial arrangements for the symbols of the transmitted constellation. Permutations of the BPSK components on the RF branches also change the shape of the resulting constellation at channel level (recall that constellation mapping can also suffer changes due to phase rotations of BPSK components). Thus, combining both mapping possibilities and g_i permutations between antennas, for any interception the eavesdropper needs to compute $M! \times (N_m! - N_m)$ combinations (if the transmitter uses Voronoi constellations, even for the smallest constellation with 16 symbols, we have 2×10^{13} combinations).

Complexity can be even increased through non-uniform spacing between antennas. Consider again the transmitter structure of Figure 9.1 with equal spaced N_a antennas, where only N_m antennas are active in each instant. Since the active antennas can be any set of N_m antennas among the N_a, we have $\frac{N_a!}{(N_a-N_m)!N_m!}$ combinations. Obviously, this situation does not reflect the real number of possibilities, since the spacing between antennas can be any real multiple of λ, leading to a phase shift between antennas $\Delta\Theta \in [0, 2\pi]$. Despite this fact, simulation results presented here are restricted to the transmitter configuration of Figure 9.1, where the distance between antennas is restricted to an integer multiple of $d = \lambda/4$ (in a real case, the spacings between antennas are not necessarily restricted to an integer multiple as shown in Figure 9.1).

Let **X** and **Y** be complex n-dimensional arrays representing the input and output of a memoryless communication channel and assume the respective alphabets $X \subseteq \mathbb{C}^n$ and $Y \subseteq \mathbb{C}^n$. For any x and y, the joint distribution given by

$$f_{\mathbf{X}|\mathbf{Y}}(x, y), \tag{9.2}$$

can be factorized as

$$f_{\mathbf{X}|\mathbf{Y}}(x, y) = f_{\mathbf{X}|}(x) f_{\mathbf{Y}|\mathbf{X}}(y|x), \tag{9.3}$$

where f_x is the input distribution determined by the modulation and constellation shape and $f_{\mathbf{Y}|\mathbf{X}}(y|x)$ is the channel distribution that aggregates the noise and distortion effects. The mutual information between **X** and **Y** is denoted as $I(\mathbf{X} : \mathbf{Y})$, and $I(\mathbf{X} : \mathbf{Y}|\mathbf{N})$ represents the conditional mutual information. Entropy and conditional entropy are denoted by $H(\mathbf{X})$ and $H(\mathbf{X}|\mathbf{N})$, respectively.

Let us consider a time memoryless channel with a user i as receiver and one transmitter, where the user tries to receive a message from the transmitter, with the input and output denoted by $\mathbf{X}_i$ and $\mathbf{Y}_i$. For a channel has only additive noise, we have

$$Y_i = X_i + N_i, \tag{9.4}$$

where X and Y are the input and the output of the channel and N is additive white Gaussian noise (AWGN) with zero mean and variance σ_N^2. For a given channel input x, channel law is given by the conditional probability density function (pdf)

$$f_{Y|X}(y|x) = \frac{1}{\sigma_N^2} f_g \left(\frac{y}{\sigma_N} \right), \tag{9.5}$$

where $f_g = (\frac{y}{\sqrt{2\pi}}) \exp(-x^2/2)$ is the zero mean unit variance Gaussian pdf (probability density function). Under these conditions, $f_{Y|X}(y|x)$ is also Gaussian for any x and the conditional entropy will be $H(Y|X) = \frac{1}{2} \log_2(2\pi e \sigma_N^2)$. For a given input distribution f_x, the output distribution f_y is given by marginalizing the joint distribution $f_{X|Y}(y, x) = f_{X|}(x) f_{Y|X}(y|x)$ and the mutual information is calculated as $I(X : Y) = H(y) - H(Y|X)$. When the receiver is aware about transmitter parameters, the input distribution f_x is known and the mutual information increases monotonic to its maximum with the increase in the transmitted power. Parameters of transmitter's configuration include the distribution of the polar components among the amplification branches and antennas and the spacing between antennas, but let restrict the analysis to the case where the receiver does not know the spacing between antennas, when it is transmitted a regular M^2-QAM constellation. As stated before, a regular M^2-QAM constellation may result as a sum of M polar components, which leads to a transmitter with $N_m = M$ amplifiers and M isotropic antennas that transmit M uncorrelated signals. For linear array arrangement with equally spaced antennas by d, as shown in Figure 9.2, the polar components $g_i e^{\theta_i} = |g_i|e^{\varphi_i}e^{\theta_i} = |g_i|e^{\beta_i}$ from each branch i are affected by a phase rotation of $\alpha_i = 2\pi(i-1)d/\lambda \cos\left(\frac{\pi}{2} + \Theta\right)$. In such conditions, the transmitted symbol is given by

$$x = \sum_{i=1}^{M} |g_i|e^{\beta_i} \exp(\alpha_i), \tag{9.6}$$

instead of $\sum_{i=0}^{M-1} g_i e^{\beta_i}$. Therefore, each polar component suffers a different rotation that depends on the antenna position in the array and sort order adopted along the M branches. Let us consider the simplest configuration with two antennas separated by d and $|g_1|e^{\beta_1}$ and $|g_2|e^{\beta_2}$. Assuming $\alpha_i = 0$ for the first antenna, the transmitted symbol is

$$\begin{aligned} x' &= |g_1|\cos(\theta_1) + |g_2|\cos(\theta_2 + \alpha_2) + j(|g_1|\sin(\theta_1) + |g_2|\sin(\theta_2 + \alpha_2)) \\ &= K_a x \exp(j\alpha_a), \end{aligned} \tag{9.7}$$

where K_a and α_a are the amplitude and phase distortion, respectively, and the symbol before the shaping performed by the array is given by

$$x = |g_1|\cos(\theta_1) + |g_2|\cos(\theta_2) + j(|g_1|\sin(\theta_1) + |g_2|\sin(\theta_2)). \tag{9.8}$$

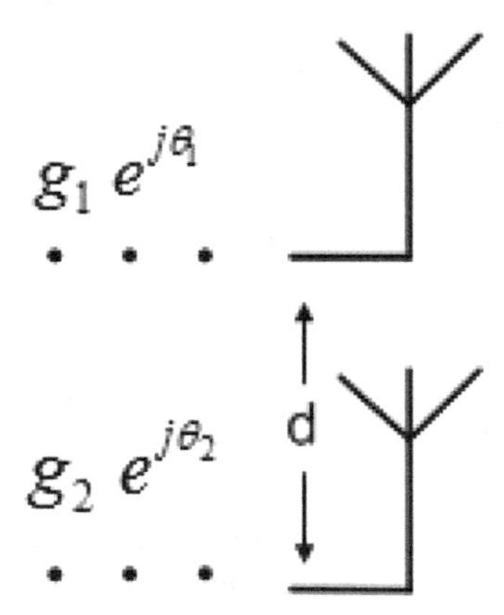

Figure 9.2 Simplest transmitter with an array of two antennas.

Obviously, any receiver without information about transmitter configuration has as reference the regular constellation unaffected by the amplitude scaling and phase rotation due to the array arrangement and distribution of the polar components over the amplification branches and the shaping of the constellation acts as a nonlinear distortion that is also unknown by the receiver. For this user, we have a channel with nonlinear distortion and the additive noise, represented as

$$Y = f_a(x) + N, \tag{9.9}$$

where $f_a(x) = k_a x e^{j\alpha_a}$. In this case, for a given channel input x, the channel law is given by the conditional pdf $f_{Y|X}(y|x) = \frac{1}{\sigma_N^2} f_g(\frac{y - f_a(x)}{\sigma_Z})$. The problem of a theoretical analysis for the previous conditional pdf relies on the definition of a closed form for $f_a(x)$ when constellation's size grows. For example, in 64-QAM, the nonlinear distortion is the result of the sum 6 phase rotations affecting the 6 different polar components (in Voronoi constellations, we must consider $M - 1$ terms). This distortion effect is shown in Figures 9.3 and 9.4,

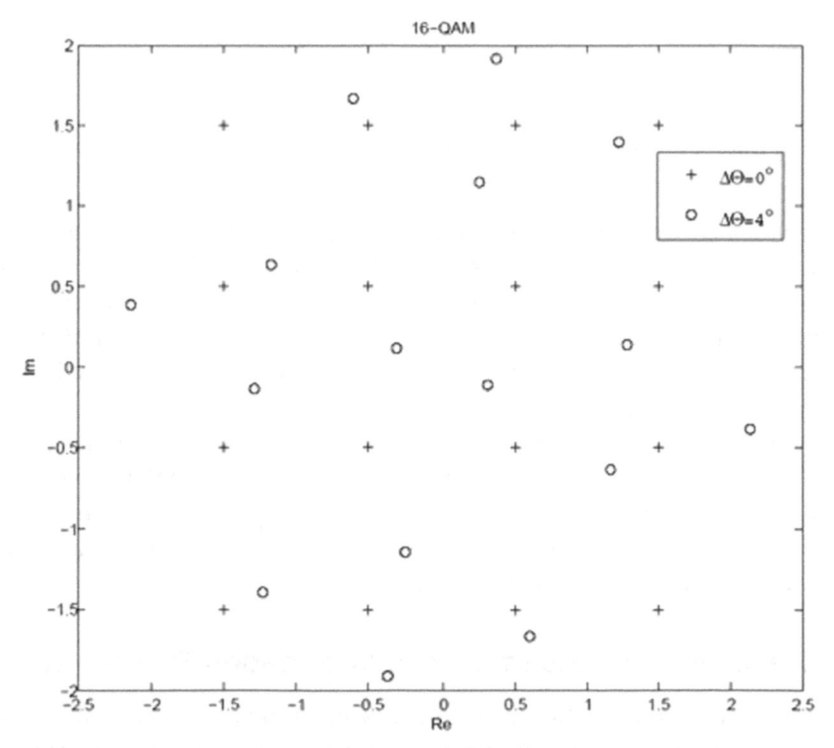

Figure 9.3 16-QAM: Effect of an error $\Delta\Theta$ in the received constellation.

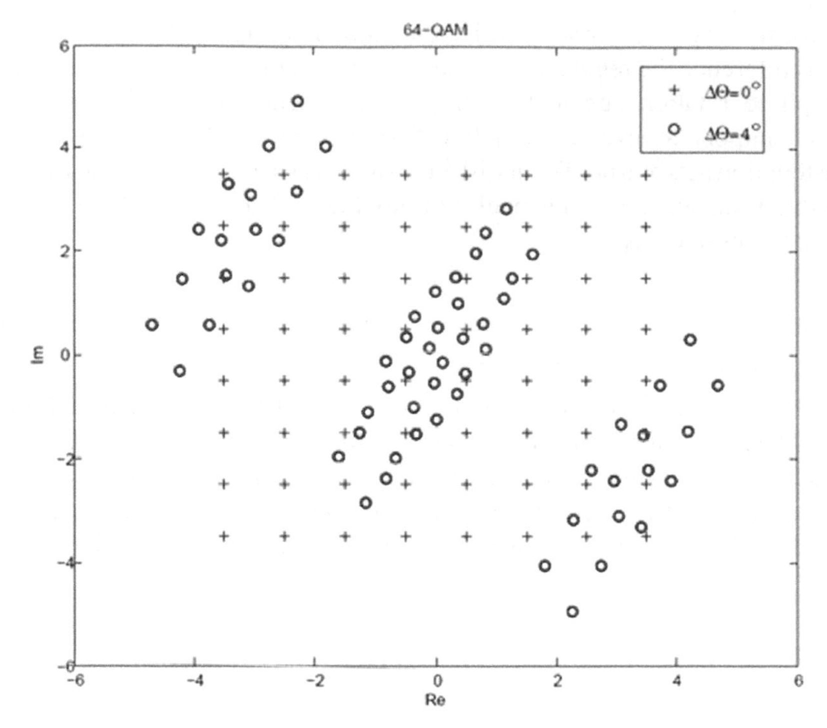

Figure 9.4 64-QAM: Effect of an error $\Delta\Theta$ in the received constellation.

where it can be seen the distortion effect in the constellation for small errors in estimation of transmitter parameters that lead to an incorrect direction in which the constellation is optimized $\Theta' = \Theta + \Delta\Theta$ (it is assumed that the constellation optimized under Θ is a regular M^2-QAM). Obviously, this effect becomes stronger when the constellation's size grows as it can be seen comparing both figures.

Having in mind these aspects, it seems obvious that shaping of the constellation achieved by this transmitter structure can be used to implement security at physical level. In the following sections, an analysis of this transmission scheme is presented with emphasis to the impact on mutual information and secrecy capacity assured by these schemes.

9.4 Receivers and the Impact of Information Directivity

The simplest communication problem of secrecy and confidentiality arises in a three-terminal system comprising a transmitter, the intended receiver (Bob), and an unauthorized receiver, referred as eavesdropper (Eve), wherein the transmitter wishes to communicate a private message to the receiver. Therefore, the receiver must know the set of N_m coefficients that define the constellation shaping in the desired direction, otherwise receives a distorted constellation. Having in mind these considerations, a question arises: What will be the tolerance against errors on the estimation of transmitter's parameters and what will be the secrecy level assured by this transmission scheme? To answer this question, we present some results related to the mutual information associated to both authorized receiver and eavesdropper, and the secrecy capacity of the proposed system.

We assume a non-degraded Gaussian wiretap channel depicted in Figure 9.5, where both channel 1 and channel 2 are AWGN channels.

Let $x(t)$ denotes the nth transmitted symbol associated to a given block

$$x(t) = s_n h_T(t - nT_S), \tag{9.10}$$

with T_S denoting the symbol duration and $h_T(t)$ denoting the adopted pulse shape. s_n belongs to a given size M constellation $\mathfrak{S}$. Under these conditions, the received signals by the authorized receiver and the eavesdropper are

$$y(t) = f_a(x(t)) + n_1(t), \tag{9.11}$$

and

$$z(t) = f_a(x(t)) + n_2(t), \tag{9.12}$$

with $n_1(t)$ and $n_2(t)$ denoting de noise terms and f_A denotes the shaping performed by the antenna array. With perfect secrecy, we have $I(X;Z) = 0$, with X the sent message and Z the received message by the eavesdropper and where $I(;)$ denotes mutual information (MI). It should be noted that the mutual information (assuming equiprobable symbols) for a given signal set $\mathfrak{S}$

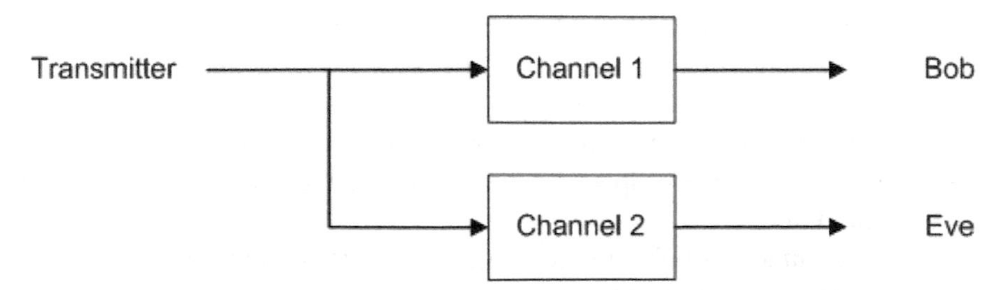

Figure 9.5 Non-degraded Gaussian wiretap channel.

gives the maximum transmission rate (in bits/channel use) at which error-free transmission is possible with such signal set [14] and can be written as

$$I(X,Y) = \log_2 M - \frac{1}{M}\sum_{s\in\mathfrak{S}} \boldsymbol{E}_n\left[\log_2\left(\sum_{x'_n\in\mathfrak{S}} \exp\left(-\frac{1}{N_0}|\sqrt{Es}(s_n - s'_n) + n|^2 - |n|^2\right)\right)\right], \quad (9.13)$$

where E denotes the expectation. Under these conditions, the secrecy capacity (SC) is given by

$$C_s = \max_{s\in F}[I(X;Y) - I(X;Z)], \quad (9.14)$$

where F is the set of all probability density functions at the channel input under power constrain at the transmitter, $I(X;Y)$ denotes the MI of the intended receiver, and $I(X;Z)$ represents the MI of eavesdropper (the capacity is always positive, since due to the absence of information regarding the transmitter configuration, the eavesdropper receives always a distorted version of the transmitted signal).

To evaluate the influence of antenna spacing on constellation shaping and the assured physical layer security degree, two scenarios are firstly considered: In first scenario, the receiver knows the set of coefficients g_i and $f_a(.)$. In the second one, the receiver is unaware about the transmitter parameters, but we admit that is able to estimate them with an error. Note that the receiver in the second scenario can be compared to the eavesdropper. We admit an AWGN channel for both scenarios. It is important to mention that perfect secrecy implies $I(MS; RE) = 0$, with MS the sent message and RE the received message by the eavesdropper.

For both scenarios, we assume two hypotheses regarding the antenna spacing. In hypothesis I, the array antennas are equally spaced by $d/\lambda = 1/4$ at the transmitter. The hypothesis II assumes non-uniform spacings between successive elements given by $d/\lambda = 1/2$, $d/\lambda = 3/4$, and $d/\lambda = 1/2$, respectively (see Table 9.1). In both hypotheses, the set of coefficients associated to the RF branches follows the arrangement of Table 9.1 and it is assumed that the eavesdropper knows somehow the spacing arrangement between antennas.

Results include MI behavior with the optimization angle θ and the MI evolution as function of signal-to-noise ratio (SNR).

9.4.1 Simulation Results

One thousand independent trials of Monte Carlo experiments are used to obtain the average results of MI. Symbols s_n are selected with equal probability from a M-QAM constellation (dimensions of $M = 16$ and $M = 64$ are considered). Linear power amplification at the transmitter and perfect synchronization are assumed. Some MI results are expressed as function of $\frac{E_b}{N_0}$, where $N_0/2$ is the noise variance and E_b is the energy of the transmitted bits.

Both hypotheses are considered in Figures 9.6 and 9.7, where it shows the MI evolution with the optimization angle θ for a fixed SNR value of 16 dB. In both figures, it is assumed that the receiver knows the set of coefficients g_i used by the transmitter. It is clear that when receiver knows the transmitter parameters, the MI is practically unaffected by the optimization angle in which the constellation is configured (it should be noted that the intended receiver acts like the "smart receiver" of [15]). This means that independent of the direction in which the constellation is optimized, the authorized receiver is able to decode with success the transmitted information. Despite this fact, the non-uniform arrangement of the transmitter's array shows more variability with Θ than uniform case, due to the higher information directivity attainable by the first one.

When the knowledge of the coefficients g_i is not perfect, the distortion effects on the transmitted constellation are comparable to nonlinear distortion introduced by a nonlinear channel with an AM/AM and an AM/PM non-null characteristic. It is important to note that the eavesdropper needs to estimate correctly the arrangement of g_i among the amplification branches and must know the antenna spacing. However, this is not a realistic supposition having in mind the previous considerations about the number of possible configurations. Despite this fact, it is important to have an idea of the impact of estimation errors on the MI for an eavesdropper, which will give an idea of exactness needed for any estimation process from the eavesdropper's side. Compare with previous results, we also include in Figures 9.8 and 9.9 the MI results for the case where the eavesdropper estimates the transmitter parameters with an precision that leads to an error of $\Delta\Theta$ in the estimate of the optimization angle Θ (we admit $\Theta = 60°$). Clearly, the MI for the eavesdropper is severely affected by any error affecting Θ. In both hypotheses, for low SNR, the overall channel (constellation distortion due to error $\Delta\Theta$ plus noise) is effectively an AWGN channel and the noise's effect predominates. In this case, the mutual information is governed by the mean value of the noise distribution.

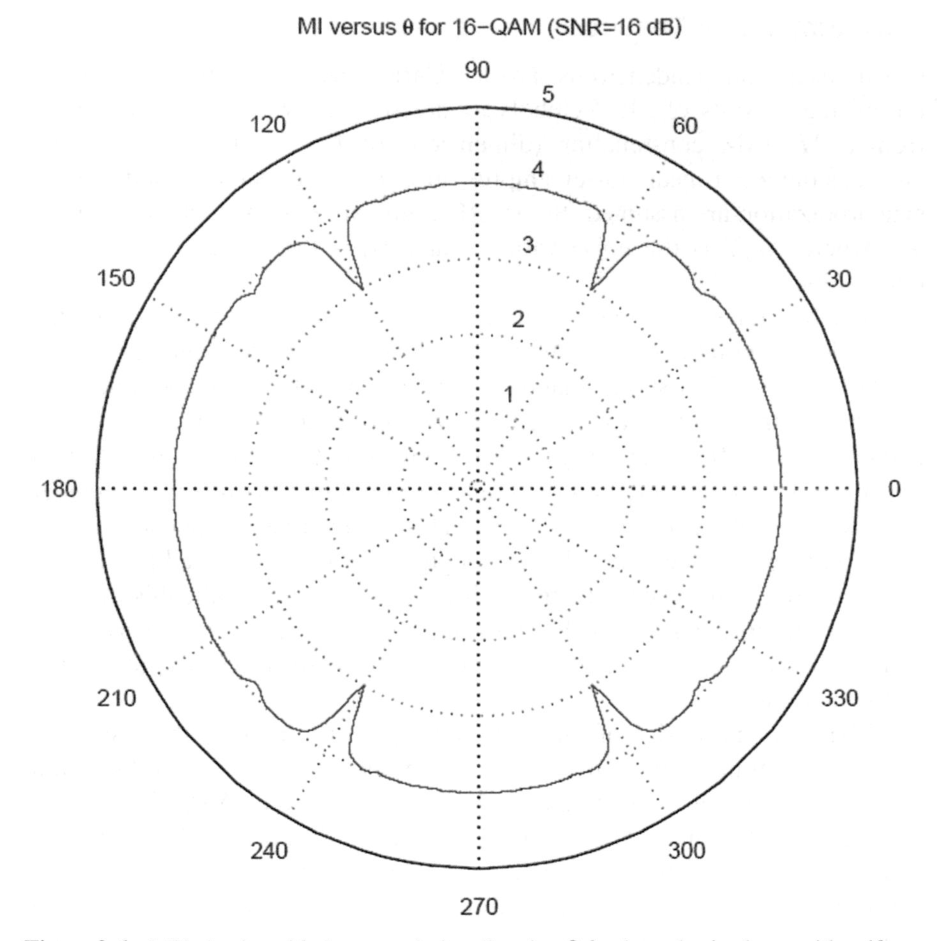

Figure 9.6 MI behavior with the transmission direction Θ for the authorized user with uniform spacing between antennas.

When the SNR is further increased, the mutual information starts to decrease which can be explained by the fact that at high enough power, almost all source samples fall in the nonlinear distortion caused by the estimation error of Θ. It should be mentioned that even a small error $\Delta\Theta$ can be the result of higher phase rotations of the BPSK components that build up each symbol of the transmitted constellation. This can be seen in MI curves of Figure 9.8 for $\Delta\Theta \geq 2°$ where the MI reaches a peak around 6 dB and decreases toward a lower value. Nevertheless, for angle errors higher than 2°, as the $\frac{E_b}{N_0}$ increases, MI curves increase toward a peak, and then, they

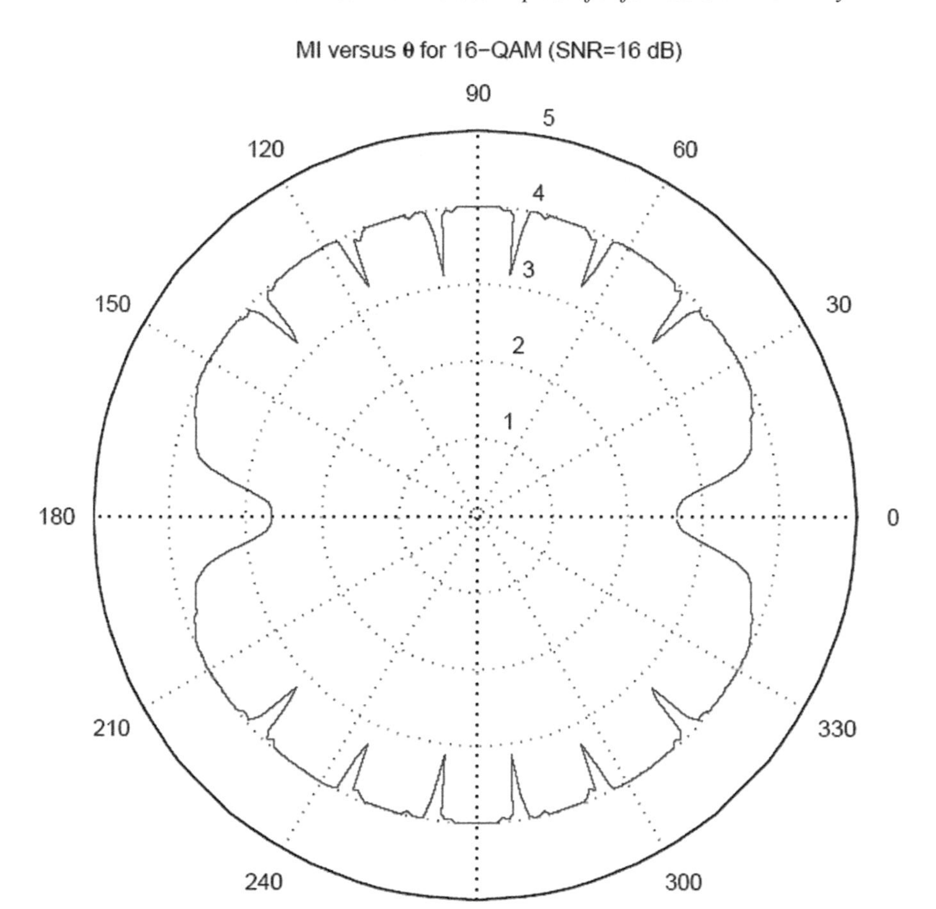

Figure 9.7 MI behavior with the transmission direction Θ for the authorized user with non-uniform spacing between antennas.

decrease again toward zero as the power is further increased. Comparing both figures, it becomes obvious that the impact of any distortion due to angle errors is more significant for hypothesis II. As expected, the mutual information results of Figure 9.9, regarding non-uniform arrays, exhibit higher sensitivity to the distortion introduced by constellation shaping. Still, for $\Delta\Theta \geq 2^{\circ}$, the MI curve has a similar behavior. However, angle errors higher than 2° affect severely the MI. It should be mentioned that we have a secrecy level of hundred percent when the information regarding the set of coefficients g_i and the array arrangement is missing (in this case, we have a null value for MI).

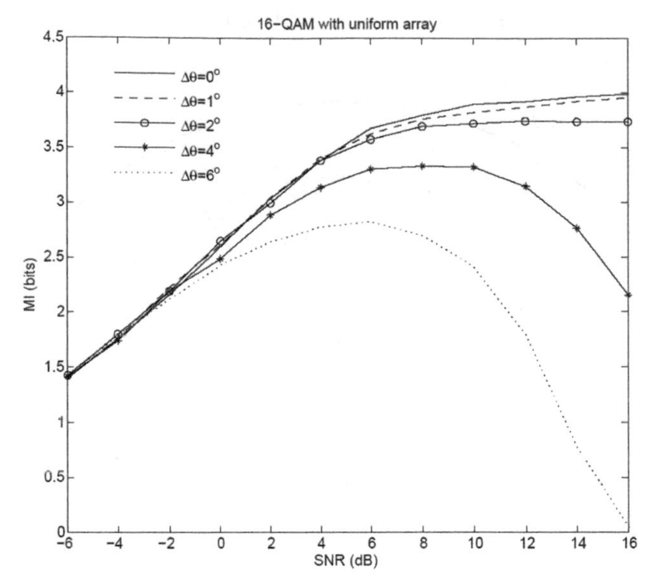

Figure 9.8 MI evolution for 16-QAM and impact of an angle error regarding the transmission direction Θ in eavesdropper's MI for uniform spacing between antennas.

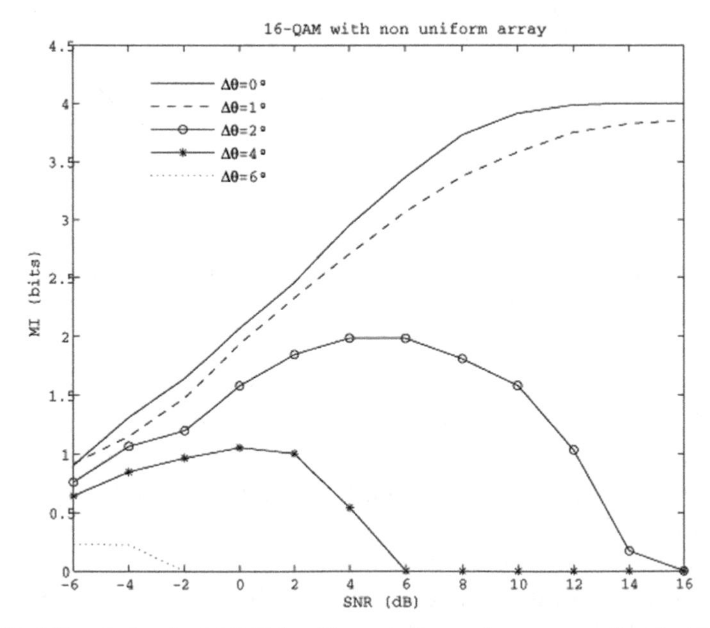

Figure 9.9 MI evolution for 16-QAM and impact of an angle error regarding the transmission direction Θ in eavesdropper's MI for non-uniform spacing between antennas.

Thus, by simple changes on array spacings, security can be improved, since the transmitted constellation's sensitivity to the transmission direction becomes higher.

9.4.2 Transmitter Configuration Effects in MI and Secrecy

Previous results have shown the high sensitivity of MI against any error on the estimation of Θ for an eavesdropper or any authorized receiver unable to estimate exactly the transmitter's parameters that lead to the constellation shape optimized for Θ. However, it is not realistic to assume that Eve knows in advance the initial configuration of the transmitter, including both coefficients g_i, the arrangement of these coefficients in the several RF branches, and the spacing between antennas. Besides that, these results were for a specific configuration of the transmitter and do not give an idea of the impact in MI and SC of transmitter configuration freedom degrees such as permutations of BPSK components between antennas and non-uniform spacing. To extend previous analysis, we consider two options regarding the transmitter configuration, namely permutations of BPSK components antennas in uniform arrays and non-uniform spacings between antennas, and their impact on the achievable secrecy rate of an arbitrary downlink transmission with authorized users and a user acting as potential eavesdropper is characterized. In first option, the antennas are equally spaced by $d/\lambda = 1/4$ at the transmitter. In non-uniform arrangements, the spacing between antennas is always an integer multiple of the distance $d/\lambda = 1/4$. All permutations were tested for 16-QAM and 64-QAM constellations as well as a set of combinations for non-uniform arrangements. Tables 9.2 and 9.3 show the combinations between coefficients g_i and antennas, for transmitters based on 16-QAM and 64-QAM that generate constellation directivity with higher effect on the MI. Spacing between antennas for uniform and non-uniform array arrangements are also presented in these tables (it should be mentioned that the values are referred to antenna 1).

Regarding Bob and Eve, two more realistic hypotheses are considered:

- I—Eve only knows the set of coefficients g_i, but do not knows the array configuration, i.e., the spacing between antennas and the direction in which the constellation is optimized and Bob has all information about the transmitter.
- II—Eve only estimates the set of coefficients g_i that lead to a small error $\Delta\Theta$ and Bob knows all information about the transmitter.

Table 9.2 16-QAM: g_i arrangements and distances between antennas for uniform and non-uniform arrays

16-QAM				
Antenna order	1	2	3	4
coefficients	2j	1	2	j
Spacing to antenna 1		3d	4d	5d
		2d	5d	8d
		3d	4d	6d
		3d	4d	7d

Table 9.3 64-QAM: g_i arrangements and distances between antennas for uniform and non-uniform arrays

64-QAM						
Antenna order	1	2	3	4	5	6
coefficients	2j	1	2	j	4	4j
Spacing to antenna 1		8d	12d	18d	20d	21d
		8d	14d	16d	17d	21d
		8d	14d	16d	20d	21d
		8d	14d	18d	19d	21d

Note that hypothesis II is still an optimistic one, since Eve must estimate the coefficients g_i and the array arrangement. As before, Monte Carlo experiments were used to obtain the average results for MI, which are expressed as function of $\frac{E_b}{N_0}$, where $N_0/2$ is the noise variance and E_b is the energy of the transmitted bits. In both options, symbols s_n are randomly selected with equal probability from a M-QAM constellation (values of $M = 16$ and $M = 64$ are considered). Linear power amplification at the transmitter and perfect synchronization for both receiver types are assumed.

In hypothesis I, all possible permutations between g_i coefficients and antennas for 16-QAM and 64-QAM were performed. It was observed that MI results for the authorized receiver (Bob) are largely unaffected by changes on antennas permutations. This finding applies equally to both constellation sizes as well as to the eavesdropper (the MI associated to Eve is always null). Having in mind this behavior, in Figure 9.10, we present the MI evolution with Θ for uniform arrays. It is important to note that the permutations between g_i and antennas presented on Tables 9.2 and 9.3 correspond to the 21st and 679th possible permutations for the transmitters based on 16-QAM and 64-QAM, respectively. It can be seen that the MI is practically unaffected by the optimization angle in which the constellation is configured (thus despite the secrecy achieved, Bob is always able to decode with success the message). Results regarding non-uniform arrays are presented in Figure 9.11.

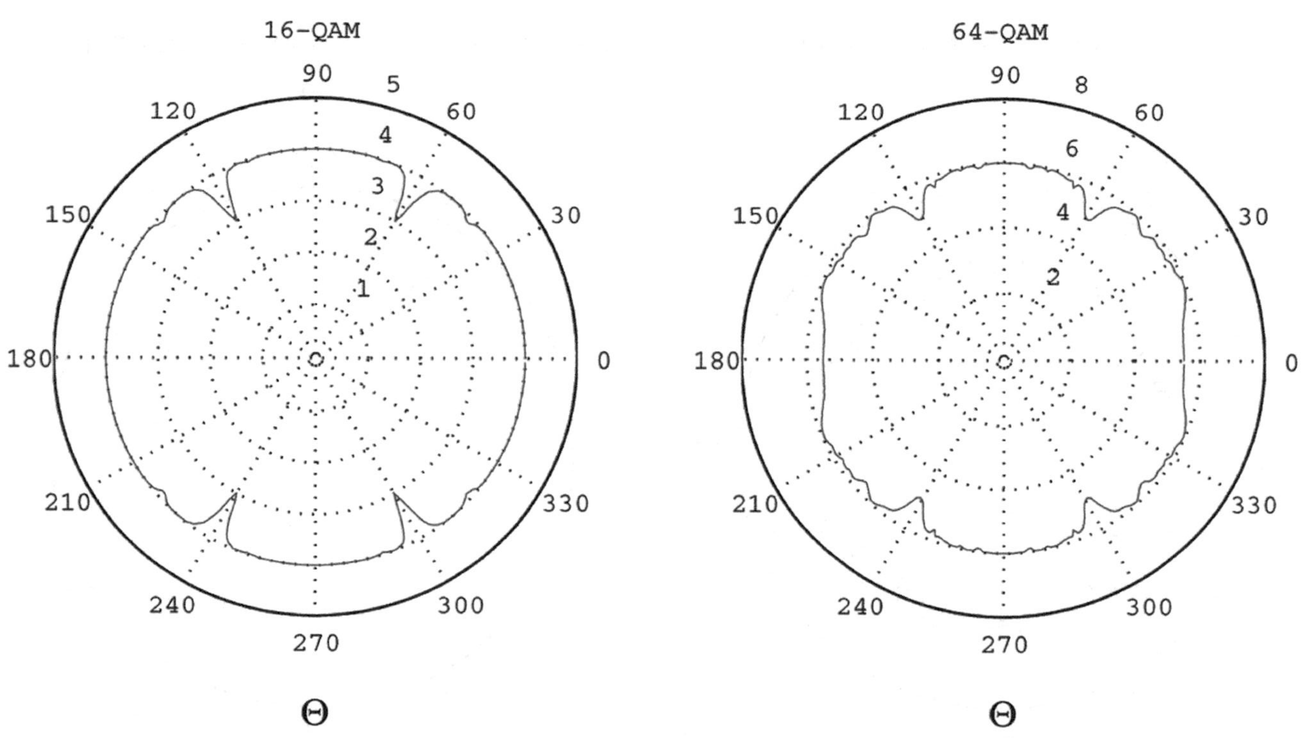

Figure 9.10 Uniform arrays: MI behavior with the angle Θ in which the transmitted constellation is optimized.

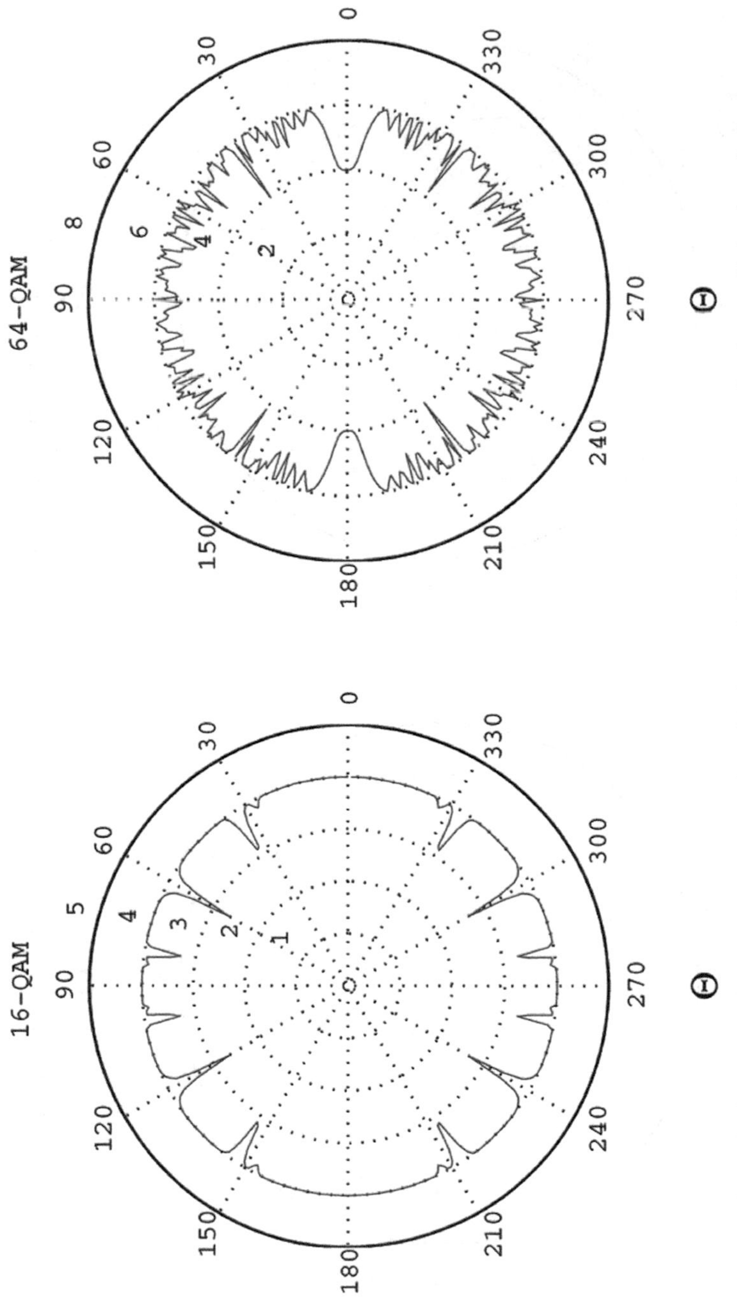

Figure 9.11 Non-uniform arrays: MI behavior with the angle Θ in which the transmitted constellation is optimized.

It is obvious that the MI is more sensitive to the direction Θ in which the constellation is optimized. Despite this higher sensitivity, the MI values are slightly affected over the range of the optimization angle Θ. Still, there are some directions where the MI is lower. The cause for that relies on the constellation shaping that may reduce significantly the separation between constellation's symbols in these cases. It is also worth to mention that the MI values for Bob are practically unaffected by changes on non-uniform arrangements between antennas for both constellations' sizes. Similar to the uniform arrays, the MI associated to Eve is always null for all the arrangements previously analyzed (so the case considered here gives a good example about the behavior of SC in non-uniform arrays). The same behavior happens even when the g_i values are known but information about the array configuration is absent. This means that the SC, represented by C_s, is equal to the mutual information of the authorized receiver.

From these results, it seems obvious that MI values for Bob and Eve are similar in both options. Despite this fact, uniform and non-uniform arrangements may have different effects on tolerance against estimation errors. Let us now consider the hypothesis II, where Bob knows the set of coefficients g_i and Eve estimates the phase shifts between antennas with an error that leads to an estimate $\Theta + \Delta\Theta$. We assume that estimation errors $\Delta\Theta$ that are limited to the set of values $6°, 10°$ in 16-QAM and to $4°, 6°$ in 64-QAM. For comparison purposes, we also include the results of hypothesis I (Figure 9.12), where Bob has perfect information about the transmitter parameters. The non-uniform array spacing configurations for 16-QAM and 64-QAM are those presented on Tables 9.2 and 9.3, respectively. Figures 9.13 and 9.14 show the secrecy capacities assured by uniform and non-uniform arrangements. Clearly, in both options, the SC increases with the number of coefficients g_i involved in the definition of the constellation. For errors on g_i estimates leading to estimation errors $\Delta\Theta$ higher than $6°$, as the $\frac{E_b}{N_0}$ increases, SC remains practically unaffected in both constellations' sizes. Only for small SNR values where the system preforms poorly, even in the absence of an eavesdropper, the secrecy rate is affected. Results of Figure 9.14, regarding the 64-QAM, exhibit practically the same behavior, but now even for small estimation errors, the SC values are close to the results of hypothesis I. However, it is important to note that SC is only affected when SNR values are below the threshold at which the system performance becomes reasonable. For instance, to achieve a BER of 10^{-5} in 16-QAM and 64-QAM, the SNR must be 14 dB and 18 dB, respectively. Above these values of SNR, the capacity reaches its maximum value and remains constant. Results also show that non-uniform arrays outperform the uniform ones, due to the lower

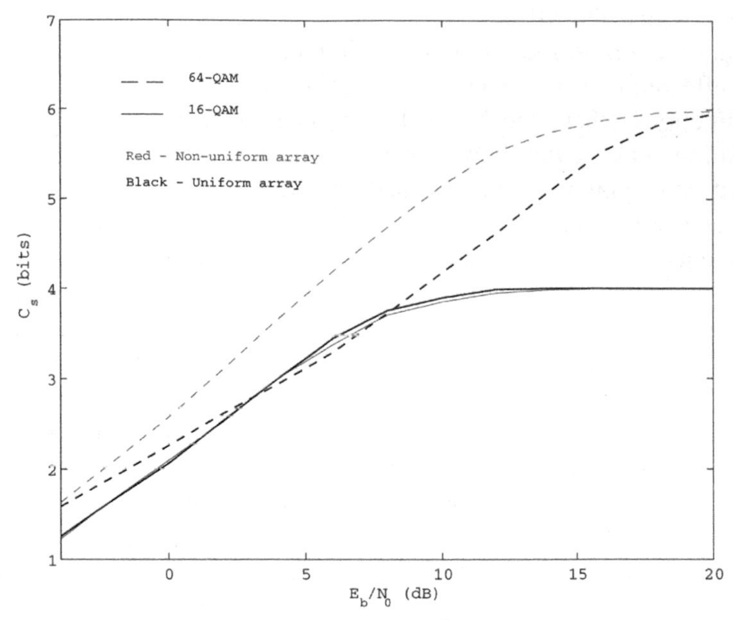

Figure 9.12 C_s evolution with $\frac{E_b}{N_0}$ for hypothesis I.

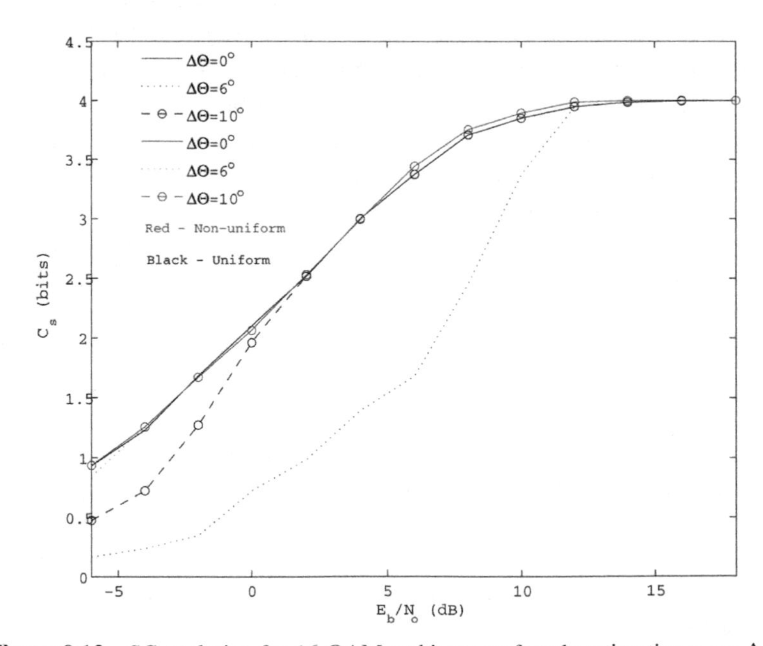

Figure 9.13 SC evolution for 16-QAM and impact of angle estimation error $\Delta\Theta$.

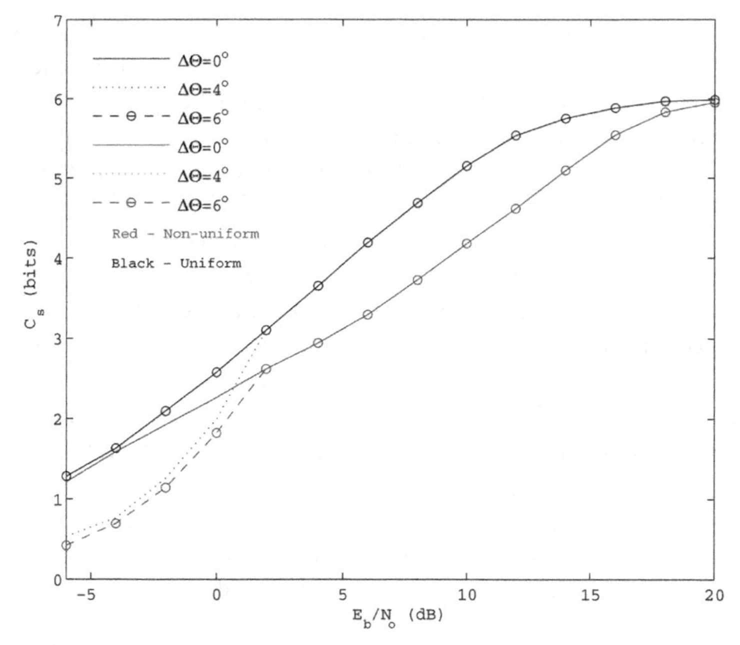

Figure 9.14 SC evolution for 64-QAM and impact of angle estimation error $\Delta\Theta$.

tolerance against any estimation error. This low tolerance, together with the complexity, means that the computational load associated to any interception by Eve can be prohibitive, due to the high number of parameters that must be estimated (the same conclusion is still valid for uniform configurations). Additional increases on security can be achieved by changing dynamically between successive transmitted blocks the configuration of coefficients g_i according to a pattern known by the transmitter and the intended user. This information directivity can be also useful to separate data streams among users in a multi-user wireless communication massive MIMO environment, where for each user, it used a different antennas' subset and consequently a different constellation shaping. Massive MIMO turns possible the association of beamforming with this scheme of information directivity to isolate data streams and eliminate residual interference among users.

9.5 Conclusions

It became obvious that a multi-antenna transmission structure with constellation shaping can be used to achieve physical layer security. The multi-antenna transmitter based on multilevel modulation decomposition as a sum of BPSK components introduces channel-independent security at the physical layer without sacrifices on spectral and power amplification efficiencies. From previous analysis, it became obvious that information directivity provided by a multi-antenna power-efficient transmission structure can assure physical layer security. The decomposition of the signal associated to a given constellation as the sum of N_m BPSK signals that are posteriorly combined at channel level assures an optimization of the transmitted constellation in a specific direction Θ that must be known by the receiver. Secrecy is assured for several configurations of the transmitter and even can be increased using non-uniform spacing between the antennas or changing other parameters. The low tolerance of MI against estimation errors of Θ implies that the set of transmitter g_i coefficients as well as the shaping of the constellation should be perfectly known by the eavesdropper, which means that the complexity involved in a real-time estimation process can be prohibitive for any unauthorized interception. Obviously, the level of security can be increased by dynamic changes on mapping rule and constellation shaping as well as through the optimization of the constellation directivity with other array configurations. Further research should analyze the influence of the array configuration and how these signals can be used together with beamforming supported by massive MIMO systems to mitigate interference among several users sharing the same channel.

Acknowledgments

This work was supported in part by CTS multi-annual funding project PEst-OE/EEI/UI0066/2011, IT UID/EEA/50008/2013 (plurianual founding and project GLANCES), GALNC EXPL/EEI-TEL/1582/2013, EnAcoMIMOCo EXPL/EEI-TEL/2408/2013, and CoPWIN PTDC/EEI-TEL/1417/2012.

References

[1] Bloch, M., Barros, J., Rodrigues, M. R. D., and McLaughlin, S. W. (2008). Wireless information-theoretic security. *IEEE Trans. Inf. Theor*, 54 (6), pp. 2515–2534, June.

[2] Massey, J. L. (1988). An introduction to contemporary cryptology. *Proc. IEEE*, 76 (5), 533–549, May.

[3] Schneier, B. (1998). Cryptographic design vulnerabilities. *IEEE Comput.*, 31 (9), 26–33, Sep.

[4] Barenghi, A., Breveglieri, L., Koren, I., and Naccache, D. (2012), Fault injection attacks on cryptographic devices: Theory, practice, and countermeasures. *Proc. IEEE*, 100 (11), 3056–3076, Nov.

[5] Harrison, W. K., Almeida, J., Bloch, M. R., McLaughlin, S. W., and Barros, J. (2013). Coding for secrecy: An overview of error-control coding techniques for physical-layer security. *IEEE Signal Process. Mag.*, 30 (5), 41–50, Sep.

[6] Gollakota, S., and Katabi, D. (2011). Physical layer wireless security made fast and channel independent. *INFOCOM, 2011 Proceedings IEEE*, April, pp. 1125–1133.

[7] Marques da Silva, M., and Monteiro, F. A. (2014). MIMO processing for 4G and beyond: fundamentals and evolution, CRC Press Auerbach Publications: ISBN: 9781466598072, FL, USA, May, (http://www.crcpress.com/product/isbn/9781466598072).

[8] Montezuma, P., and Gusmão, A. (1999). Design of TC-OQAM schemes using a generalised nonlinear OQPSK-type Format. *IEEE Elect. Letters*, 35 (11), 860–861, May.

[9] Astucia, V., Montezuma, P., Dinis, R., and Beko, M. (2013). On the use of Multiple grossly Nonlinear amplifiers for Highly Efficient Linear amplification of multilevel constellations. *Proc. IEEE VTC2013-Fall*, Las Vegas, NV, US, September 2013.

[10] Montezuma, P., and Dinis, R. (2015). Implementing physical layer security using transmitters with constellation shaping. *Proc. IEEE ICCN 2015*, Las Vegas, NV, US, August.

[11] Tse, D. N. C., and Viswanath, P. (2005). Fundamentals of wireless communications. Cambridge, UK: Cambridge University Press.

[12] Dinis, R., Montezuma, P., Souto, N., and Silva, J. (2010). Iterative frequency-domain equalization for general constellations. *IEEE Sarnoff Symposium*, Princeton, USA, Apr.

[13] Amoroso, F., and Kivett, J. (1977). Simplified MSK signalling technique. *IEEE Trans. Comm.*, 25, Apr.

[14] Caire, G., Taricco, G., and Biglieri, E. (1998). Bit-interleaved coded modulation. *IEEE Trans. Inf. Theory*, 44, 927–947, May.

[15] Montezuma, P., and Dinis, R., Marques, D., and Beko, M. (2014). Robust frequency-domain receivers for a transmission technique with directivity at the constellation level. *VTC2014-Fall*, 14–17 September 2014, Vancouver, Canada.

PART VI

Reliable System Design

10

Active Sub-Areas-Based Multi-Copy Routing in VDTNs

Bo Wu[1], Haiying Shen[1] and Kang Chen[2]

[1]Electrical and Computer Engineering, Clemson University,
Clemson, SC 29634, USA
[2]Electrical and Computer Engineering,
Southern Illinois University Carbondale, Carbondale, IL 62901, USA

Abstract

In vehicle delay-tolerant networks (VDTNs), current routing algorithms select relay vehicles based on either vehicle encounter history or predicted future locations. The former method may fail to find relays that can encounter the target vehicle in a large-scale VDTN, while the latter method may not provide accurate location prediction due to traffic variance. Therefore, these methods cannot achieve high performance in terms of routing success rate and delay. In this chapter, we aim to improve the routing performance in VDTNs. We first analyze vehicle network traces and observe that (i) each vehicle has only a few active sub-areas that it frequently visits and (ii) two frequently encountered vehicles usually encounter each other in their active sub-areas. We then propose active area-based routing (AAR) method which consists of two steps based on the two observations correspondingly. AAR first distributes a packet copy to each active sub-area of the target vehicle using a traffic-considered shortest path spreading algorithm, and then in each sub-area, each packet carrier tries to forward the packet to a vehicle that has high encounter frequency with the target vehicle. Furthermore, we propose a distributed AAR (DAAR) to improve the performance of AAR. Extensive trace-driven simulation demonstrates that AAR produces higher success rates and shorter delay in comparison with the state-of-the-art routing algorithms in VDTNs. Also, DAAR has a higher success rate and a lower average delay compared with AAR since information of dynamic active sub-areas tends to be updated from time to time, while the information of static active sub-areas may be outdated due to the change of vehicles' behaviors.

Keywords: VDTN, Active area, Routing algorithm.

10.1 Introduction

Recently, the problem of providing data communications in vehicle networks (VNETs) has attracted a lot of attention. Vehicle delay-tolerant networks (VDTNs) create a communication infrastructure composed by vehicle nodes, which offers a low-cost communication solution without relying on base stations. In this chapter, we focus on routing algorithms in VDTNs.

Current routing algorithms in VDTNs can be classified into three categories: contact [1–4]-, centrality [5–7]-, and location [8–11]-based routing algorithms. Based on the fact that vehicles which encountered frequently in the past tend to encounter frequently in the future, contact-based routing algorithms relay packets according to the encounter history. In centrality-based algorithms, a packet carrier forwards the packet to the vehicle with the highest centrality, i.e., the vehicle that can encounter more vehicles. However, in the contact- and centrality-based algorithms, a packet carrier may fail to find relays that can encounter the target vehicle in a large-scale VDTN with thousands of vehicles on a very large area, leading to low routing efficiency.

Location-based routing algorithms predict the future locations of vehicles, find the shortest path from the source vehicle to the target vehicle, and select the vehicles with trajectories on the shortest path as relay vehicles. These algorithms require highly accurate prediction so that relay vehicles and the packet carrier will be close to each other in a certain short distance (e.g., less than 100 meters). However, it is difficult to achieve accurate prediction because vehicles have high mobility and vehicle trajectories are greatly influenced by many random factors such as the traffic and speed of vehicles. Also, since the shortest path is determined without considering the traffic, there may be few vehicles on the path. Therefore, if the selected relay vehicle is missed due to low prediction accuracy, it is difficult to find other relay candidates, which leads to low routing efficiency.

Therefore, current routing algorithms cannot achieve high performance in terms of routing success rate and delay. In this chapter, we aim to improve the routing performance in VDTNs. We first analyze real vehicle network (VNET)

Roma [12] and *SanF* [13] traces and gain the following two observations: (i) Each vehicle has only a few active sub-areas in the entire VDTN area that it frequently visits and (ii) two frequently encountered vehicles usually have high probability to encounter each other in their active sub-areas, while have very low probability to encounter each other in the rest area on the entire VDTN area. We then propose active area-based routing method (AAR) which consists of two phases based on the two observations correspondingly.

As shown in Figure 10.1, unlike the contact- and centrality-based routing algorithms that search the target vehicle in the entire VDTN area, AAR constrains the searching areas to the active sub-areas of the target vehicle, which greatly improves the routing efficiency. AAR first distributes a packet copy to each active sub-area of the target vehicle, and then in each sub-area, each packet carrier tries to forward the packet to a vehicle that has high encounter frequency with the target vehicle. Specifically, AAR consists of the following two algorithms for these two steps.

Traffic-considered shortest path spreading algorithm. It jointly considers traffic and path length in order to ensure there are many relay candidates in the identified short paths to efficiently distribute multiple packet copies. Figure 10.2 shows an example of the basic idea of our traffic-considered shortest path spreading algorithm. Current location-based routing algorithms relay the packet from road intersection a to b through the shortest path (i.e., the dotted line) but fail to consider whether there are enough relay vehicles in the path. If the shortest path only consists of small roads with less traffic, it leads to a long time for a packet to reach the target sub-area. In our spreading algorithm, the packet is routed along the circuitous path (i.e., the solid line) which consists of main roads that are full of traffic. Then, the packet can easily find next hop relay vehicle and reach b faster in spite of the longer length of the path.

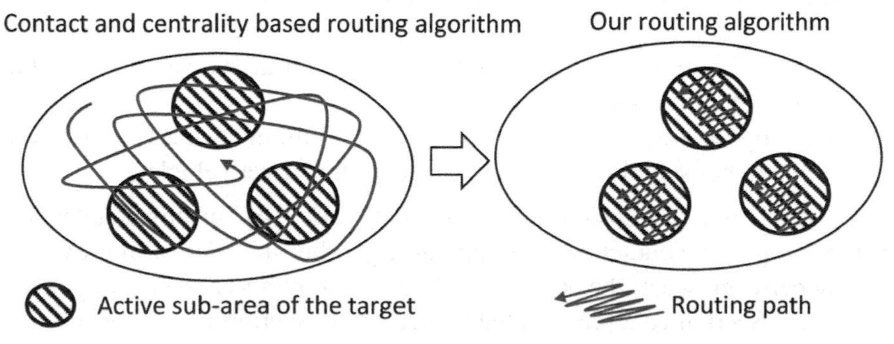

Figure 10.1 Current routing algorithm vs. AAR.

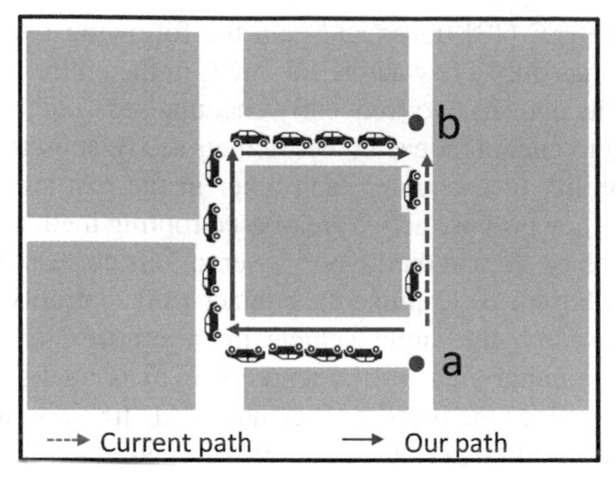

Figure 10.2 An example of the traffic-considered shortest path spreading algorithm.

Contact-based scanning algorithm. It restricts each packet copy in its responsible active sub-area to find relay vehicles with high encounter frequencies with the target vehicle. Specifically, the packet copy is forwarded to vehicles traveling in different road sections so that it can evenly scan the sub-area.

By avoiding searching the non-active sub-areas of the target vehicle as in the contact- and centrality-based routing algorithms, AAR greatly improves routing efficiency. Instead of pursuing the target vehicle as in the location-based routing algorithms, each packet copy in an active sub-area of the target vehicle is relayed by vehicles with high encounter frequency with the target vehicle, thus bypassing the insufficiently accurate location prediction problem in location-based routing algorithms.

To sum up, the main contributions of this chapter are as follows:

1. We analyze two real VNET *Roma* and *SanF* traces, which serve as the foundation for our proposed routing algorithm for VNETs.
2. We propose a traffic-considered shortest path spreading algorithm to spread different copies of a packet to different active sub-areas of the target vehicle efficiently.
3. We propose a contact-based scanning algorithm in each active sub-area of the target vehicle to relay the packet to the target vehicle.

This chapter is an extension of our original work [14]. In this chapter, we further propose a contact-based scanning algorithm in order to improve the routing performance of the original work. The rest of this chapter is organized as follows. Section 10.2 presents the related work. Section 10.4 measures and analyzes the pattern of vehicles' trajectories and the distance among different encounter locations of pairs of vehicles in two real VNET traces. Section 10.5 introduces the detailed design of AAR. In Section 10.6.3, the performance of AAR is evaluated by trace-driven experiments in comparison with the state-of-the-art routing algorithms. Section 10.7 summarizes the chapter with remarks on our future work.

10.2 Related Work

Current routing algorithms in VDTNs can be classified into three categories: contact-based [1–4], centrality-based [5–7, 15], and location-based [8–11] routing algorithms.

In the category of contact-based routing algorithms, PROPHET [1] simply selects vehicles with higher encounter frequency with target vehicles for relaying packets. PROPHET is improved by MaxProp [2] with the consideration of the successful deliveries history. R3 [3] considers not only the encounter frequency history, but also the history of delays among encounters to decrease the routing delay performance. Zhu *et al.* [4] found that two consecutive encounter opportunities drop exponentially and, based on the observation, improved the prediction of encounter opportunity by Markov chain to design the routing algorithm in vehicle networks.

In the category of centrality-based routing algorithms, PeopleRank [5] is inspired by the PageRank algorithm, which calculates the rank of vehicles and forwards packets to the vehicles with higher ranks. SimBet [6] identifies some bridge nodes as relay nodes which can better connect the VNETs by centrality characteristics to relay packets. Instead of directly forwarding packets to target nodes, Bubble [7] clusters the nodes to different communities based on encounter history and still utilizes the bridge nodes to forward packets to the destination community. However, though vehicles with high centrality can encounter more vehicles, they may not have a high probability of encountering the target vehicle. Also, the main problem in both contact- and centrality-based routing algorithm is that packets may hardly encounter suitable relay vehicle due to the low encounter frequencies among vehicles in a large-scale VDTN.

In the category of location-based routing algorithms, GeOpps [8] directly obtains the future location of the target vehicle from GPS data and spreads packets to certain geographical locations for routing opportunities through shortest paths. GeoDTN [9] encodes historical geographical movement information in a vector to predict the possibility that two vehicles become neighbors. Wu *et al.* [10] exploited the correlation between location and time in vehicle mobility when they used trajectory history to predict the future location of the target vehicle in order to improve the prediction accuracy. Instead of predicting exact future location, DTN-FLOW [11] divides the map to different areas and predicts the future visiting area of vehicles, which improves the routing performance since it is much easier to predict the future visiting areas than exact future locations. However, as indicated previously, the location-based algorithms may lead to low routing efficiency due to insufficiently accurate location prediction due to traffic and vehicle speed variance.

A number of multi-copy routing algorithms have been proposed. Spyropoulos *et al.* [16] introduced a "spray" family of routing schemes that directly replicate a few copies by source vehicle into the network and forward each copy independently toward the target vehicle. R3 [3] simply adopts the "spray" routing schemes based on its own single-copy routing. Bian *et al.* [17] proposed a scheme for controlling the number of copies per packet by adding an encounter counter for each packet carrier. If the counter reaches the threshold, then the packet will be discarded by the packet carrier. Uddin *et al.* [18] minimized the energy efficiency by studying how to control the number of copies in a disaster-response application, where energy is a vital resource. However, in current multi-copy routing algorithms, different copies of each packet may search the same area on the entire VDTN area, which decreases routing efficiency. AAR spreads different copies of each packet to different active sub-areas of the target vehicle.

10.3 Identification of Each Vehicle's Active Sub-areas

Current routing algorithms search the target vehicle in the entire VDTN map, which leads to low routing efficiency since some routing paths may be outside of the active sub-areas of the target vehicle. Using multi-copies to search in different active sub-areas of the target vehicle can improve routing efficiency. Because our trace measurement uses the concept of each vehicle's active sub-areas, we first introduce our method of identification of each vehicle's active sub-areas in this section before we introduce our trace measurement.

In the entire VDTN map, a *road section* is the road part that does not contain any intersections and it is denoted by the two IDs of intersections on its two ends such as ab in Figure 10.2. Then, instead of searching target vehicle v on the entire VDTN map, we only direct packets to search in the road sections where the target vehicle v visits frequently. We call these road sections the *active road sections* of vehicle v. To be more specific, we define the set of active road sections of vehicle v (denoted by S_v) by

$$S_v = \{\forall s \in S | f(s, v) > r\} \tag{10.1}$$

where S is the set of all road sections, $f(s, v)$ is the frequency that vehicle v visits road section s, and r is a visit frequency threshold. A smaller threshold r leads to more road sections in the S_v and a larger routing area and vice versa. In this chapter, we set $r = 7$ and $r = 5$ in *Roma* and *SanF* traces, respectively.

Sending a packet copy to each active road section of the target vehicle generates many packet copies and high overhead. Actually, sending a packet copy to a set of connected active road sections is sufficient because the copy can be forwarded to vehicles traveling along all these road sections to search the target vehicle. We define an active sub-area of a vehicle as a set of connected active road sections of the vehicle. We propose a method to create the active sub-areas of each vehicle by following rules:

1. Each sub-area of a vehicle consists of connected road sections of the vehicle so that a packet copy can scan the entire sub-area for the target vehicle without the need of traveling on the inactive road sections.
2. Each sub-area of a vehicle should have similar number of road sections so that the load balance on the size of scanning sub-areas of multiple copies can be guaranteed.

Specifically, our active sub-area identification algorithm works as follows:

1. First, we transform the entire VDTN map to a graph. We consider each road section as a node and connect two nodes if the corresponding two active road sections share the same road intersection. Also, we tag each node with weight 1. Then, the areas' division problem is translated to a graph partition problem.
2. Next, as shown in Figure 10.3, we continually select a directly connected node with the smallest sum of weights, remove the edges between these two nodes, merge them to one node, and set its weight to the sum. If all pairs of directly connected nodes have equal sum of weights, we randomly select a node pair to merge.
3. We repeat step (2) until the number of nodes equals the number of active sub-areas required. Then, the corresponding road sections in one node constitute an active sub-area.

In the above process, two disconnected nodes cannot be merged, which guarantees that the road sections in each active sub-area are connected. Also, since the weight of a node represents the number of road sections corresponding to the node, merging two nodes with the smallest sum of weights

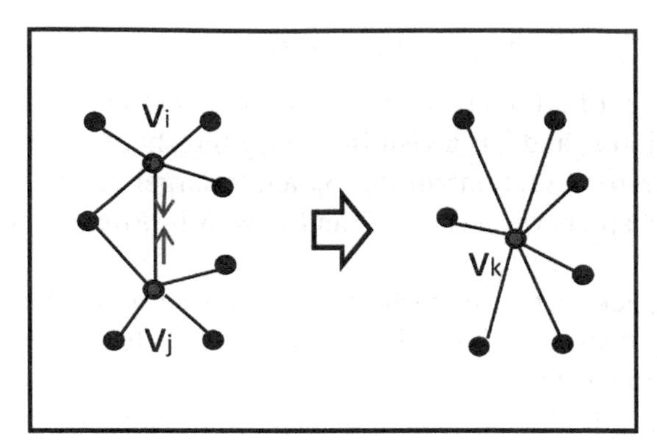

Figure 10.3 Active sub-area identification.

can constrain the difference between the number of road sections in different active sub-areas. As a result, the above two rules are followed, which facilitates the execution of our proposed routing algorithm. The active sub-areas of each vehicle and the entire VDTN map are stored in each vehicle in VDTN. When a vehicle joins in the VDTN, it receives this information.

10.4 Trace Measurement

In order to design a new routing algorithm to improve the performance of current routing algorithms, we first need to better understand the pattern of vehicles' trajectories and the relationship between vehicle contact and location. Therefore, we analyze the real-world VNET *Roma* and *SanF* traces gathered by taxi GPS in different cities, referred to as *Roma* [12] and *SanF* [13]. The *Roma* trace contains mobility trajectories of 320 taxies in the center of Roma from February 1 to March 2, 2014. The *SanF* trace contains mobility trajectories of approximately 500 taxies collected over 30 days in San Francisco Bay Area. Our analysis focuses on following two aspects:

1. *Vehicle mobility pattern.* We expect to find out whether the movement of each vehicle exhibits a certain pattern. If each vehicle frequently visits a few sub-areas in the entire VDTN area, then our routing algorithm only needs to search these sub-areas of a target vehicle in order to improve routing efficiency.
2. *Relationship between contact and location.* Contact- and centrality-based routing algorithms search vehicles that have high encounter frequency

with the target vehicle in the entire VDTN area. If we can identify the locations that the vehicles frequently meet the target vehicle, we can reduce the relay search area to improve the routing efficiency. Therefore, we expect to find out whether such locations can be identified.

10.4.1 Vehicle Mobility Pattern

In order to measure the pattern of vehicles' mobility on the entire VDTN area, we normalize the total driving time of each vehicle to 100 hours and normalize its real visiting time on each road section by

$$\bar{t}(s_i, v_i) = \frac{100 \times t(s_i, v_i)}{t_{v_i}} \tag{10.2}$$

where s_i denotes road section s_i, v_i denotes vehicle i, $\bar{t}(s_i, v_i)$ is the normalized visiting time of vehicle v_i on road section s_i, t_{v_i} is the real total driving time of vehicle v_i, and $t(s_i, v_i)$ is the real visiting time of vehicle v_i on road section s_i. We then calculate the deviation of visiting time of vehicle v_i (D_{v_i}) by

$$D_{v_i} = \frac{1}{|S|} \sum_{i \in S} \left(\bar{t}(s_i, v_i) - \bar{t}_{v_i}\right)^2 \tag{10.3}$$

where S is the set of all the road sections in the entire VDTN map and $\bar{t}_{v_i}$ is the average visiting time of vehicle v_i per road section. Since the total visiting time of each vehicle is normalized to 100 hours and S is fixed, $\bar{t}_{v_i}$ is a fixed value $\frac{100}{|S|}$ for any vehicle v_i. Figure 10.4 shows the distributions of the deviation of visiting time of vehicles in the *Roma* and *SanF* traces. The high deviations of most vehicles indicate that these vehicles' trajectories are unevenly distributed among road sections, and they frequently visit a few road sections.

Next, we measure the percentage of time of vehicles spent on their active sub-areas. Figure 10.5 shows the distribution of the percentage of time of vehicles spent on active area. As we can see, most vehicles spent more than 90% of time on their active sub-areas. Also, as shown in Figure 10.6, our measurement shows that usually the total size of active sub-areas of each vehicle is smaller than 10% of the size of the entire VDTN area. From Figures 10.4 and 10.5, we conclude our first observation (**O1**) as follows:

O1: *Each vehicle has its own active sub-areas which are usually very small comparing to the entire VDTN map.*

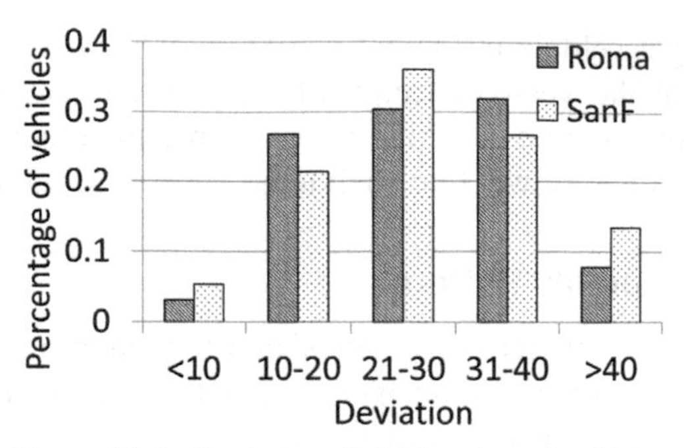

Figure 10.4 Deviation of visiting time of vehicles.

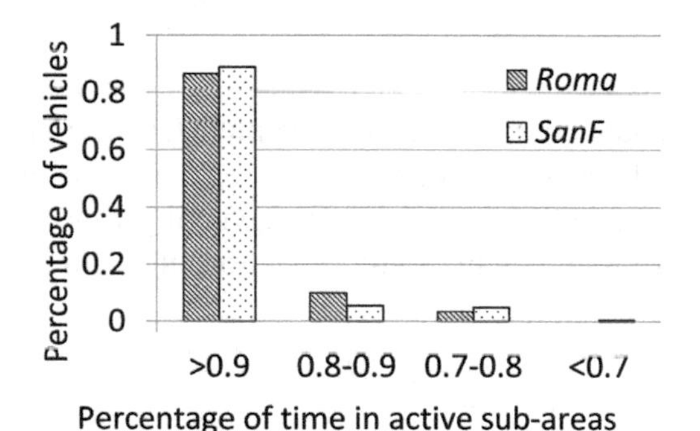

Figure 10.5 Percentage of time spent on active sub-areas.

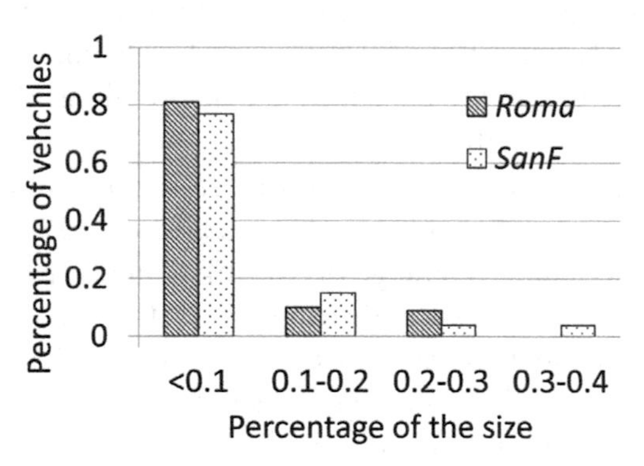

Figure 10.6 Percentage of size of the entire map.

Based on this observation, we can constrain the areas of searching the target vehicle to its active sub-areas. Then, routing of packets on inactive areas of the target vehicle can be avoided and the routing efficiency can be improved.

10.4.2 Relationship between Contact and Location

First, we define a pair of vehicles as frequently encountered pair of vehicles if they encounter more than 10 times. Then, for any frequently encountered pair of vehicles v_i and v_j that frequently meet each other, we calculate the average distance $\overline{d(v_i, v_j)}$ by

$$\overline{d(v_i, v_j)} = \frac{\sum_{i=1}^{n} d_i(v_i, v_j)}{|n|} \tag{10.4}$$

where $d_i(v_i, v_j)$ is the shortest distance between the ith encounter location and the shared active sub-areas of vehicles v_i and v_j, and n is the number of encounters happened between these two vehicles. Figure 10.7 shows the distribution of the average distances of pairs of vehicles that encountered frequently in the *Roma* and *SanF* traces. We find that for most pairs of vehicles, their encounter locations are nearby their active sub-areas. Actually, most encounters are happened in their active sub-areas (i.e., encounter locations with average distance 0). Therefore, we conclude our second observation (**O2**) as follows:

O2: *The frequently encountered vehicles usually encounter each other in their active sub-areas.*

Based on this observation, we can use contact-based routing in each active sub-area of the target vehicle rather than the entire VDTN map, which will greatly improve the routing efficiency.

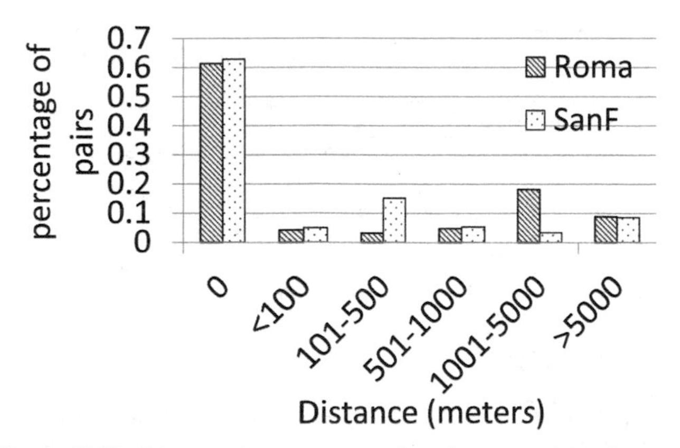

Figure 10.7 Distance from encounter locations to active sub-areas.

10.5 Active Area-based Routing Method

Before introducing the detailed design of AAR, we first give an overview of the routing process for a packet in AAR. The active sub-areas of each vehicle and the entire VDTN map are stored in each vehicle in the VDTN statically. When a vehicle joins the VDTN, it receives this information.

1. In the traffic-considered shortest path spreading algorithm, the source vehicle spreads different copies of the packet to the target vehicles' active sub-areas through paths with short distance and more traffic, as shown in the left part of Figure 10.8. Different from current location-based routing algorithms, we identify the spreading paths with the consideration of not only the physical distance of the paths but also the traffic condition in order to have enough relay vehicle candidates in spreading, which improves the spreading efficiency.
2. In the contact-based scanning algorithm, a packet copy in an active sub-area continually scans the sub-area until it encounters the target vehicle or a vehicle that can encounter the target vehicle more frequently as a relay vehicle, as shown in the right part of Figure 10.8. Since the scanning focuses on the active sub-areas of the target vehicle that it visits frequently and also encounters its frequently encountered vehicles, the routing efficiency is improved.

In the following, we introduce these two algorithms. As the work in [19], we assume that each road intersection is installed with a road unit. The road unit can send information to and receive information from nearby vehicles

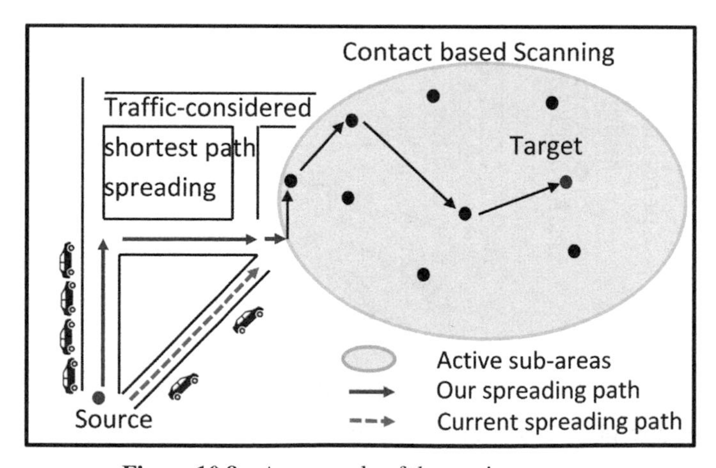

Figure 10.8 An example of the routing process.

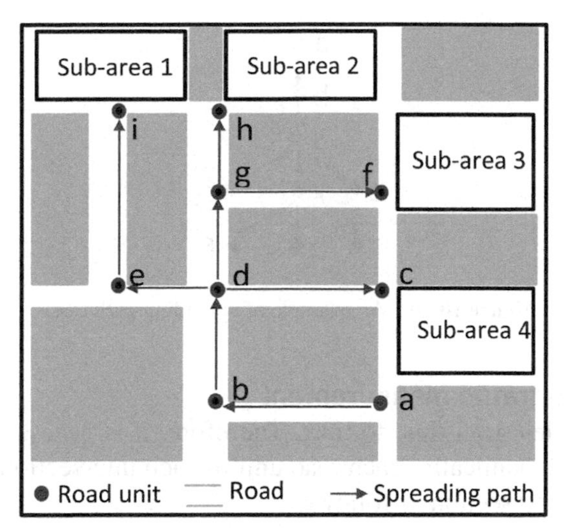

Figure 10.9 The shortest path to different sub-areas.

and store information. The road units help to calculate traffic and receive and forward packets.

10.5.1 Traffic-Considered Shortest Path Spreading

To spread the copies of a packet to different active sub-areas of the target vehicle, as we indicated previously, if we directly calculate the shortest path only based on the distance, the identified path may have few vehicles to function as relays, which leads to low routing efficiency. Therefore, our traffic-considered shortest path spreading algorithm jointly considers distance and traffic in selecting a spreading path. To spread the multiple packet copies, the source vehicle can send a copy to each sub-area individually, which however generates high overhead. To handle this problem, our spreading algorithm builds the spreading path tree that combines common paths of different copies in copy spreading. For example, Figure 10.10 shows an example of such a spreading path tree to spread packet copies to active sub-areas as shown in Figure 10.10, where letters $a - i$ represent the road units. Then, the copies of a packet are relayed to their responsible active sub-areas one road unit by one road unit through vehicles. Each source vehicle needs the traffic information in determining the spreading path. Below, we first introduce how the traffic is calculated and dynamically updated in each vehicle in Section 10.5.1.1. We then introduce the details of our spreading algorithm in Section 10.5.2.

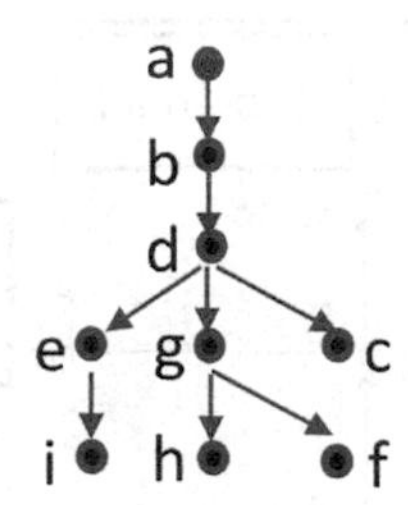

Figure 10.10 An example of spreading path tree.

10.5.1.1 Road traffic measurement

Road traffic varies from time to time. Therefore, it is necessary to measure the road traffic dynamically. Each road unit in each intersection measures the traffic of each road section as follows:

1. When a vehicle v passes intersection a, vehicle v sends ID of the previous road unit it passed to road unit in intersection a (denoted by u_a).
2. Road unit u_a periodically updates the traffic in road section ab by

$$T_{ba}^{t} = \alpha T_{ba}^{t-1} + (1 - \alpha) N_{ba}^{t} \tag{10.5}$$

where T_{ba}^{t} is the traffic from intersection b to a at time t and N_{ba}^{t} is the number of vehicles that have pass through intersection b to a during the time period.

Also, each vehicle updates its road traffic information via two ways as follows:

1. When a vehicle v_a passes road unit u_a, road unit u_a sends its stored traffic information to vehicle v.
2. When vehicles v_a and v_b encounter each other, they exchange their stored traffic information. Then, the vehicles compare and update the traffic information of each road section with updated information.

10.5.1.2 Building traffic-considered shortest path tree

Based on the traffic information, we introduce our algorithm for each vehicle to build the traffic-considered shortest path tree to spread packet copies to different active sub-areas.

1. First, we introduce a metric called *traffic-considered distance* that jointly considers the distance and traffic of a road section:

$$D_{ba} = \frac{d_{ba}}{T_{ba}} \tag{10.6}$$

where T_{ba} is the updated traffic from intersection b to a, d_{ba} is the physical distance length between b and a, and D_{ba} is traffic-considered distance from b to a. It is not necessary that $D_{ba} = D_{ab}$. Using this metric in selecting spreading path, we can find path with shorter distance and higher traffic (i.e., more relay candidates), which can improve the routing efficiency.

2. Recall that an active sub-area of a target vehicle consists of several connected road sections. In order to successfully send a packet copy to a sub-area, a source vehicle must send the copy to an intersection of one road section in the sub-area. To find the path with minimum traffic-considered distance for a sub-area, the source vehicle first builds a graph, in which the nodes are the intersection it will pass and all the intersections of the road sections in the sub-area, two nodes are connected if their corresponding intersections are connected by a road section, and the weight of each edge equals the traffic-considered distance. It then finds the shortest path to each road intersection using the Dijkstra algorithm. Among these paths, it further picks up the shortest path as the shortest path to the sub-area.
3. After the source vehicle calculates the shortest paths to all the active sub-areas of the target vehicle, it combines the common paths in these shortest paths to build the spreading path tree. For example, paths $a \rightarrow b \rightarrow d \rightarrow c$, $a \rightarrow b \rightarrow d \rightarrow e$, $a \rightarrow b \rightarrow d \rightarrow g \rightarrow h$, $a \rightarrow b \rightarrow d \rightarrow g \rightarrow f$, and $a \rightarrow b \rightarrow d \rightarrow e \rightarrow i$ are combined to a tree by merging the same road intersections on different paths (as shown in Figures 10.9 and 10.10), so that copies can be spread efficiently.

When the source vehicle arrives at the next road unit, i.e., u_a, it drops the packet to u_a. When a vehicle traveling to road unit u_b passes u_a, u_a sends a packet copy to the vehicle, which will drop the packet to u_b. u_b will send a copy to u_d through a vehicle traveling to u_d. Then, u_d sends packet copies to three vehicles traveling to u_e, u_g, and u_c, respectively. This process repeats until all road units in the spreading path tree receive a packet copy.

10.5.2 Contact-based Scanning in Each Active Sub-area

After a packet copy arrives at an active sub-area of the target vehicle, the packet carriers (i.e., road units and vehicles) use the contact-based scanning algorithm in the sub-area to forward the packet to the target vehicle. As in current contact-based routing algorithms, each vehicle records its contact frequency to others and exchange such information upon entering. Therefore, a packet carrier can judge whether its encountered vehicle is a better packet carrier, i.e., has a higher encounter frequency with the target vehicle. In our contact-based scanning algorithm, the packet is being forwarded to vehicles traveling in different road sections in order to evenly scan the sub-area to meet the target vehicle. During the scanning process, if the packet carrier meets a vehicle which is a better packet carrier or can lead to more even scanning, it forwards the packet to this entered vehicle. Once a packet carrier is about to leave the active sub-area, it drops the copy to the boundary road unit of the sub-area, which will forward the packet to the vehicle whose traveling direction is the road section that should be scanned.

10.5.2.1 Maintaining scanning history table

In order to ensure that the entire active sub-area can be scanned by a packet, each packet maintains a scanning history table. The scanning history table records the scanning history of the packet. For example, as shown in Table 10.1, each road section in the sub-area has a time stamp which is its last scanning time. A road unit chooses the road section that has the oldest time stamp among the reachable road sections as the next scanning road section. For example, as shown in Figure 10.11, from intersection c, the road sections that can be scanned are road sections ac, cd, and cf, while road section ef cannot be scanned. Since road section cd has the oldest time stamp, it is the next scanning road section. Once a packet finishes scanning a road section, the time stamp of this road section in its scanning history table is updated with the current time.

10.5.2.2 Routing algorithm in a sub-area

We adopt the method in [1] to measure the encounter frequency of each pair of vehicles. Specifically, the contact utility is calculated every time when once two vehicles encounter by

$$C(v_i, v_j) = C_{old}(v_i, v_j) + (1 - C_{old}(v_i, v_j)) \times C_{init}(v_i, v_j) \tag{10.7}$$

Table 10.1 An example of active sub-area information table

Road Section	Time Stamp	Road Section	Time Stamp
ac	1:01	fg	1:14
bd	1:12	fh	1:03
cd	0:00	gi	1:09
cf	1:02	hi	1:06
dg	1:13	hj	1:05
ef	1:16	ik	1:08

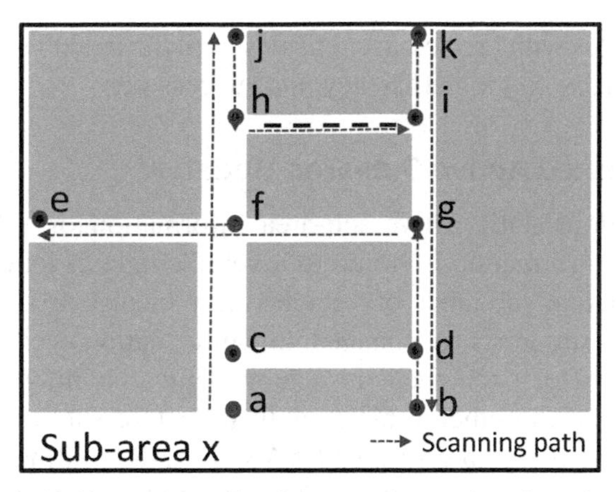

Figure 10.11 An example of the scanning road section selection.

where $C(v_i, v_j)$ is the updated encounter frequency utility; $C_{old}(v_i, v_j)$ is the old encounter frequency utility; and $C_{init}(v_i, v_j)$ is the initial value of contact utility of all the pairs of vehicles, which is set to a value selected from $(0, 1)$. This definition ensures that the two vehicles with a high encounter frequency have a larger encounter frequency utility.

Below, we explain the contact-based scanning algorithm. Recall that the traffic-considered shortest path algorithm sends a packet copy to a road unit in an active sub-area of the target vehicle. Then, the contact-based scanning algorithm is executed. First, the road unit determines the road section that the packet should scan, which is the road section that has the oldest scan time stamp among the reachable road sections, as explained previously. Then, the road unit will forward the packet to the passing vehicle, say v_i, with the direction to the selected road section. When v_i travels along the road section, for each of its encountered vehicle v_j, if v_j has a higher contact utility to the target vehicle or v_j's direction has a smaller time stamp, then v_i's direction in the scanning history table of the packet among all the reached road sections, v_i forwards the packet to v_j. After a vehicle finishes scanning a road section, it drops the packet to the road unit on the intersection in the end of this road section. If a vehicle is leaving the active sub-area, it also drops the packet to the boundary road unit. Then, the road unit decides the next scanning road section and the process repeats until the packet meets the target vehicle.

As shown in Section 10.4, the target vehicle spends most of traveling time in its active sub-areas and also it meets its frequently encountered vehicle's active sub-areas. Therefore, scanning the target vehicle's active sub-areas and relying on vehicles with high contact utilities with the target vehicle in routing can greatly improve routing efficiency and success rate.

10.5.3 Distributed Active Sub-area Updates

In AAR, we adopt static active sub-area information of vehicles for the routing and the information is stored to a vehicle once it joins the network. However, the active sub-areas of vehicles may change from time to time. Once a vehicle's sub-areas are changed, this information needs to be updated in the system. Also, once there is a new vehicle joining the VDTN, it needs to notify all the other vehicles with its active sub-area information. In this section, we design a distributed AAR (DAAR) routing method which enables vehicles to dynamically update the information of active sub-areas of vehicles in a distributed manner. In this method, each node maintains a table recording the sub-area information of all other vehicles. Once two vehicles meet each other, they update the tables. Therefore, instead of adopting the static information from the beginning when vehicles participate in the network, vehicles can dynamically maintain and update the active sub-areas information of other vehicles they contacted frequently from time to time distributedly. We introduce the detailed process for building and maintaining the active sub-area information table as follows.

10.5.3.1 Building the active sub-area information table

In order to share the information of the dynamic active sub-areas of vehicles in a distributed manner, in the DAAR method, each vehicle maintains an active sub-area information table by itself. For example, as shown in Table 10.2, the active sub-area information table stores the active road sections, visited frequencies, and updating times of each corresponding vehicle. Based on the visiting frequency stored in its active sub-area information table, a source vehicle can select a number of top frequently visited active sub-areas of the target vehicle to send packets. For example, if a packet is allowed to have three copies in routing, then the source vehicle selects the active sub-areas with top three visiting frequencies.

Table 10.2 An example of active sub-area information table

Location	Visited Frequency	Updated Time
Road a	1	1:00
Road b	1	5:00
Road c	2	2:00
Road d	1	14:00

10.5.3.2 Maintaining the active sub-area information table

After vehicle v_i has updated its own active sub-area information, it needs to notify other vehicles about this update. Below, we introduce how another vehicle, say vehicle v_j, builds and maintains the active sub-area information of vehicle v_i.

1. When vehicle v_i sends a routing packet to target vehicle v_j, vehicle v_i piggybacks its active sub-area information on the packet to vehicle v_j.
2. Once vehicle v_j receives the packet and the active sub-area information from vehicle v_i, vehicle v_j updates the active sub-area information of vehicle v_i by the new active sub-area information and changes the corresponding updating time to the current time stamp.

Besides obtaining the active sub-area information of vehicle v_i from vehicle v_i itself, vehicle v_j can also update the active sub-area information of vehicle v_i from other vehicles as follows.

1. Once vehicle v_j encounters a vehicle, say vehicle v_k. If v_k is v_i, go to Step (4); otherwise, go to Step (2).
2. Vehicle v_j checks the corresponding updating time of vehicle v_i's active sub-area information in the active sub-areas information table of vehicle v_k. If the updating time is later than the updating time of vehicle v_i in v_j's active sub-areas information table, go to Step (3); otherwise, go to Step (4).
3. Vehicle v_j updates vehicle v_i's active sub-area information and the corresponding updating time by the information in the active sub-areas information table of vehicle v_k.
4. Vehicle v_j updates the active sub-area information of vehicle v_i by the new active sub-area information and changes the corresponding updating time to the current time stamp.

Based on the active sub-area information table, we can dynamically update the active sub-area information. At the same time, vehicles' frequently contacted active sub-areas' information is more likely to be updated.

10.6 Performance Evaluation

In order to evaluate the performance of AAR, we conduct the trace-driven experiments on both the *Roma* and *SanF* traces in comparison with DTN-FLOW [11], PeopleRank [5], and PROPHET [1] algorithms. DTN-FLOW represents location-based routing algorithms, PeopleRank represents centrality-based routing algorithms, and PROPHET represents contact-based routing algorithms. The details of the algorithms are introduced in Section 10.2. We measure the following metrics:

1. *Success rate:* The percentage of packets that successfully arrive at their destination vehicles.
2. *Average delay:* The average time per packet for successfully delivered packets to reach their destination vehicles.
3. *Average number of hops:* The average number of hops per packet routing for the packets successfully delivered to their destination vehicles.

In our experiments, the number of active sub-areas of the target vehicle depends on the number of multiple copies of the packet. To be more specific, we spread one copy of a packet to each active sub-area and, therefore, the number of active sub-areas equals the number of multiple copies.

10.6.1 Performance with Different Number of Copies

Since our algorithm is designed for multi-copy routing, we compare AAR with the other three algorithms with multiple copies of each packet replicated by the spray and wait multi-copy routing algorithm [16] for fair comparisons. Figures 10.12(a) and 10.13(a) show the success rates with different numbers of copies per packet in the *Roma* and *SanF* traces, respectively. Generally, the success rate of AAR is higher than the success rate of DTN-FLOW, the success rate of DTN-FLOW is higher than the success rate of PeopleRank, and the success rate of PeopleRank is higher than the success rate of PROPHET. The performance of DTN-FLOW is better than PeopleRank since DTN-FLOW divides the very large area to sub-areas and avoid to search the target vehicles on a very large area. AAR performs better than DTN-FLOW since AAR considers the encounter history. PROPHET performs the worst, since it is difficult to encounter a vehicle that encounters the target vehicle frequently

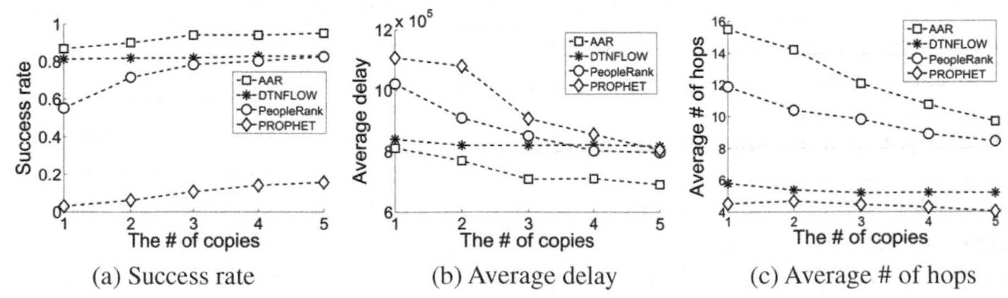

(a) Success rate (b) Average delay (c) Average # of hops

Figure 10.12 The performance with different number of copies on *Roma* trace.

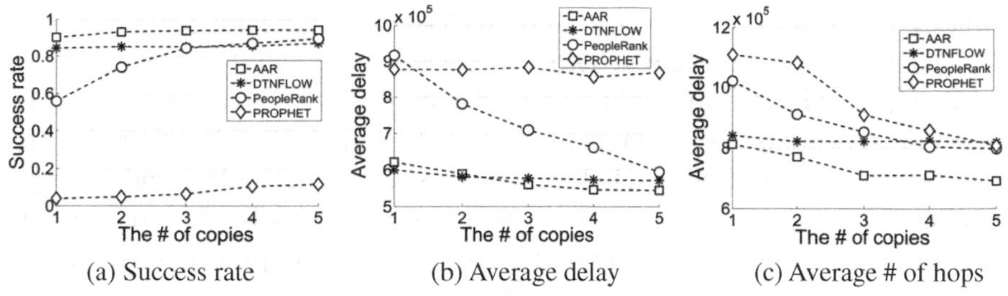

(a) Success rate (b) Average delay (c) Average # of hops

Figure 10.13 The performance with different number of copies on *SanF* trace.

in the very large area. Figures 10.12(b) and 10.13(b) show the average delays with different numbers of copies per packet. Generally, the average delay of PROPHET is higher than the average delay of PeopleRank, the average delay of PeopleRank is higher than the average delay of DTN-FLOW, and the average delay of DTN-FLOW is higher than the average delay of AAR. The delay of PROPHET is the largest, since the copies of a packet waste most time outside of active sub-areas where target vehicle barely visits brought by relay vehicles, as shown in the left part of Figure 10.1. The delay of DTN-FLOW is smaller than PROPHET since DTN-FLOW limits the routing paths in certain sub-areas. The delay of AAR is the smallest, since AAR not only spreads each copy to its responsible active sub-area efficiently by traffic-considered paths, but also scans different active sub-areas with the help of vehicles that encounter target vehicles frequently simultaneously.

Figures 10.12(c) and 10.13(c) show the average number of hops with different numbers of copies per packet. Generally, the average number of hops of AAR is larger than the average number of hops of PeopleRank, the average number of hops of PeopleRank is larger than the average number of hops of DTN-FLOW, and the average number of hops of DTN-FLOW is higher than the average number of hops of PROPHET. The number of hops of PROPHET is the smallest since the packets are directly forwarded to the vehicles with high probability to encounter the target vehicles. However, PROPHET has very low success rate due to such a routing strategy. The number of hops of DTNFLOW is also very small since the packets wait in the landmarks most of the time. The number of hops of PeopleRank is larger than PROPHET and DTNFLOW since the packets are forwarded only by the PeopleRank value without any reachability information to different vehicles. AAR uses the most number of hops than other methods since AAR keeps scanning vehicles' active sub-areas by different vehicles' one road section by one road section.

Although AAR uses most number of hops, AAR has highest success rates and shortest average delays than PROPHET, DTNFLOW, and PeopleRank.

Then, we analyze the influence of the number of copies per packet to different algorithms. As shown in Figures 10.12 and 10.13, when there is only 1 copy, the performance (include success rate, average delay, and average number of hops) of AAR is a little worse than PeopleRank and SimBet, since AAR is designed for multi-copy only and each copy can search in its community only before it encounters the destination community. However, when the number of copies is slightly increased, the performance of AAR is improved significantly and exceeds the other three algorithms. This is because our weak tie multi-copy-based routing algorithm carefully allocates the different copies and fully utilizes each of the copies.

10.6.2 Performance with Different Memory Sizes

Besides the number of copies per packet, the memory size of each vehicle also influences the performance. Therefore, we analyze the influence of memory size to different algorithms. Figures 10.14 and 10.15 show the success rates, average delays, and average numbers of hops with different memory sizes, where we suppose that 1 unit memory (horizontal axis) can store 1 packet.

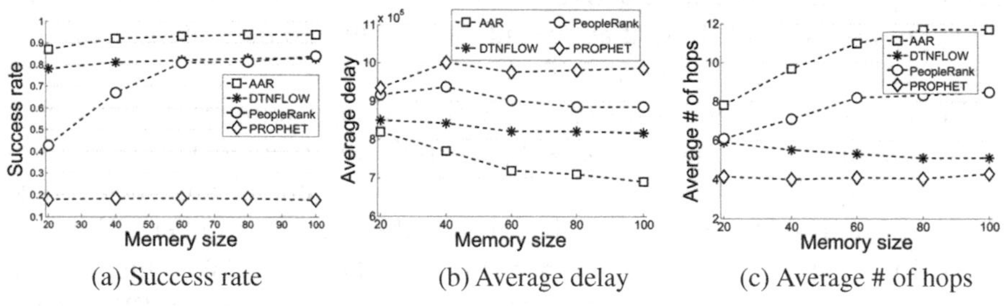

(a) Success rate (b) Average delay (c) Average # of hops

Figure 10.14 The performance with different memory sizes on *Roma* trace.

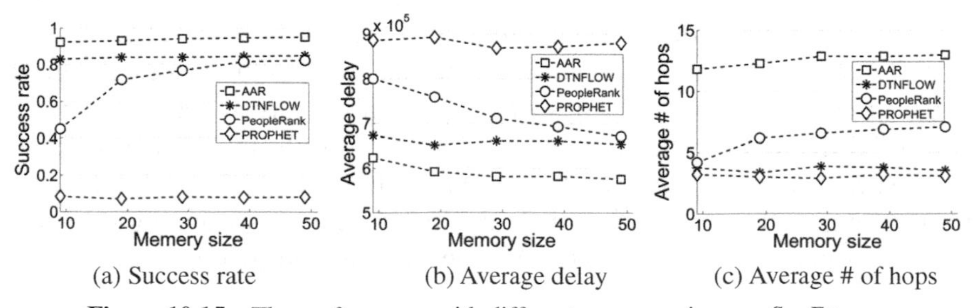

(a) Success rate (b) Average delay (c) Average # of hops

Figure 10.15 The performance with different memory sizes on *SanF* trace.

Generally, PeopleRank is more sensitive to the memory size than AAR, AAR is more sensitive to the memory size than DTN-FLOW, and DTN-FLOW is more sensitive to the memory size than PROPHET. The performance of PeopleRank is very sensitive to the memory size, since all the packets tend to be forwarded to few vehicles with very high PeopleRank values and the limited memory size can significantly influence the routing process negatively. PROPHET is insensitive to the memory size, since the packets only tend to find those specific vehicles with high probability to encounter the target vehicles, which guarantees load balance. However, PROPHET generates low success rate and long delay due to the reasons we mentioned in Section 10.6.1. DTNFLOW is also not sensitive to the memory size since each packet is relayed in limited times from one landmark to another landmark. The performance success rate and average delay of AAR are slightly improved with the increasing memory size since a larger memory size allows packets to scan sub-areas more frequently.

To sum up, AAR has the highest success rate and the lowest average delay. However, AAR is a little sensitive to the number of copies and the memory size. DTNFLOW and PeopleRank have the medium success rate and average delay. However, PeopleRank is very sensitive to the number of copies and the memory size. DTNFLOW and PROPHET are not sensitive to the number of copies and the memory size. However, PROPHET has very low success rate and high average delay. To sum up, considering memory size and limited number of copies is not a main concern in VDTN routing, and AAR performs best in the four routing algorithms.

10.6.3 Performance of Distributed AAR (DAAR)

In this section, we compare AAR and DAAR with different number of copies per packet and different memory sizes of each vehicle. In DAAR, we first calculate the initial active sub-areas information of each vehicle using the first half of data in the data trace. Later on, each time when a vehicle's active sub-areas information is requested by an encountering vehicle or it generates a new packet, it updates its active sub-areas information by counting its recently visited sub-areas.

Figures 10.16 and 10.17 show the success rates, average delays, and average numbers of hops with different number of copies per packet. Generally, the success rate of DAAR is higher than the success rate of AAR, the average delay of DAAR is lower than the average delay of AAR, and the average number of hops of DAAR is smaller than the average number of hops of AAR.

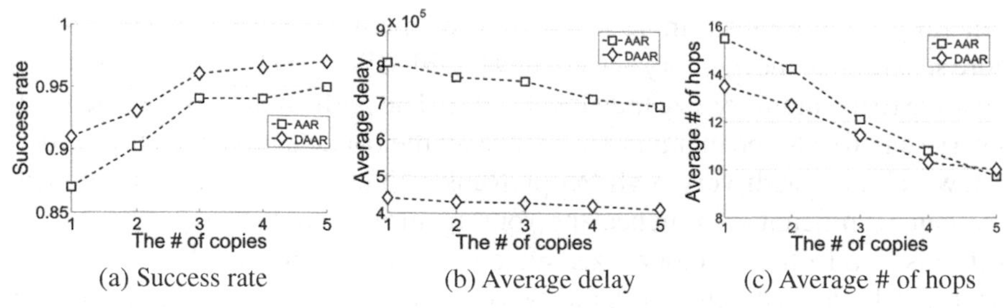

(a) Success rate (b) Average delay (c) Average # of hops

Figure 10.16 Performance comparison between AAR and DAAR with different number of copies on *Roma* trace.

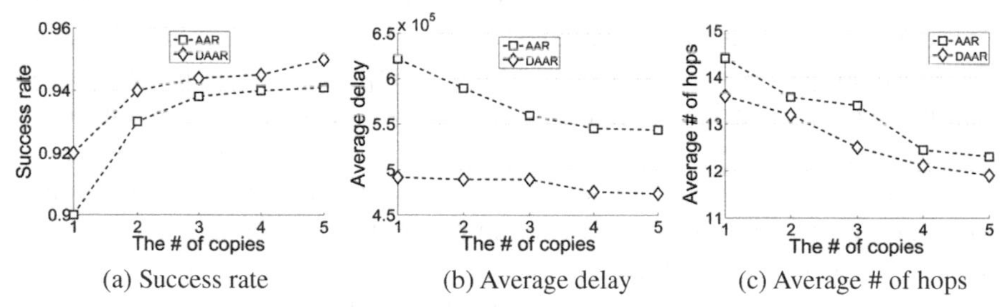

(a) Success rate (b) Average delay (c) Average # of hops

Figure 10.17 Performance comparison between AAR and DAAR with different number of copies per packet on *SanF* trace.

The improvement of performance success rate is not so significant since AAR already has a very good performance on success rate comparing to other algorithms. AAR generates relatively high average delay and much higher numbers of hops due to more relays in scanning. In DAAR, since the packet copy scans on the road unit which are most likely to be visited by the target vehicle at the current time for the most time during the routing period, the number of relays in scanning is reduced, and hence, the performance on average delay and the average number of hops is improved.

Figures 10.18 and 10.19 show the success rates, average delays, and the average numbers of hops with different memory sizes of each vehicle. When a node's memory is full, it drops some packets. We can have the same observations as the previous figures due to the same reasons. Also, AAR is more sensitive to the memory size than DAAR; that is, AAR increases or decreases faster than DAAR as the memory size increases generally. This is because the active sub-areas' information of DAAR can more accurately reflect the sub-areas where the target vehicle is active currently since the active sub-areas' information of DAAR is updated from time to time.

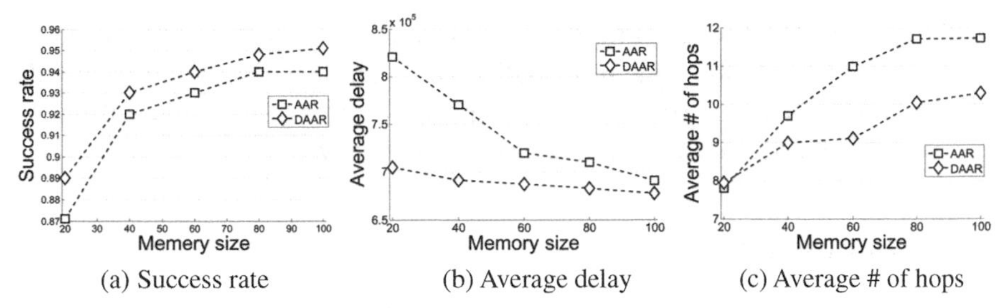

(a) Success rate (b) Average delay (c) Average # of hops

Figure 10.18 Performance comparison between AAR and DAAR with different memory sizes on *Roma* trace.

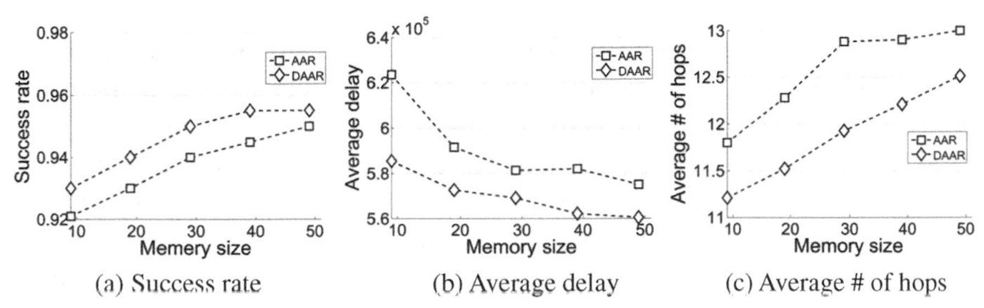

(a) Success rate (b) Average delay (c) Average # of hops

Figure 10.19 Performance comparison between AAR and DAAR with different memory sizes on *SanF* trace.

Therefore, DAAR can use fewer copies of a packet for the delivery and the failure probability of each copy is decreased. Therefore, even though some copies are lost due to the small memory size, the performances are not significantly influenced since fewer copies are needed for the delivery. On the contrary, since the active sub-areas' information of AAR is not updated, the active sub-areas' information of AAR can only reflect the general active sub-areas of vehicles in a long term rather than the exact active sub-areas where vehicles are active currently. Therefore, AAR needs more memory to spread enough copies of packets to each possible active sub-areas in order to find the sub-area where the target vehicle is active currently. Hence, the probability of failure of each copy is increased and the loss of copies may decrease the probability to reach the current active sub-areas of the target vehicle. The performances of success rate and average delay of both AAR and DAAR are improved with the increasing memory size since a larger memory size allows packets to scan sub-areas more frequently. The average numbers of hops of both AAR and DAAR increase with the increasing memory size since a larger memory size can lead to more times of relays.

To sum up, DAAR has a higher success rate and a lower average delay compared with AAR since the information of dynamic active sub-areas tends to be updated from time to time, while the information of static active sub-areas may be outdated due to the change of vehicles' behaviors.

10.7 Conclusion

In this chapter, we first measured the pattern of vehicles' mobility and the relationship between contact and location for each pair of vehicles. Then, by taking advantage of the observations, we proposed active area-based routing (AAR) method. Instead of pursuing the target vehicle on the entire VDTN area, AAR spreads copies of a packet to the active sub-areas of the target vehicle where it visits frequently and restricts each copy in its responsible sub-area to search the target vehicle based on contact frequency. Furthermore, we proposed a distributed AAR (DAAR) to improve the performance of AAR. The trace-driven simulation demonstrates that AAR has the highest success rate and lowest average delay in comparison with other algorithms. Also, DAAR has the higher success rate and the lower average delay compared with AAR since information of dynamic active sub-areas tends to be updated from time to time, while the information of static active sub-areas may be outdated due to the change of vehicles' behaviors. In our future work, we will discuss the possibility of routing in VDTNs without the help of road units.

Acknowledgments

This research was supported in part by U.S. NSF grants NSF-1404981, IIS-1354123, CNS-1254006, CNS-1249603, and Microsoft Research Faculty Fellowship 8300751.

References

[1] Lindgren, A., Doria, A., and Scheln, O. (2003). Probabilistic routing in intermittently connected networks. *Mobile Comput. Commun. Rev.*, 7, 19–20.

[2] Burgess, J., Gallagher, B., Jensen, D., and Levine, B. N. (2006). Maxprop: routing for vehicle-based disruption-tolerant networks. In *Proceedings of the INFOCOM*, IEEE.

[3] Symington, A., and Trigoni, N. (2012). Encounter based sensor tracking. In *Proceedings of the MobiHoc*, ACM, pp. 15–24.
[4] Zhu, H., Chang, S., Li, M., Naik, K., and Shen, S. X. (2011). Exploiting temporal dependency for opportunistic forwarding in urban vehicular networks. In *Proceedings of INFOCOM*, IEEE.
[5] Mtibaa, A., May, M., Diot, C., and Ammar, M. H. (2010). Peoplerank: Social opportunistic forwarding. In *Proceedings of the INFOCOM*, IEEE, pp. 111–115.
[6] Daly, E. M., and Haahr, M. (2007). Social network analysis for routing in disconnected delay-tolerant manets. In *Proceedings of MobiHoc*, pp. 32–40, ACM.
[7] Hui, P., Crowcroft, J., and Yoneki, E. (2011). Bubble rap: Social-based forwarding in delay-tolerant networks. *IEEE Trans. Mob. Comput.*, 10 (11), 1576–1589.
[8] Leontiadis, I., and Mascolo, C. (2007). Geopps: Geographical opportunistic routing for vehicular networks. In *Proceedings of WOWMOM*, IEEE, pp. 1–6.
[9] Link, J. g. B., Schmitz, D., and Wehrle, K. (2011). Geodtn: Geographic routing in disruption tolerant networks. In *Proceedings of GLOBECOM*, IEEE, pp. 1–5.
[10] Y. Wu, Y. Zhu, and B. L. 0001, “Trajectory improves data delivery in vehicular networks.,” in *Proc. of Infocom*, pp. 2183–2191, IEEE, 2011.
[11] K. Chen and H. Shen, “Dtn-flow: Inter-landmark data flow for high-throughput routing in dtns,” in *Proc. of IPDPS*, pp. 726–737, IEEE, 2013.
[12] B. Lorenzo, B. Marco, L. Pierpaolo, B. Giuseppe, A. Raul, and R. Antonello, “CRAWDAD data set roma/taxi (v. 2014-07-17).” Downloaded from http://crawdad.org/roma/taxi/, July 2014.
[13] Piorkowski, M., Sarafijanovic-Djukic, N., and Grossglauser, M. (2009). CRAWDAD data set epfl/mobility (v. 2009-02-24). Downloaded from http://crawdad.org/epfl/mobility/, Feb. 2009.
[14] Wu, B., Shen, H., and Chen, K. (2015). Exploiting active sub-areas for multi-copy routing in vdtns. In *Proceedings of ICCCN*, IEEE, pp. 1–8.
[15] Hossmann, T., Spyropoulos, T., and Legendre, F. (2010). Know thy neighbor: Towards optimal mapping of contacts to social graphs for DTN routing. In *Proceedings of INFOCOM*, IEEE. pp. 866–874.
[16] Spyropoulos, T., Psounis, K., and Raghavendra, C. S. (2008). Efficient routing in intermittently connected mobile networks: the multiple-copy case. *IEEE/ACM Trans. Netw*, 16 (1), 77–90.

[17] Hong, B., and Haizheng, Y. (2010). An efficient control method of multi-copy routing in DTN. In *Proceedings of NSWCTC*, IEEE. pp. 153–156.

[18] Uddin, M. Y. S., Ahmadi, H., Abdelzaher, T. F., and Kravets, R. (2009). A low-energy, multi-copy inter-contact routing protocol for disaster response networks. In *Proceedings of SECON*, IEEE pp. 1–9.

[19] Song, C., Liu, M., Wen, Y., Cao, J., and Chen, G. (2011). Buffer and switch: An efficient road-to-road routing scheme for vanets. In *Proceedings of MSN*, IEEE. pp. 310–317.

11

RobustGeo: A Disruption-Tolerant Geo-Routing Protocol

Ruolin Fan[1], Yu-Ting Yu[2] and Mario Gerla[1]

[1]Department of Computer Science, University of California, Los Angeles, CA, USA
[2]Qualcomm Research, Bridgewater, NJ, USA

Abstract

While geo-routing is a promising routing algorithm in urban vehicular ad-hoc networks (VANETs), there is still much work to be done for it to become truly usable for such environments. One of the biggest obstacles to this goal is the intermittent nature of VANETs due to mobility. Traditional geo-routing algorithms do not perform well in these conditions because they drop packets whenever they cannot find an immediate forwarder for the packet. In this chapter, we propose RobustGeo, a routing protocol that combines the simplicity and efficiency of the greedy forwarding technique in geo-routing algorithms, with the robustness of delay-tolerant networks in the face of disruptions in network connectivity. When there exists a good connection between the source and destination, RobustGeo can route the packet like the traditional geo-routing algorithms, and when the network faces disruptions, RobustGeo relies upon vehicle mobility and packet replication to explore multiple geo-route paths and quickly recover the packet back to greedy forwarding. We show that for a highly intermittent scenario, RobustGeo has a delivery ratio of over 20% (compared to a pure geo-routing protocol's delivery ratio of 0), and it reduces the delay by over 40% as compared to a more pure delay-tolerant routing solution.

Keywords: Delay-tolerant networks, geo-routing, VANET, ad-hoc networks, mobility.

11.1 Introduction

Location-based routing algorithms are a class of routing algorithms that use locations of forwarding nodes to route packets. Unlike the more traditional link-state (LS) or distance-vector (DV) algorithms that rely on each node having some overarching information on the network topology [1], location-based routing generally only requires each node to have information on its immediate neighbors. This is a desirable property to have because it obviates the need for nodes to periodically flood the network with routing-related messages. Because of their scalability in this respect, these algorithms offer a very promising solution to the difficult problem of routing in vehicular ad-hoc networks (VANETs), where the network topology changes rapidly and bandwidth is limited [2].

One of the first and the most well known of these routing protocols is the Greedy Perimeter Stateless Routing (GPSR) protocol [3], which sets the foundation for almost all subsequent location-based routing algorithms. In GPSR, each node in the network keeps a list of its immediate neighbors and their locations. Packets are thus routed based on *greedy forwarding* under normal circumstances, where the forwarding node transmits packets to the neighbor who is the closest geographically to the packets' destination. When no node closer to the destination than the current one can be found, the packet is said to have encountered a *local maximum* and must be routed around the perimeter via *perimeter forwarding*, which requires a careful mapping of the nodes to form planar graphs so that routing loops can be avoided.

While the greedy forwarding method proposed by GPSR offers a great solution to routing in VANETs, the perimeter routing approach as a solution to the local maximum problem is inadequate [4], as the highly mercurial network topology common to VANETs quickly outdates any planar graphs that may have been built from the nodes, much like routing tables in the DV and LS algorithms. GPCR (Greedy Perimeter Coordinator Routing) [5], another location-based routing protocol that focuses more on urban VANETs, takes advantage of the urban layout that naturally forms a planar graph for its *recovery strategy* from a local maximum and proposes the *restricted* greedy forwarding method, which prioritizes forwarding packets to urban cross-streets for better routing decision making.

To the end of avoiding local maximums and getting out of them in urban VANETs, many proposals are made [6–9]. However, all of them make the assumption that despite mobility and the urban landscape, the network is always connected. Unfortunately, this is not the case for VANETs in reality, where **the network is often partitioned or disrupted due to mobility, giving no instant end-to-end routes**. According to [10], network connections in VANETs are highly dependent upon inter-vehicle spacing, which is never constant, and that on freeways network disconnectivity generally heals on the order of seconds. In such situations, conventional geo-routing algorithms will cause large numbers of packets to be dropped. To address this problem, we propose RobustGeo, a location-based routing algorithm that takes the *store and forward* and *replication* approaches of delay-tolerant networks to explore the possibility of multiple paths to overcome such disruptions in VANET connectivity.

Although delay-tolerant networks (DTNs) were originally proposed for networks that experience very frequent and long periods of disconnectivity [11], the DTN methods, with modification, can be used to deal with the shorter intermittencies in geo-routing with great benefits. One such method is the Data MULE (Mobile Ubiquitous LAN Extension) [12] that exploits node mobility to deliver packets. While this used by itself translates to unacceptable delay lengths in more conventional networks, we can, in the same spirit, rely on node mobility to recover from a disconnected scenario by saving the packet and actively looking for greedy forwarding routes at the same time. Another important aspect of DTNs is packet *replication*, as seen in Epidemic Routing [13] and Spray and Wait [14]. The drawback to these methods is mainly in bandwidth overhead when too many packet replications occur. However, by using similar methods while limiting the number of replications, we can increase the packet delivery ratio and minimize delay.

With the combination of traditional geo-routing schemes and DTN approaches, RobustGeo indeed offers a more *robust* solution for geo-routing in urban VANETs. When the network connectivity is good, RobustGeo simply acts like any other state-of-the-art location-based routing algorithm. When the connectivity is intermittent, RobustGeo is not only able to withstand these disruptions and continue forwarding packets when the network heals, but also make use of the mobility of nearby neighboring nodes to even increase the healing rate (in the perspective of the packet stored by the node).

The rest of this chapter is organized as follows. Section 11.3 describes the general design of RobustGeo, while Section 11.4 analyzes the potential effects and consequences of packet replication. In Section 11.5, we evaluate RobustGeo by comparing it with three other geo-routing and DTN approaches and conclude with Section 11.7.

11.2 Background

11.2.1 Location-based Routing Algorithms

Location-based routing algorithms are a class of routing algorithms that uses locations of forwarding nodes to route packets, rather than using the more traditional link-state or distance-vector algorithms. One of the first and the most well known of these routing protocols is Greedy Perimeter Stateless Routing (GPSR) protocol [3]. GPSR improves upon the link-state and distance-vector algorithms because it employs a much more scalable approach that obviates the need for each node to have information about the network over more than a single hop. This means that it is not bogged down as much by the trade-off between having stale routes leading to routing loops, or overwhelming the network with update or status exchange messages.

In GPSR, each node in the network is able to route packets simply by keeping a list of its immediate neighbors and their locations. Under normal circumstances, GPSR routes packets based on what it calls "*greedy* forwarding". With greedy forwarding, a node greedily routes the packet to its neighbor who is the closest geographically to the packet destination. With this approach, the packet is routed hop by hop and any beaconing is done purely in a single hop basis, preventing the network from becoming congested with routing messages.

However, greedy forwarding does not work completely on its own, since situations sometimes arise where a packet arrives at what is called a "local maximum" node, where another closer node cannot be found. Rather, the packet would temporarily need to be routed *further away* from the destination before greedy forwarding can be resumed. GPSR's solution to this problem is perimeter routing, which routes the packet around the perimeter of the void area. To avoid crosslinks that can cause routing loops, a planar graph must first be constructed. The packet can then be routed along the planar graph using right hand rule to the destination. Two ways to construct such a planar graph are the "Relative Neighborhood Graph" (RNG) or "Gabrel Graph" (GG). If the destination is unreachable, the packet would be routed back to the local maximum node and dropped.

While the greedy forwarding method proposed by GPSR offers a great solution to routing in VANETS, the perimeter routing approach to solve the local maximum problem is not much help. This is because in an urban setting with VANETs, mobility causes rapid changes in the network topology that quickly outdates any planar graphs that may have been built from the nodes. Moreover, physical buildings and radio interference common in these environments cause the local maximum phenomenon to occur frequently, especially if pure greedy forwarding was used. As highlighted in [5], a packet may be greedily forwarded to a location exactly at the opposite side of a building from the destination, prompting perimeter routing to be used for further forwarding of the packet. Lochert et al.'s solution to this problem is Greedy Perimeter Coordinator Routing (GPCR), which proposes the idea of *restricted greedy forwarding*, where the packet is greedily forwarded to *junction nodes* rather than as close to the destination as possible. In GPCR, junction nodes are nodes located at the block intersections in the city, which makes better decisions on where the packet should be forwarded next because it can help the packet to take a turning path. GPCR also presents a form of perimeter routing of its own, of which it terms a "recovery strategy". In its perimeter routing, GPCR does not construct a planar graph from scratch as suggested by GPSR, but takes advantage of the urban layout with the junction nodes, which naturally forms a planar graph on which the packet can traverse.

The problem of how to most effectively route packets using positional-based routing techniques in an urban setting is an interesting one, because the urban grid has a more stable topology, and it is very easy for packets to get stuck in a local maxima because of buildings. To this end, many further enhancements were made. References [6, 7] propose to propagate the beacons by an extra hop for nodes to get a better view of the geographic topology and find the junction nodes more efficiently. References [6, 9] both suggest an *anchoring* system where packets are routed along anchored streets, with information from static maps or GPS with live traffic information. When a packet reaches the local maximum, a new anchor is computed from the map and the packet is forwarded along the new street.

All of these contributions are either suggesting ways to avoid the local maximum situation by some new greedy forwarding scheme, or proposing some new way of recovering from a local maximum situation. RobustGeo is compatible with most of the suggested ways of avoiding the local maximum situation, but in terms of recovery, it utilizes a solution in the spirit of the delay-tolerant networks.

11.2.2 Delay-Tolerant Networks

As suggested by its name, delay-tolerant networks are a class of networks that experience very frequent and long periods of disconnectivity [11]. In these networks, delivery rate and power consumption take on a higher priority than the speed of delivery, though higher speeds of dissemination are still preferred if possible. Almost all DTN routing algorithms use some variations of store and forward, where mobile nodes store the packet and physically deliver it to the destination, to other nodes that eventually forwards it to the destination, or to a more connected network that will guarantee delivery.

Data Mobile Ubiquitous LAN Extensions (MULEs) [12] and message ferries [15] are two examples of such algorithms. While they both exploit the higher chances of delivery in mobile nodes, the two approaches differ in that MULEs are *reactive*, whereas ferries are *proactive*. This means that MULEs passes data along to any mobile nodes that comes into the range of the less mobile nodes and hope that it travels to somewhere with connectivity for delivery. Ferries, on the other hand, employ mobile nodes that move in a non-random fashion with the purpose of picking up packets and delivering them to their destinations. Another more extreme approach is epidemic routing [13], where any node that meets any other nodes would pass along the message so that it can eventually reach its destination. While this approach results in a high chance of delivery rate given enough time (eventually 100%), it also introduces a lot of overhead and higher congestion into the system, especially where node storage capacity is an issue. The spray and wait routing algorithm [14] seeks to alleviate such transmission overhead concerns by limiting the overall number of copies of the packet that can be in the network. Spray and wait operates in two phases: in the spray phase, L copies of the message are disseminated into the network, and in the wait phase, each node with a copy of the message move around until it can be delivered to the destination.

As seen from above, almost all DTN routing approaches are replication-based, that is, multiple copies of the message are disseminated in the network to increase the chances of data delivery. They are much more preferred over single copy approaches because of the high frequency and long periods of network disconnectivity, which poses a great obstacle in finding a route that is needed in a single copy routing approach. However, it must be noted at this junction that the scenarios for which these protocols are designed are quite different from the urban VANET scenario. While urban VANETs do experience high frequencies of network partitioning due to environment and mobility, the period for which the network is disconnected is usually short. At the same time, though the message delivery speeds in the proposed VANET scenarios are more tolerant of delays than the Internet, delivery times on the order of hours [11] are unacceptable. In addition, transmission overheads caused by overzealous replication of packets can cause more harm than good in an environment where radio interference itself is a cause for network intermittency. For these reasons, directly using a DTN approach to route in a VANET would generally not work well simply due to tighter delay constraints and network bandwidth availability. However, a hybrid approach that uses ideas from DTN, such as data MULEs in [12] and limited replication of messages in [14], stabilizes a VANET scenario that is otherwise bogged down by frequent interruptions to connectivity.

11.3 Design

RobustGeo combines the quick speed of geo-routing with the robustness of delay-tolerant networking. Under normal circumstances, greedy forwarding is used to quickly route packets to its destination. When, unavoidably, a local maximum situation arises, it can use a good recovery algorithm like the one suggested in [8] to route around the perimeter while at the same time safeguard against complete disconnectivity by employing the delay-tolerant networking routing approach. In this approach, the nodes keep the packets in their disruption tolerant queue (DTQ), where the packets are opportunistically routed as available greedy forwarding neighbors are found. Periodically, nodes broadcast the packets inside their DTQ to their one-hop neighbors to explore other possible paths and increase the chances of finding a geo-routed path toward the destination.

11.3.1 Geo-Routing

The geo-routing component of RobustGeo is compatible with all types of greedy-forwarding mechanism based on the one proposed in [3]. Generally, each node keeps a list of its immediate neighbors and looks for the neighbor geographically closest to the packet destination. Following these neighbors, the packets are forwarded greedily step by step until it reaches its destination. To combat the challenges posed by urban grid layouts and high urban error rates, restrictive forwarding approaches as described in GPCR [5], GpsrJ + [7], and TO-GO [8] can be employed.

Likewise, RobustGeo can utilize any of the previously existing solutions for perimeter forwarding to attempt local maxima recovery.

11.3.2 Disruption Tolerance

The DTN component of RobustGeo exploits the mobile nature of VANETS. In such a highly mobile environment, it is beneficial to not give up on a local maximum node so quickly since topology can evolve quite rapidly. In many cases, a local maximum at one moment in time may very well no longer be a local maximum in the next simply because of mobility. Therefore, unlike conventional geo-routing algorithms, when the perimeter forwarding phase is entered, RobustGeo routes a *replica* of the packet around the perimeter rather than the original packet itself. It then pushes the original packet into the node's DTQ, essentially turning the node into an "intelligent" data MULE [12] that looks for greedy forwarding neighbors and at the same time further packet replication by periodically broadcasting to its one-hop neighbors. In this way, RobustGeo recovers from a local maximum situation by entering DTN mode in addition to perimeter forwarding.

11.3.2.1 Perimeter forwarding with packet replication

When a node encounters a local maximum, it sends out a packet replica for perimeter forwarding, stores the original in its DTQ, and continues to actively try and look for nodes that are closer to the destination than the local maximum. If found, the node has successfully recovered greedy forwarding via DTN means.

Meanwhile, as the replicated packet is routed around the perimeter it records each node it visits as breadcrumbs. If a greedy path is recovered, a receipt can be sent back to the originator using the breadcrumb path so that the originator can remove the packet from its DTQ and abort the broadcasting countdown time. Otherwise, the perimeter-routed packet can continue to be routed by the right-hand rule until either a greedy route is found, or ends up back at the originator or reaches its TTL. In both of the latter cases the packet is discarded. Note that packet replication due to perimeter forwarding happens only once at the original local maximum node. This prevents the packet from replicating out of control as it is forwarded around the perimeter.

If the original packet and its replicas both manage to find a greedy forwarding path, two cases should be considered. In the first case, as illustrated in Figure 11.1, multiple different paths to the destination are found. This arguably improves the scenario, since having multiple paths to the destination causes the intermittent network to be more robust. In the second case (Figure 11.2), both packets are forwarded greedily onto the same path. While on the surface this may seem problematic, it is not a great cause for concern since in the unlikely event that this happens, the receiving node can simply drop one of the duplicates and continue to forward the packet onwards. However, this does mean that the packets should be uniquely identified so that identical packets can be easily discovered. A way to realize this is to use a large sequence number in combination with the sender's identifier (like an IP address).

The decision to replicate the packet when doing perimeter forwarding is due to the fact that in urban scenarios, the local maximum situation often occurs because of complete network partitioning. When this is the case, no amount of perimeter forwarding can bring the packet to its destination. Fortunately, these situations are generally not permanent because of mobile

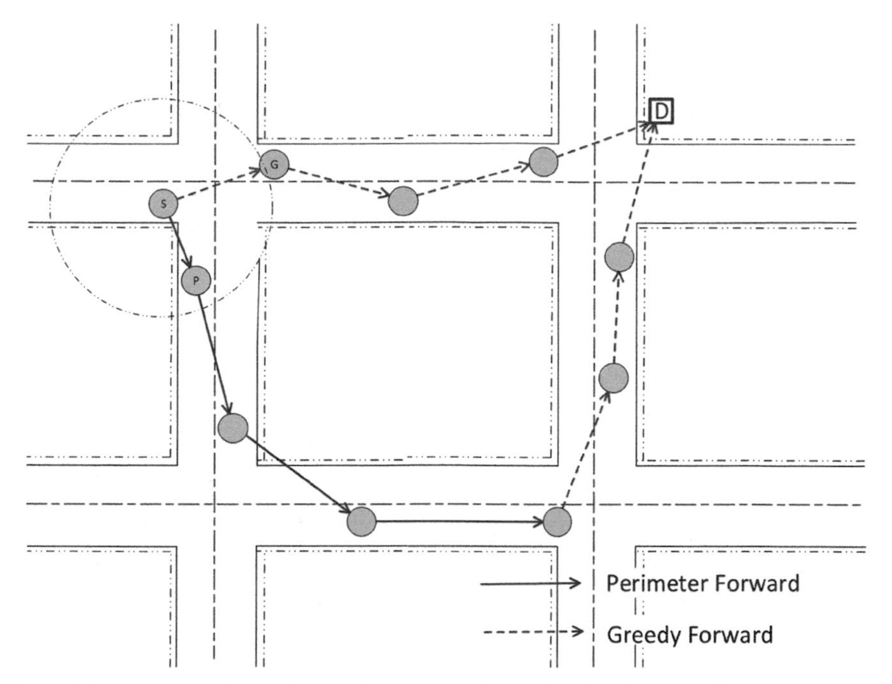

Figure 11.1 Greedy forwarding and perimeter forwarding finding two different paths. The greedy forwarding recipient node G moves into the source node S's range after S has perimeter forwarded the packet to node P.

nature of VANETs. An example of such an occurrence is when there are gaps in vehicular traffic caused by something as simple as a long red light. Traditional perimeter forwarding solutions in these situations will cause the packet to be dropped when it either reaches its TTL or ends up back at its originator [3, 16]. With RobustGeo, however, because the local maximum node keeps the perimeter forwarded packets, connectivity that is reestablished a little later can allow greedy forwarding to resume. Although this will cause the network to experience delays, the receiver is still able to receive the message eventually rather than experiencing a complete disconnectivity.

11.3.2.2 Single-hop broadcasting to explore multiple paths

Packets that are not replicas of any previous broadcasts in the DTQ are scheduled to be one-hop broadcasted periodically to explore multiple paths, which potentially increases the delivery rate and decreases latency. Although the node is still continuing to look for greedy routes for the packet, once the packet was broadcasted, it essentially switches into "full DTN mode", relying more on node mobility to recover the next greedy routing path.

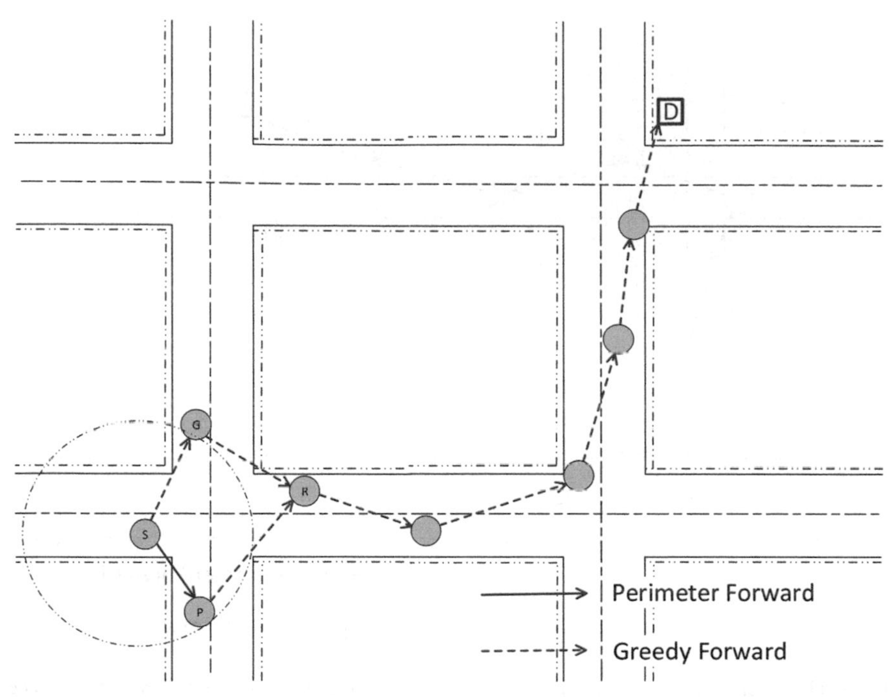

Figure 11.2 Greedy forwarding and perimeter forwarding finding the same path. Once again, the greedy forwarding recipient node G moves into the source node S's range a moment after perimeter routing is done by S. The two identical packets both pass through node R and the latter one is dropped.

When a node receives a broadcasted packet, it first makes sure that it does not already have the packet in its DTQ. Then it adds the packet into its DTQ, where the node can repeatedly attempt to greedy forward it.

The single-hop broadcasting technique is beneficial in situations where node density is sparser, taking advantage of the replication effects in delay-tolerant networking [17]. In RobustGeo, we choose to replicate the packets using single-hop broadcast because it has a high replication-to-transmission ratio in that a packet can reach all the neighboring nodes with just a single transmission.

Since packet replication increases bandwidth usage, it is important to consider the length of the period in between broadcasts. Set too long, the node can lose opportunities to send to neighbors who are good data MULE candidates; set too short, the network would be saturated with replicated packets. With this in mind, we configure the period to be 6 sec, which we believe is short enough to not miss most of the potential neighbors, yet long enough that it does not overload the network (see Section 11.4). We believe that most of the potential neighbors would not be missed because vehicles rarely travel faster than 20 m/s (about 45 mph) in an urban area. Taking incoming traffic into account, the relative speed can be up to 40 m/s. Therefore, 6 sec is equivalent to 240 m, still within most broadcasting ranges. Additionally, we set this as an adjustable parameter depending on needs. For example, if the target scenario is a very sparse network, then it would be beneficial to make this period shorter. In all, due to the relatively long broadcasting period, momentary interruptions in the network will not trigger one-hop broadcasts.

To further limit the number of packet replicas, a packet that was once received via broadcasting can never be broadcasted again. This is easily done by marking a single bit in the packet header and checking that bit whenever a packet is about to be put into broadcast mode. Finally, when the first copy of the packet, be it a replica or the original, reaches its destination, the destination can send out an *active receipt* [18] to clean up all of the packet replicas that are still in the network.

11.3.2.3 Scheduling

When a local maximum situation occurs, the node stores the packet into its DTQ, and packets bound for other destinations with available next-hop nodes continue to be forwarded normally. Every time the node receives a new beacon, be it from any of its existing neighbors or from a new neighbor, the node would check its DTQ to see if any of the packets can now be routed.

The DTQ is similar in function to a normal FIFO queue, in that packets which are enqueued earlier generally get dequeued earlier as well. However, the DTQ is not necessarily dequeued from the head all the time. Whenever a node receives a new beacon from its neighbor, potentially marking a new route, RobustGeo attempts to dequeue from the DTQ by attempting to greedily route to the destination every packet in the queue, beginning with the packet at the head. When a route is found for a packet, that packet is dequeued to be sent out. This speeds up average packet delivery time and keeps the queue length short.

To prevent a large number of packets in the DTQ from suddenly flooding the network, RobustGeo allows only one packet to be sent out for a set period of time. For example, it may be decided that the backlogged packets should only take up about 1 Mbps of the overall bandwidth, so a maximum of 1 packet can be sent out every millisecond. However, since the beaconing period is much longer than that, a dispatch buffer is introduced to manage the send rate. Therefore, the DTQ dequeues all eligible packets into the dispatch buffer up to a maximum of $send\ rate * beaconing\ period$ packets each beaconing period. Then at every $\frac{1}{send\ rate}$ period, RobustGeo sends out a packet from the dispatch buffer.

Packets marked for one-hop broadcasting are not sent out immediately either. Rather, RobustGeo adds the packet into a separate broadcasting queue and continues to look through its DTQ for other packets to be greedily forwarded. Additionally, the packet marked for broadcasting still remains in the DTQ for more future attempts at greedy forwarding recovery. If the recovery attempt is successful, the packet is removed from both the DTQ and the broadcasting queue. On the other hand, if the packet was broadcasted, it is removed from the broadcast queue only but kept in the DTQ, with the broadcasting timer reset to begin a new period of broadcasting.

Having two queues with different priorities, we employ our dispatch buffer to enforce these priorities. The broadcasting packets are never added to the dispatch buffer. Instead, the dequeuing method for the dispatch buffer dequeues a packet from the broadcast queue only if the dispatch buffer is empty. As a result, packets are only broadcasted when there are no more geo-routed packets to be sent out in the beaconing period. Figure 11.3 gives an overall view of the components of RobustGeo working together inside a node upon a packet arrival.

11.4 Analysis

Packet replication is a tricky parameter in ad hoc networks. In a lossy and especially delay-tolerant environment, replication can provide substantial benefits in that it increases the delivery rate and decreases delay [17]. Unfortunately, replication is a double-edged sword, as it also introduces bandwidth overhead into the network. Therefore, as a protocol that stresses replication, it is important to see just how much overhead replication brings into the network.

In RobustGeo, packets are only replicated when a local maximum is reached. When the node is attempting to route the packet via perimeter routing, it introduces a single packet replica. If the recovery time is long, the packet can be replicated many times due to periodic broadcasting. The overall number of packets replicated as the result of a single local maximum is therefore

$$rep_1 = mn \left\lfloor \frac{t}{\pi} \right\rfloor + 1, \tag{11.1}$$

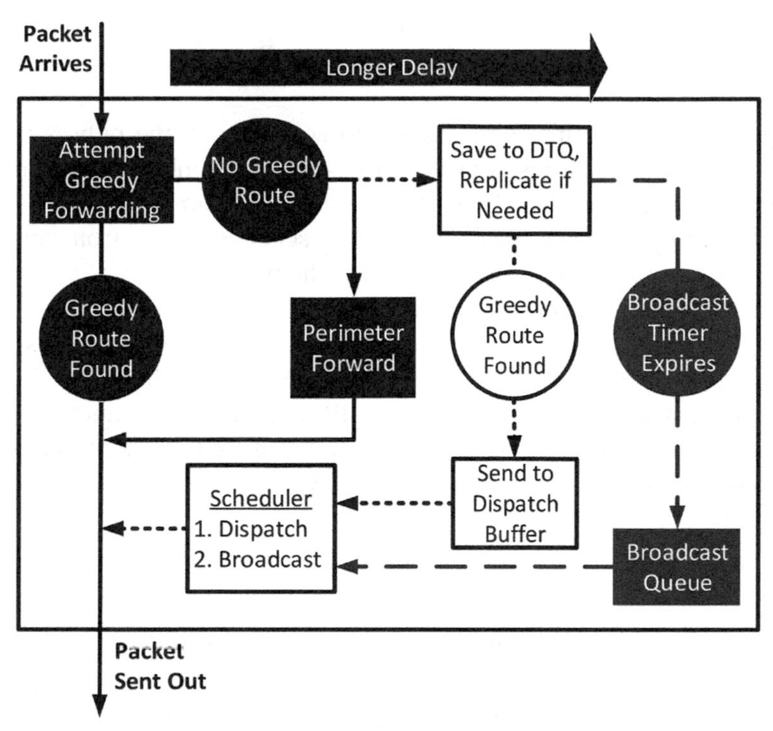

Figure 11.3 RobustGeo processing a packet. Normal packets use all paths as necessary; Broadcasted packet replicas do not traverse the long-dashed grey path; perimeter forwarded packet replicas only stay on the solid black path.

where m is the number of packets in the network that experience local maximum recovery exactly once, t is the average local maximum recovery time in seconds; π is the broadcasting period of each node, and n is the number of neighbors that receive the broadcasted packet for the first time.

However, not every replicated packet ends up finding a greedy route. Packets that do not find alternate paths are ultimately dropped as they time out, and perimeter-routed replicas that find a greedy route cancels out its original. For this reason, we assign probabilities to the replicated packets to indicate that they actually remain in the network.

Suppose that each broadcasted packet finds an alternative route with probability P_b, and the perimeter-routed packet finds an alternative route without being able to inform its originator with probability P_p, then the number of replicated packets produced from a single local maximum recovery with these probabilities take into account is

$$rep_1 = mn \left\lfloor \frac{t}{\pi} \right\rfloor P_b + P_p. \tag{11.2}$$

Multiplying P_b to the first term takes into account that the only replicated packets that remain are those who are forwarded to neighbors that eventually find a greedy geo-forwarded route. The value 1 from Equation (11.1) is replaced with P_p to take into account that the packet replicated from perimeter routing ends up remaining in the network at the probability P_p.

If the distance between the sender and receiver is long, a packet may experience multiple cases of local maximum. Each time it happens, more packets are replicated. In RobustGeo, no packet replicas from broadcasting can be broadcasted again, but they may still replicate via perimeter-routing. We now calculate the number of replicas produced from the second time that m packets encounter local maximum recovery:

$$rep_2 = mn \left\lfloor \frac{t}{\pi} \right\rfloor P_b + P_p \left(mn \left\lfloor \frac{t}{\pi} \right\rfloor P_b \right) + P_p + P_p^2. \tag{11.3}$$

The first term, $mn[\frac{t}{\pi}]P_b$, is the number of packet produced from broadcasting the original packet. The second term, $P_p(mn[\frac{t}{\pi}]P_b)$, is the probabilistic number of packets produced from broadcasting the perimeter-routed packet. P_p is the replica from perimeter-routing the current original packet, and P_p^2 is the replica from perimeter-routing the replicated packet produced from the first time that perimeter-routing was performed.

Since P_p is a probability with value between 0 and 1, as it gets higher powers it becomes more negligible and is disregarded. Further, each time the packet is perimeter-routed, a new *broadcastable* packet is replicated from perimeter routing with probability P_p. The next time that the local maximum is encountered, all of these replicated packets are eligible for broadcasting. Therefore, the total number of replica packets in the network as m original packets are routed through K local maximums is approximately

$$rep \approx \sum_{k=1}^{K} mn \left\lfloor \frac{t}{\pi} \right\rfloor P_b(1 + (k-1)P_p) + P_p. \tag{11.4}$$

Figure 11.4 shows the number of replicas a single packet accumulates as it is routed in the network. It describes an urban scenario with short but frequent

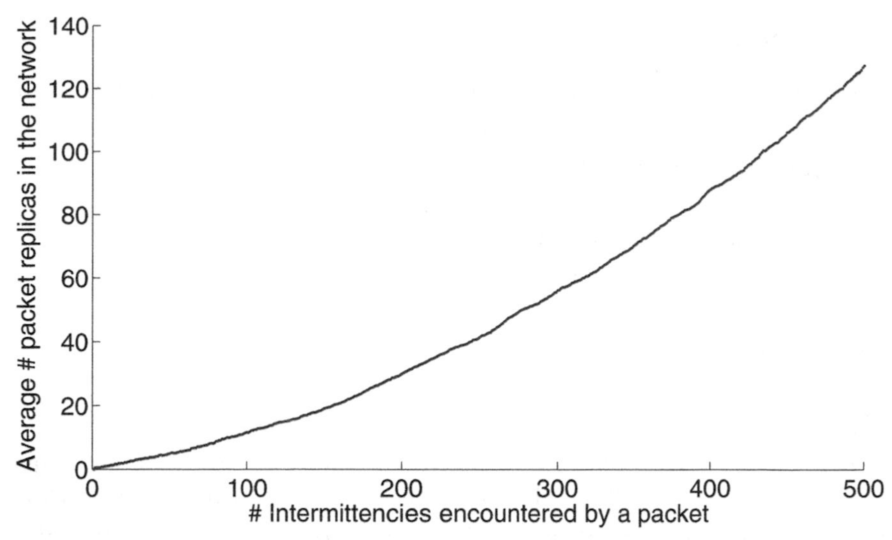

Figure 11.4 The average number of packet replicas in the network vs. the number of intermittencies encountered by a packet, where the intermittency length t is modeled as a Poisson random variable with $\lambda = 3$. All cases of intermittencies are independent of one another.

intermittencies. In this scenario, the recovery time is assumed to take 3 sec on average, and the broadcasting period is 6 sec. We also assume that the node has on average 10 new neighbors to broadcast to each time that it broadcasts. We also set P_p to be 0.01, a very low number because we assume that for the majority of times, the local maximum occurs due to brief network partitions and so it fails. For a similar reason, we set P_b to be 0.2 because we assume that most of the paths found through broadcasting are not unique. Because the broadcast rate is a floor function, we assume that the length of recovery is roughly a Poisson distribution with $\lambda = 3$[1] (for the 3 sec of average recovery time) so that the results are not just summations of P_p.

We see that the number of packet replicates increase slowly with the number of intermittencies it encounters. At 100 intermittencies, the average number of replicas is less than 5 per packet. Thus, packet replication is manageable.

[1]we can use a discrete pmf in this case since the output values we are concerned with are discrete.

11.5 Evaluation

We evaluate RobustGeo's performance by simulating three scenarios using the NS3 simulator, with the following parameters:

- PhyMode: OFDM 36 Mbps
- Propagation Delay Model: Constant Speed
- Propagation Loss Model: Friis
- WiFi Standard: 802.11a
- Wifi Mac Type: Adhoc Wifi Mac
- Transport Protocol: UDP
- Application: CBR Client-Server Pair (20 Kbps)
- Simulation Time: 200 s

We begin with an artificially set-up scenario mainly to showcase the major features of RobustGeo, followed by two progressively more realistic scenarios, first with a realistically simulated mobility model running on an actual map of Washington D.C and then with an actual trace of taxicabs in the San Francisco area. We believe that these three scenarios give a clear view of the superiority of RobustGeo over the other three competing routing algorithms.

To highlight the effectiveness of RobustGeo in intermittent situations, we treat all cases of local maximum in the simulation as network partitioning by not allowing perimeter routing in any of the algorithms we evaluate. For each scenario, we compare RobustGeo against *pure geo-routing*, which simply drops the packet that it cannot immediately route, *geo-routing with a DTQ* that improves delay tolerance but without packet replication, similar in behavior to GeoDTN+Nav [16], and *geo-routing with controlled flooding*, where nodes can only pass along the packet to 5 unique neighbors before dropping the packet (Epidemic Routing [13] with restrictions).

We base our comparisons on the following metrics: packet delivery rate, average delay, and overall traffic per packet received. While packet delivery rates and average delay are obvious metrics to measure, we believe that the overall traffic per packet received is equally important because packet replication is a major feature of RobustGeo. Finally, using data gathered from the San Francisco cab trace, we analyze the number of intermittencies packets generally encounter in the network, as well as the number of times each packet is broadcasted, to further cement our claim that RobustGeo is scalable in terms of packet replication.

We begin by testing a synthetic scenario as illustrated in Figure 11.5. This 5-part system consists of groups of three static nodes and two groups of mobile nodes, with a total of 11 nodes. The two clusters of static nodes on the left form

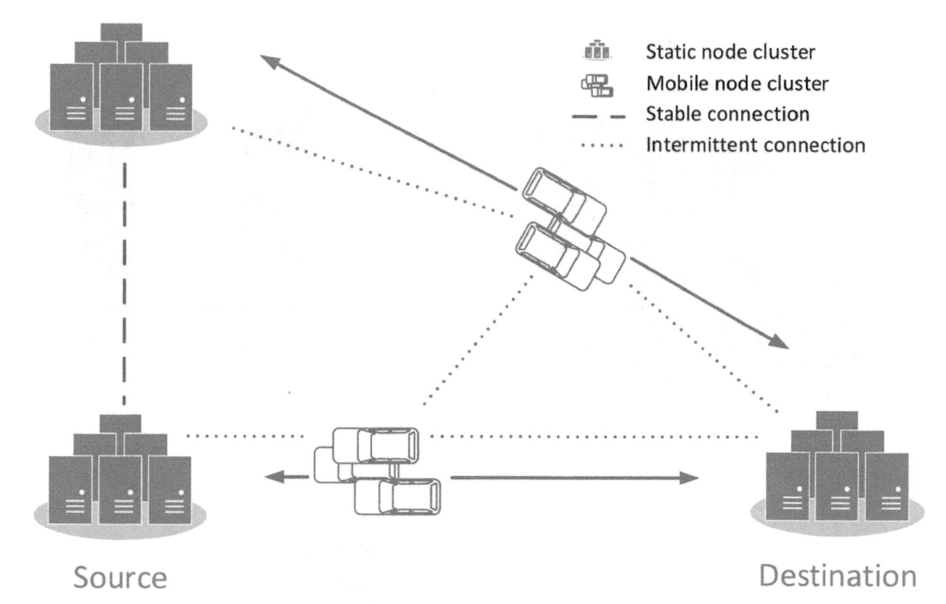

Figure 11.5 5 groups of nodes in an intermittent situation; arrows denote node mobility, and dotted lines denote network connection.

a stable connection, while the single group of static nodes on the right relies on the two mobile groups to receive messages. As the mobile nodes move around, there are short intervals and they form a bridge of network connection between the source and the destination. Most of the time, however, the network on the left side is completely partitioned off from the right. We believe that this is a scenario best fit for RobustGeo since the source is able to reach its stably connected neighbor via periodic broadcasting and thus make use of both routes as each become available. Such a path cannot be realized with perimeter routing because when the mobile node on the slant is connected to the static relay, it is not connected with anything else, so all of the perimeter-routed packets would be dropped.

We repeat our simulation 10 times with different seeds for each of the routing algorithms in this scenario. In Figure 11.6, we see that as expected, the pure geo-routing method performs poorly because the sender can only transmit to the receiver when the two mobile clusters line up to form the transmission bridge. For the other three schemes, while RobustGeo performs marginally better than the rest in terms of packet delivery, it generated the most packets. We attribute this to the small number of nodes in the network. Controlled flooding only sends packets to another node if the packet is brand

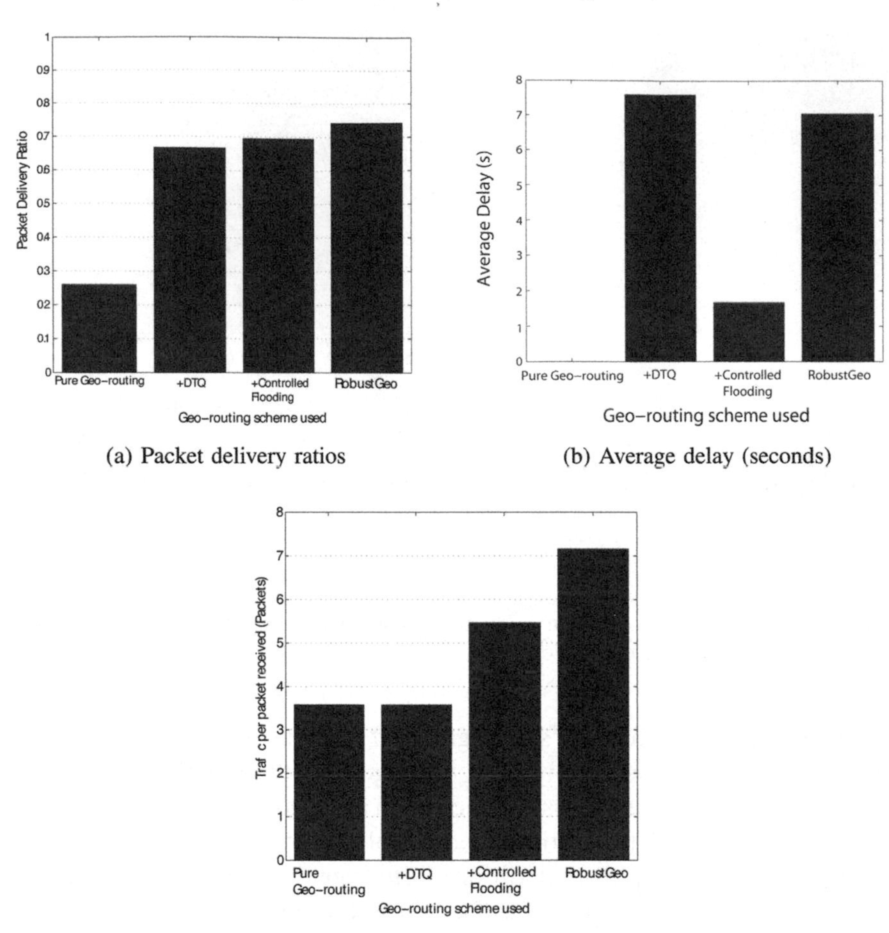

(a) Packet delivery ratios

(b) Average delay (seconds)

(c) Total traffic generated per packet received

Figure 11.6 Results for the 5-group artificial scenario.

new to the node. With only 11 nodes in a highly partitioned system, there is not as much chance for sending packets in such a fashion. RobustGeo, on the other hand, broadcasts periodically regardless of who is around.

We also simulated a more realistic environment by downloading a 2,000 m by 2,000 m map of the Washington, DC area made available by the US Census Bureau's TIGER database, and simulating mobility on the map using the Intelligent Driver Model with Intersection Management by VanetMobisim [19]. This model is complete with intersection and stop light rules to simulate realistic vehicular traffic on the streets provided by the

Washington, DC map. The vehicles in the scenarios have maximum and minimum speeds of 10 m/s and 20 m/s, respectively. We generate 4 scenarios by varying the number of nodes between 50 and 125. With each scenario, we randomly place a static node on the map as the destination (server) and choose a random mobile node as the source (client).

As Figure 11.7 shows, in the very sparse scenario of only 50 nodes, pure geo-routing completely breaks down and has a packet delivery ratio of 0. Having the DTQ and using RobustGeo marginally increase the delivery ratio, but it is still under 10%. As the nodes become more dense, the protocols generally trend upwards in their packet delivery ratios. Something quite unexpected is that geo-routing + controlled flooding did considerably worse than pure geo-routing when the number of nodes increased to 125. We attribute this to the fact that each time a node cannot find a greedy forwarding path for a packet, the packet is sent out at least 5 times by that single node alone and multiplied by each recipient 5 more times. This generates a large amount of traffic, taking up available bandwidth and causing interference for nearby nodes. This suspicion is confirmed by the traffic line plot in Figure 11.7. In all, from the VanetMobisim scenarios we see that RobustGeo is indeed the most robust of all four routing schemes compared, having the highest delivery ratio for every situation. In terms of delay, geo-routing + DTQ did better in the 75-node case, but RobustGeo did better in all other cases. Additionally, we see that as predicted in Section 11.4, the replication overhead of RobustGeo is a manageable one, since it has only a slightly higher traffic to packet received ratio (about 10% or less) than the geo-routing + DTQ scheme that does not have replication.

Finally, we used real traces of an extremely intermittent network to further compare the four geo-routing schemes. In this scenario, we use actual mobility traces of taxicabs in San Francisco downloaded from Crawdad [20]. We run the simulation in a 5,700 m by 6,600 m area with 116 nodes moving for 200 sec and again place a static node on the map to be the receiver. In this scenario, the nodes are extremely sparse because only cab movements are recorded in the tracefile. This means that the network's period of disconnection is likely longer than its connectivity.

As Figure 11.8 shows, the pure geo-routing approach completely breaks down, with a receiving ratio of 0. Geo-routing + DTQ and RobustGeo do much better, with RobustGeo almost achieving a 30% packet delivery ratio. Once again, the geo-routing + controlled flooding approach fails, with a less than 1% delivery rate. We again attribute this to its high delay time. This means that had the simulation time been longer, the geo-routing + controlled

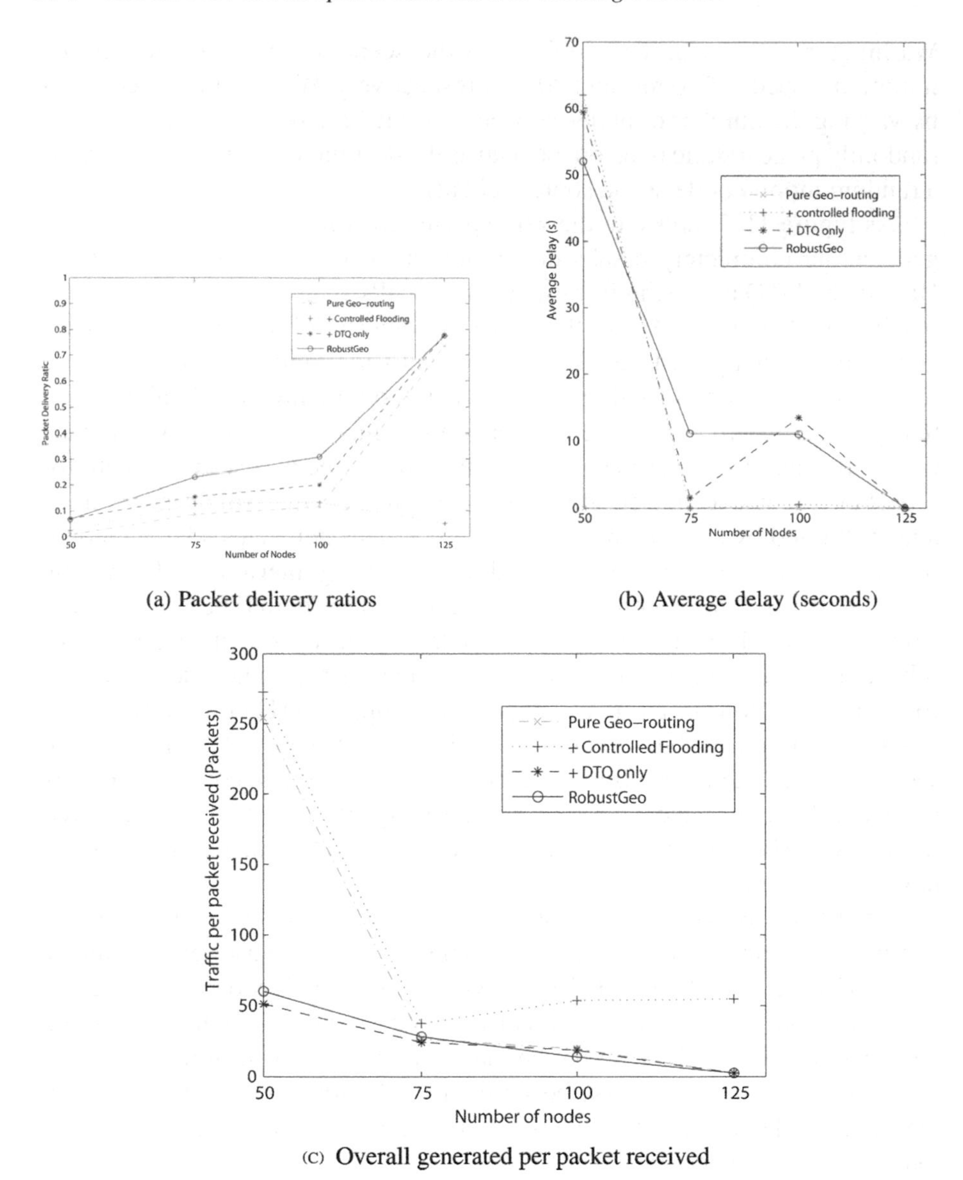

(a) Packet delivery ratios

(b) Average delay (seconds)

(c) Overall generated per packet received

Figure 11.7 Results of the simulation using realistically simulated mobility patterns generated with VanetMobisim.

flooding approach can likely a more acceptable packet delivery rate at the cost of even higher delay times. We choose to show the traffic per packet delivered metric for only pure geo-routing + DTQ and RobustGeo because the other

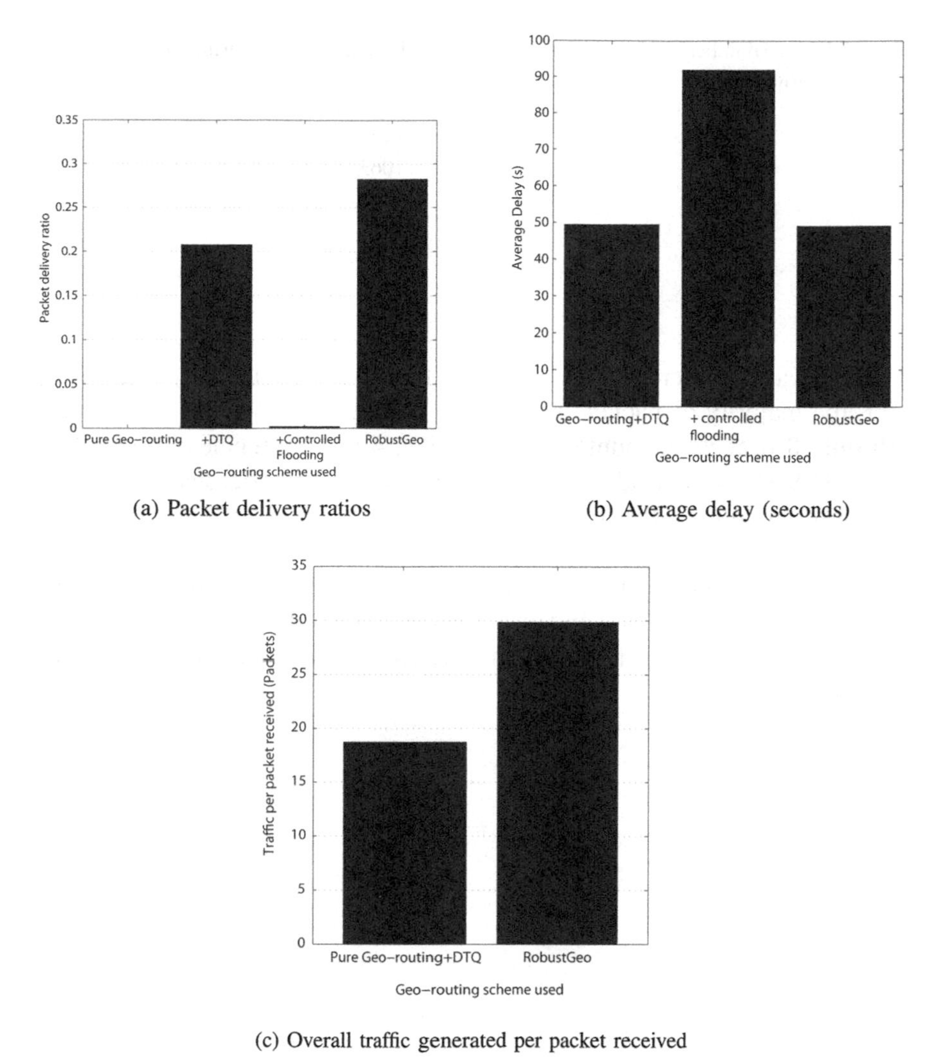

(a) Packet delivery ratios

(b) Average delay (seconds)

(c) Overall traffic generated per packet received

Figure 11.8 Results of the simulation using real life San Francisco taxicab traces.

two approaches yield incredibly high numbers (infinity and about 2500). As expected, RobustGeo generated 58% more traffic per packet delivered than geo-routing + DTQ. At the same time, it also has a 36% higher delivery ratio with a very slightly lower average delay. We believe that this is a good trade-off as packet delivery ratios should be prioritized over bandwidth conservation.

Since the San Francisco cab scenario is very realistic and at the same time gives an extremely intermittent network, we use it to analyze disconnectivity

Table 11.1 The number of packets that encounter each frequency of intermittencies in the SF taxicab scenario

# Intermittencies	# Packets
1	1434
2	1063
3	981
4	214
5	3
6	5

and broadcast frequencies. Table 11.1 shows that in the given environment, the majority of packets experience disconnectivity 3 or fewer times, with only 8 packets encountering more than 4 intermittencies. Meanwhile, Figure 11.9 shows that about 80% of packets get broadcasted 6 times or fewer. This confirms the fact that packet replication will not spiral out of control.

We also postulate that the majority of packets experiencing higher frequencies of broadcast do not end up at the destination. If true, this can be used as a good metric to decide how long the packet should remain in a node's DTQ before being dropped.

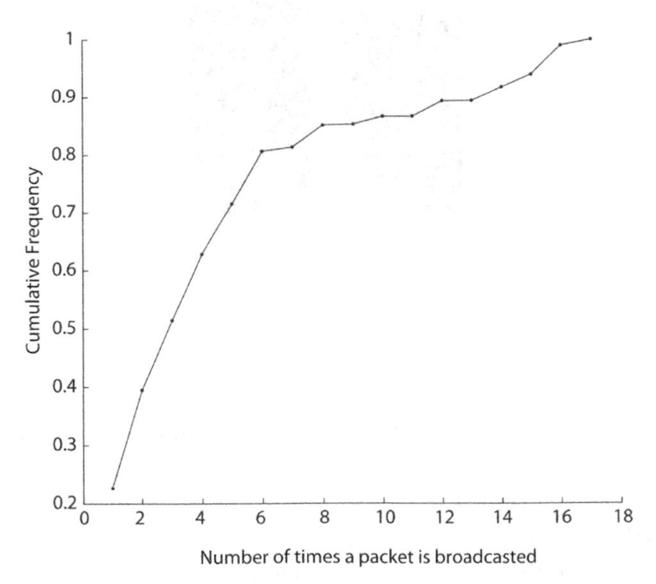

Figure 11.9 The CDF of packet broadcast rate in the SF taxicab scenario.

11.6 Related Work

Several other works have also explored the possibility of using delay-tolerant networks that exploits location knowledge in VANETs. GeOpps [21] is one such DTN protocol. GeOpps uses the vehicle's navigation systems to find the *Nearest Point* (NP) to the message's destination that the vehicle will travel to, assuming that the vehicle has a predetermined destination of its own. While on the way, the vehicle will periodically broadcast the message's destination to its one-hop neighbors. The neighbors that receive the destination location calculates their Minimum Estimated Time of Delivery (METD) for the message based on their own calculated NP and replies to the message carrier. The message carrier makes the decision to pass along to its neighbor that has the shortest METD, and the process is repeated. While GeOpps relying on a "smarter" way of choosing what are essentially Data MULEs can increase the data dissemination speed, its high dependence on vehicular movements to carry the message to its destination causes it to experience much higher delays than greedy forwarding, since network propagation is much faster than physical node movements.

GeoSpray [22] is another VANET routing solution that combines geographical knowledge with DTN ideas. Inspired by GeOpps, GeoSpray works in a similar fashion in that it takes into account the neighbors' METD and forwards the message to the ones with small METDs. However, they differ in that GeoSpray utilizes the concept of the spray phase in the binary Spray and Wait [14] method where it utilizes controlled packet replication to explore multiple paths and decrease delay. This is similar to RobustGeo's singe hop broadcast phase, which also makes use of replication. However, just like GeOpps, GeoSpray suffers in a similar fashion in that it cannot take advantage of nodes who can otherwise be valid relay nodes in greedy forwarding, simply because they have a low METD value.

The work that is the most similar to RobustGeo is GeoDTN + Nav [16]. GeoDTN + Nav fully incorporates the DTN paradigm into existing works on location-based routing protocols by operating with three modes: *restricted greedy forwarding*, *perimeter forwarding*, and *DTN*. The added DTN mode of GeoDTN + Nav allows packets to still be routed to the destination with a DTN approach in the event that the network is segmented due to sparse node topologies. The default greedy forward strategy for GeoDTN + Nav is the same as the one used in GPCR, and the default recovery mode is the same as that of VCLCR [23], a location-based routing protocol that can detect routing loops in perimeter forwarding otherwise missed by GPCR. Finally, it uses the simple store-and-forward approach in DTN mode until greedy forwarding can be recovered.

GeoDTN + Nav switches from perimeter forwarding mode into DTN mode by having each node on the perimeter path calculate the *switching scores* of its neighbors to see if they are good candidates for carry-forward MULEs. These scores are based on three metrics: the probability of network disconnectivity, the neighbor's delivery quality in DTN mode, and the neighbors' travel direction. If the score is high enough, the packet is passed to the node's neighbor and DTN mode commences. In the DTN mode, the receiving node stores the packet and relies upon its own mobility to carry the packet until restricted greedy forwarding can once again be used to forward the packet to its destination.

GeoDTN + Nav differs from RobustGeo mainly in that the DTN method employed in GeoDTN + Nav uses a single-forwarding based algorithm, which generally has low reliability in packet delivery [17]. For example, GeoDTN uses the neighbor's current travelling direction to determine its switching score. Unfortunately, vehicles' travelling directions are not constant. A change in direction by the vehicle after receiving the packet means that the packet would be actually brought *away* from the destination and thus increasing delays. Moreover, if no "good" neighbors are found within the packet's TTL, it is dropped. RobustGeo performs better in these situations because it explores the possibility of limited message replication by having the local maximum node keep a copy of the packet in the recovery phase, allowing multiple nodes to look for greedy paths toward the destination. Additionally, GeoDTN + Nav is very much focused upon the method with which to switch into the DTN mode from perimeter mode, whereas RobustGeo is designed so that unless the packet is immediately forwarded, it is, in some ways, always both in geo-routing mode *and* in DTN mode. RobustGeo is able to do this because it includes a simple scheduling system to dispatch packets opportunistically. Having such a feature allows RobustGeo to have much greater flexibility and thus be more robust.

11.7 Conclusion and Future Work

In this chapter we presented another location-based routing algorithm, RobustGeo, that is designed to withstand network intermittencies common to urban VANETs. RobustGeo achieves this result by taking advantage of both the store and forward and packet replication strategies found in delay-tolerant networks. With RobustGeo, we were willing to make the trade-off of introducing extra bandwidth overhead into the network with packet replications in favor of more reliable delivery. We showed in both Sections 11.4 and 11.5 that the trade-off is a viable one. As future work, we would like to analyze the relationship between a packet's frequency of broadcasts and its journey completion likelihood to get a better idea of how long the packet should remain in the DTQ before timing out and being dropped. This can be very helpful in RobustGeo as it helps reduce node *storage overhead* in addition to bandwidth overhead. In all, RobustGeo is a hybrid solution that allows for connections in both reliable and intermittent networks, which are attributes of VANETs.

References

[1] Huhtonen. A. (2004). Comparing AODV and OLSR routing protocols. *Telecommun. Software Multimedia*, 1–9.

[2] Yousefi, S., Mousavi, M. S., and Fathy, M. (2006). Vehicular ad hoc networks (vanets): challenges and perspectives. In *2006 6th International Conference on ITS Telecommunications Proceedings*, pp. 761–766. IEEE.

[3] Karp, B., and Kung, H.-T. (2000). GPSR: Greedy perimeter stateless routing for wireless networks. In *Proceedings of the 6th Annual International Conference on Mobile Computing and Networking*, pp. 243–254. ACM.

[4] Lee. K. C. (2010). *Geographic routing in vehicular ad hoc networks*. University of California, Los Angeles.

[5] Lochert, C., Mauve, M., Füßler, H., and Hartenstein, H. (2005). Geographic routing in city scenarios. *ACM SIGMOBILE Mobile Comput. Commun. Rev.*, 9(1), 69–72.

[6] Jerbi, M., Senouci, S.-M., Meraihi, R., and Ghamri-Doudane, Y. (2007). An improved vehicular ad hoc routing protocol for city environments. In *IEEE International Conference on Communications, 2007. ICC'07*, IEEE. pp. 3972–3979.

[7] Lee, K. C., Haerri, J., Lee, U., and Gerla. M. (2007). Enhanced perimeter routing for geographic forwarding protocols in urban vehicular scenarios. In *Globecom Workshops, 2007 IEEE*, pp. 1–10. IEEE.

[8] Lee, K. C., Lee, U., and Gerla, M. (2009). To-go: topology-assist geo-opportunistic routing in urban vehicular grids. In *Sixth International Conference on Wireless On-Demand Network Systems and Services, 2009. WONS 2009*, IEEE. pp, 11–18.

[9] Seet, B.-C., Liu, G., Lee, B.-S., Foh, C.-H., Wong, K.-J., and Lee, K.-K. (2004). A-star: A mobile ad hoc routing strategy for metropolis vehicular communications. In *Performance of Computer and Communication Networks; Mobile and Wireless Communications, NETWORKING 2004. Networking Technologies, Services, and Protocols*. Pp. 989–999. Springer, Berlin.

[10] Wisitpongphan, N., Bai, F., Mudalige, P., and Tonguz, O. (2007). On the routing problem in disconnected vehicular ad-hoc networks. In *International Conference on Computer Communications, INFOCOM 2007. 26th IEEE.* Pp. 2291–2295, May.

[11] Jain, S., Fall, K., and Patra, R. (2004). Routing in a delay tolerant network. In *Proceedings of the 2004 Conference on Applications, Technologies, Architectures, and Protocols for Computer Communications*, SIGCOMM '04, pp. 145–158, New York, NY, USA. ACM.

[12] Shah, R. C., Roy, S., Jain, S., and Brunette, W. (2003). Data mules: modeling and analysis of a three-tier architecture for sparse sensor networks. *Ad Hoc Networks*, 1 (2), 215–233.

[13] Vahdat, A., and Becker, D. et al. (2000). Epidemic routing for partially connected ad hoc networks. Technical report, Technical Report CS-200006, Duke University.

[14] Spyropoulos, T., Psounis, K., and Raghavendra, C. S. (2005). Spray and wait: an efficient routing scheme for intermittently connected mobile networks. In *Proceedings of the 2005 ACM SIGCOMM Workshop on Delay-tolerant Networking*, pp. 252–259. ACM.

[15] Zhao, W., Ammar, M., and Zegura, E. (2004). A message ferrying approach for data delivery in sparse mobile ad hoc networks. In *Proceedings of the 5th ACM International Symposium on Mobile Ad Hoc Networking and Computing*, MobiHoc'04, pp. 187–198, New York, NY, USA. ACM.

[16] Cheng, P.-C., Lee, K. C., Gerla, M., and Härri, J. (2010). GeoDTN + NAV: Geographic dtn routing with navigator prediction for urban vehicular environments. *Mob. Netw. Appl.*, 15 (1), 61–82, Feb.

[17] Zhang, Z. (2006). Routing in intermittently connected mobile ad hoc networks and delay tolerant networks: overview and challenges. *Commun. Surveys Tutorials, IEEE*, 8 (1), 24–37.

[18] Harras, K. A., and Almeroth, K. C. (2006). Transport layer issues in delay tolerant mobile networks. In *NETWORKING 2006. Networking Technologies, Services, and Protocols; Performance of Computer*

and Communication Networks; Mobile and Wireless Communications Systems, pp. 463–475. Springer, Berlin.

[19] Härri, J., Filali, F., Bonnet, C., and Fiore, M. (2006). Vanetmobisim: Generating realistic mobility patterns for vanets. In *Proceedings of the 3rd International Workshop on Vehicular Ad Hoc Networks*, VANET '06, pp. 96–97, New York, NY, USA. ACM.

[20] Piorkowski, M., Sarafijanovic-Djukic, N., and Grossglauser, M. (2009). CRAWDAD data set epfl/mobility (v. 2009-02-24). Downloaded from http:// crawdad.org/epfl/mobility/, Feb.

[21] Leontiadis, I., and Mascolo, C. (2007). Geopps: Geographical opportunistic routing for vehicular networks. In *IEEE International Symposium on a World of Wireless, Mobile and Multimedia Networks, 2007. WoWMoM 2007*, pp. 1–6. IEEE.

[22] Soares, V. N., Rodrigues, J. J., and Farahmand, F. (2014). Geospray: A geographic routing protocol for vehicular delay-tolerant networks. *Information Fusion*, 15:102–113.

[23] Lee, K. C., Cheng, P.-C., Weng, J.-T., Tung, L.-C., and Gerla, M. (2008). VCLCR: a practical geographic routing protocol in urban scenarios. *UCLA Computer Science Department, Tech. Rep. TR080009*.

12

Social Similarity-based Multicast Framework in Opportunistic Mobile Social Networks

Xiao Chen[1], Yuan Xu[1] and Suho Oh[2]

[1]Department of Computer Science, Texas State University,
San Marcos, TX 78666, USA
[2]Department of Mathematics, Texas State University,
San Marcos, TX 78666, USA

Abstract

With the proliferation of mobile devices, opportunistic mobile social networks (OMSNs) where the communication takes place on the fly by the opportunistic contacts among mobile users when they gather together at events have become increasingly popular. Multicast is an important routing service which supports the dissemination of messages to a group of users. Some existing multicast algorithms are designed by taking advantage of the internal social features of nodes in the network. This approach is motivated by the fact that nodes come in contact more frequently if they have more social features in common. These social features are obtained from nodes' profiles and thus static. Different from these multicast protocols that utilize static social features, in this chapter, we adopt dynamic social features to more accurately capture node contact behavior and thereafter propose a novel social similarity-based multicast framework using the dynamic social features and a compare-split scheme to improve multicast efficiency in OMSNs. We instantiate the framework with two multicast algorithms named Multi-SoSim and E-Multi-SoSim that adopt the dynamic and enhanced dynamic social features, respectively. A detailed analysis of the proposed algorithms is given, and simulations are conducted to evaluate our proposed algorithms by comparing them with their variations and the existing one using static social features.

Keywords: Multicast, Multicast tree, Opportunistic mobile social networks, Social features, Social similarity.

12.1 Introduction

With the proliferation of smartphones, PDAs, and laptops, *opportunistic mobile social networks* (OMSNs) formed by people moving around carrying these mobile devices have become popular in recent years [1–5]. Unlike the popular online social networks such as Facebook and LinkedIn, the OMSNs we discuss here are a special kind of delay tolerant networks (DTNs) [6] where the communication takes place on the fly by the opportunistic contacts among mobile users when they gather together at conferences, social events, rescue sites, campus activities, and so on. This type of communication relies on a lightweight mechanism via local wireless bandwidth such as Bluetooth or Wi-fi without a network infrastructure [2, 7, 8]. In OMSNs, node connections are usually short term, time dependent, and unstable as people come and go at events.

Multicast, a service where a source node sends messages to multiple destinations, widely occurs in OMSNs. For example, in a conference, presentations are delivered to inform the participants about the newest technology; in an emergency scenario, information regarding the local conditions and hazard levels is disseminated to the rescue workers; and in campus life, school information is sent to a group of student mobile users over their wireless interfaces.

Due to the uncertainty and time-dependent nature of OMSNs, there does not guarantee a path from a source to the destinations at any time, which poses special challenges to routing, either unicast or multicast. Nodes in OMSNs can only communicate in a store–carry–forward fashion: When two nodes move within each other's transmission range, they *meet* each other and can communicate directly, and when they are out of the range, their contact is lost. The message to be delivered needs to be stored in the local buffer until a contact occurs in the next hop.

Most existing multicast algorithms focus on DTNs [9–13] without considering social factors. Recently, a few algorithms propose to take advantage of the social features in user profiles to facilitate routing [2, 14, 15]. Among these algorithms, the one proposed by Deng et al. [14] addresses multicast. Specifically, the researchers found, through the study of the INFOCOM 2006 trace, that the social features in user profiles could effectively reflect nodes' contact behavior and developed a social profile-based multicast (SPM) scheme based on the two most important social features: *Affiliation* and *Language*. In their scheme, social features F_i can refer to non-private user attributes such as *Nationality*, *City*, *Language*, *and Affiliation* and these social features can take different values f_i. For example, a social feature can be *Language* and its value can be *English*. The intuition is that nodes having more common social features come to meet more often. Thus, the nodes having more common social features with the destination are better forwarders to deliver the message to it. We believe, in the dynamic environment of OMSNs, the multicast algorithm can be further improved because the static social features may not always capture nodes' dynamic contact behavior. For example, a student who puts *New York* as his *state* in his profile may actually attend a conference in Texas. In that case, the static information in his profile can not reflect his behavior in Texas. The information that is helpful in making multicast decisions can only be gathered from the nodes' contact behavior at the conference. Therefore, in this chapter, we extend static social features to *dynamic social features* to better reflect nodes' contact behavior and thereafter develop a new multicast algorithm specifically for OMSNs based on the dynamic social features.

In dynamic social features, we want to embed information that can reflect users' dynamic behavior to facilitate routing and that can be easy to obtain and inexpensive to maintain in OMSNs. Thus, we not only record if a node has the same social feature value with the destination, but also record the frequency this node has met other nodes that have the same social feature value during the time interval we observe. For example, we not only record if node A, same as the destination, is a *New Yorker* but also record that it has met *New Yorkers* 90% of the time during the observation interval. Unlike the static social features from user profiles, dynamic social features are time-related. So they change as user contact behavior changes over time. So we can have a more accurate way to choose the best forwarders in multicast. In this chapter, we first apply the frequency-based dynamic social features and then the enhanced dynamic social features to multicast to improve its performance.

In multicast, a message holder is expected to forward a message to multiple destinations. To reduce the overhead and forwarding cost, the destinations should share the routing path as much as possible until the point that they have to be separated. Thus, the overall multicast process results in a tree structure. A compare-split scheme to determine the separation point is critical to the efficiency of a multicast. In our multicast, if a message holder x meets another node y, the scheme of compare-split is based on the social similarity of each of the destinations with x and y using dynamic social features. That is, whichever, either x or y, is more socially close to the destination will have a higher chance to deliver the message and thus should relay the message to that destination.

Based on the notions of dynamic social features and the scheme of compare-split, we propose a novel *social similarity-based multicast* framework for OMSNs. Two algorithms instantiate this framework: the *social similarity-based multicast* (Multi-SoSim) algorithm which utilizes dynamic social features to capture node contact behavior and a compare-split scheme to select the best relay node for each destination in each hop to improve multicast efficiency and the *enhanced social similarity-based multicast* (E-Multi-SoSim) algorithm which upgrades the dynamic social features in Multi-SoSim to enhance dynamic social features to further improve multicast efficiency. To evaluate the performance of our algorithms, we conduct an analysis and compare them with the existing algorithm that uses static social features and some variations of the proposed algorithms. Simulation results conclude that using dynamic social features can make better multicast routing decisions than using the static ones, letting destinations share the paths longer can reduce the cost, and separating destinations and allocating them to better forwarders can reduce latency.

The rest of the chapter is organized as follows: Section 12.2 references the related works; Section 12.3 gives the definitions of dynamic social features and the calculation of social similarity; Section 12.4 presents our multicast algorithms; Section 12.5 gives the analysis of the algorithms; Section 12.6 shows the simulation results; and the conclusion is in Section 12.7.

12.2 Related Works

The multicast algorithm in mobile social networks (MSNs) can be implemented using rudimentary approaches such as flooding [16], but it has inevitable high forwarding cost. Most of the existing multicast algorithms are designed for DTNs where social factors are not considered. Zhao et al. [13] introduce some new semantic models for multicast and conclude that the group-based strategy is suitable for multicast in DTNs. Lee et al. [9] study the scalability property of multicast in DTNs and introduce RelayCast to improve the throughput bound of multicast using mobility-assist routing algorithm. By utilizing mobility features of DTNs, Xi et al. [12] present an encounter-based multicast routing, and Chuah et al. [17] develop a context-aware adaptive multicast routing scheme. Mongiovi et al. [10] use graph indexing to minimize the remote communication cost of multicast. And Wang et al. [11] exploit the contact state information and use a compare-split scheme to construct a multicast tree with a small number of relay nodes.

There are a few multicast papers that involve social factors. Gao et al. [18] propose a community-based multicast routing scheme by exploiting node centrality and social community structures. This approach is based on the fact that "social relations among mobile users are more likely to be long-term characteristics and less volatile than node mobility" [18] in MSNs. Hu et al. [19, 20] put forward multicast algorithms to disseminate data in MSNs. In [19], the content owners multicast to their social contacts which are defined by the geographic social strength between nodes and in [20], node centrality in the social contact graph extracted from node contact trace is adopted to select the initial receiver set [20]. Deng et al. [14] propose a social profile-based multicast (SPM) algorithm that uses social features in user profiles to guide the multicast routing in MSNs. This approach has the advantage of not having to record node contact history, but the static social features may not catch people's dynamic contact behavior in the OMSNs. So the multicast algorithm for OMSNs can be further improved by catching the dynamic features of the network.

12.3 Preliminary

In this section, we first introduce static social features used in the existing papers [14], then define dynamic social features and its enhanced version, and then give the formula to calculate nodes' social similarities which will be used in the compare-split scheme in our multicast algorithms.

12.3.1 Definition of Static Social Features

Suppose we consider m social features $\langle F_1, F_2, \ldots, F_m \rangle$ in an OMSN. Note that the choice of which social features to include in a network depends on the situation and the nodes can agree on the selection through message exchange or manually when the network was first set up. A node x's static social feature is a vector in the form of $\langle x_1, x_2, \ldots, x_m \rangle$, where x_i is the social feature value for F_i obtained from the user's profile.

12.3.2 Definitions of Dynamic Social Features

In dynamic social features, we define x_i as follows based on nodes' encounter history to capture nodes' contact behavior.

12.3.2.1 Dynamic social features

One definition of x_i is the frequency of node x meeting nodes with the same f_i out of all of the nodes it has met in the history we observe. That is,

$$x_i = \frac{M_i}{M_{total}} \tag{12.1}$$

In Definition (12.1), M_i is the number of times that x has met nodes with the same f_i in the history we observe and M_{total} is all of the nodes that x has met in that interval. For example, if f_i refers to $Student$ and if x has met 20 $Students$ out of a total of 100 people, then $x_i = 20/100 = 0.2$.

Nevertheless, one problem with the frequency definition of x_i is that if node x has met one $Student$ out of two people it has met in total in the history we observe and node y has met five $Students$ out of ten people it has met in total, using Definition (12.1), both of their frequencies are 0.5 in meeting $Students$. So which one is more likely to meet $Students$ in future? From the intuition, node y should be given a higher priority because it is more active in meeting people. There are many formulas we can design to favor y. In the following, we present one formula, which will be proved in the later Analysis section, that can break the tie and favor the more active node.

12.3.2.2 Enhanced dynamic social features

In this enhanced definition of dynamic social features, x_i is calculated as follows:

$$x_i = \left(\frac{M_i+1}{M_{total}+1}\right)^{p_i} \left(\frac{M_i}{M_{total}+1}\right)^{1-p_i} = (M_i+1)^{p_i}\frac{M_i^{1-p_i}}{M_{total}+1} \tag{12.2}$$

In Definition (12.2), $p_i = M_i/M_{total}$. This definition predicts x_i by looking at the next meeting probability of node x with another node having the same f_i. In the next time, the total meeting times will be $M_{total} + 1$. The first part $\left(\frac{M_i+1}{M_{total}+1}\right)^{p_i}$ means that there will be p_i probability that x will have a "good" meeting with another node having the same social feature value f_i next time. In this case, M_i will also be incremented by 1. The second part $\left(\frac{M_i}{M_{total}+1}\right)^{1-p_i}$ means that there will be $1 - p_i$ probability for x not to meet a node with the same f_i next time. In that case, M_i will remain the same. The definition of x_i then takes the geometric mean of the two parts.

With Definition (12.2), we can break the tie in the example above. For node x, $M_i = 1, M_{total} = 2, p_i = 0.5$; for node y, $M_i = 5, M_{total} = 10$, $p_i = 0.5$. Using Definition (12.2), $x_i = (1+1)^{0.5} * \frac{1^{(1-0.5)}}{2+1} = 0.4714$ and

$y_i = (5+1)^{0.5} * \frac{5^{(1-0.5)}}{10+1} = 0.4979$. These two results are close, reflecting that the two nodes had the same frequency using Definition (12.1), yet they tell us that y is better because it is more active meeting nodes.

Dynamic social features, as shown in the definitions, not only record if a node has certain social features, but also predict the probability of this node meeting other nodes with the same social features. Unlike the static social features, dynamic social features change as user activities change over time. So they can better reflect users' dynamic contact behavior in OMSNs.

12.3.3 Calculation of Social Similarity

With nodes' dynamic social features defined, we can use the similarity metrics such as Tanimoto [21], cosine [22], Euclidean [23], and weighted Euclidean [24] derived from data mining [25] to calculate the social similarity $S(x, y)$ of two nodes x and y. We finally decide to use the Euclidean similarity metric because it does not require the calculation of additional weighting values and performs slightly better than Tanimoto and cosine in terms of latency when these metrics are compared in our simulations [24].

Euclidean Similarity Metric

After normalizing the original definition of the Euclidean similarity [23] in data mining to the range of $[0, 1]$ and subtracting it from 1, it is now defined as follows:

$$S(x, y) = 1 - \frac{\sqrt{\sum_{i=1}^{m} (y_i - x_i)^2}}{\sqrt{m}}.$$

Using the Euclidean similarity metric, if two nodes x and y have the same dynamic social features, e.g., $x_i = y_i$, then $S(x, y) = 1$. In other words, they have 0 *social similarity gap*. So the social similarity gap of two nodes is defined as $1 - S(x, y)$.

Here is how the metric is used in our algorithms. Suppose we consider three social features $\langle City, Language, Position \rangle$ of the nodes in the network. Assume destination d has social feature values $\langle NewYork, English, Student \rangle$. The vector of d is set to $\langle 1, 1, 1 \rangle$ because this is our target. Suppose there are two relay candidates x and y. We want to decide which is a better one to deliver the message to the destination. From the history of observation, node x has met people from New York 70% of the time, people who speak English 93% of the time, and students 41% of the time. If we use Definition (12.1) of the dynamic social features, node x has a vector of $x = \langle 0.7, 0.93, 0.41 \rangle$. Suppose y's vector is $y = \langle 0.23, 0.81, 0.5 \rangle$. Using the Euclidean social similarity, $S(x, d) = 0.62$ and $S(y, d) = 0.46$. So x has a social similarity gap of $1 - 0.62 = 0.38$ to d and y has a social similarity gap of 0.54 to d. Thus, x is more socially similar to d and therefore is more likely to deliver the message to the destination. Definition (12.2) of the dynamic social features can be used in a similar way.

12.4 Multicast Routing Protocols

In this section, we propose a social similarity based multicast framework that selects the best forwarding nodes depending on the social similarity of nodes using dynamic social features and a compare-split scheme. Here, we assume that there is one multicast source. If there are multiple multicast sources, then the framework can be used by each individual source to multicast messages. In the framework, when the social similarities of the nodes are calculated using dynamic social feature Definitions (12.1) and (12.2), the resulting multicast algorithms are called Multi-SoSim and E-Multi-SoSim, respectively.

12.4.1 Social Similarity-based Multicast Framework

Our multicast framework is shown in Figure 12.1. In the beginning, suppose a source node s has a message to send to a set of destinations which we refer to as its destination set $D_s = \{d_1, d_2, \ldots, d_k\}$. The destination sets of all of the other nodes are initially empty. The message holder is denoted as x. Initially, x is the source node s.

If the message holder x meets a node y, we first check whether y is one of the destinations. If it is, x will deliver the message to y and remove it from its destination set. Next, we combine the destination sets of x and y into D_{xy} and make the destination sets D_x and D_y empty. Then, we use a compare-split scheme to split the destinations in D_{xy} to D_x and D_y by comparing the social similarity of each of the destinations d_i with x and y. The social similarity $S(x, y)$ of x and y is calculated either by dynamic social feature Definition (12.1) or by (12.2). If y is more socially similar to d_i, then d_i will be placed into D_y, meaning that y will be the next forwarder for the message destined for d_i; otherwise, d_i will be put into D_x. After this, nodes x and y will become new message holders and the process will repeat until all of the destinations have received the message.

Starting from the source node s and through the splits in the middle, the multicast process naturally forms a tree. It follows the cost reduction intuition that the destinations should share the paths on the tree as long as possible until

Multicast Framework: social-similarity-based multicast framework

Require: The source node s and its destination set $D_s = \{d_1, d_2, \ldots, d_k\}$; the destination sets of all of the nodes except s are empty; the initial message holder x is s

1: /* On contact between a message holder x and node y: */
2: **if** $y \in D_x$ **then**
3: x forwards the message to destination y and removes y from D_x
4: **end if**
5: /* Combine the destination sets of x and y */
6: Let $D_{xy} = D_x \cup D_y$ and $D_x = D_y = \emptyset$
7: /* Compare node social similarities and split the destinations in D_{xy} to D_x and D_y */
8: **for** each of the destinations $d_i \in D_{xy}$ **do**
9: **if** $S(x, d_i) < S(y, d_i)$ **then**
10: add d_i to D_y, and x forwards the message to y if y does not have it
11: **else**
12: add d_i to D_x
13: **end if**
14: **end for**

Figure 12.1 Our multicast framework.

a better node appears to carry over some of the destinations. This idea can be clearly presented in the example shown in Figure 12.2. In the figure, the label in a solid circle represents an intermediate relay and the label in a dashed circle represents a destination. Initially, the source node or message holder x has a message to deliver to the destination set $D_x = \{d_1, d_2, d_3, d_4, d_5\}$. When x meets a node y, if destinations d_1, d_3, d_5 are more socially similar to x than y, they will be allocated to D_x, and d_2, d_4 will be allocated to D_y if they are more socially similar to y. The notation "$S(x, d_i : d_j : d_k) > S(y, d_i : d_j : d_k)$" is a shortened form of "$S(x, d_i) > S(y, d_i)$ and $S(x, d_j) > S(y, d_j)$ and $S(x, d_k) > S(y, d_k)$." Later, when x meets node a and a meets node b, they will make decisions following the same rule. The multicast tree continues expanding until all of the destinations are reached.

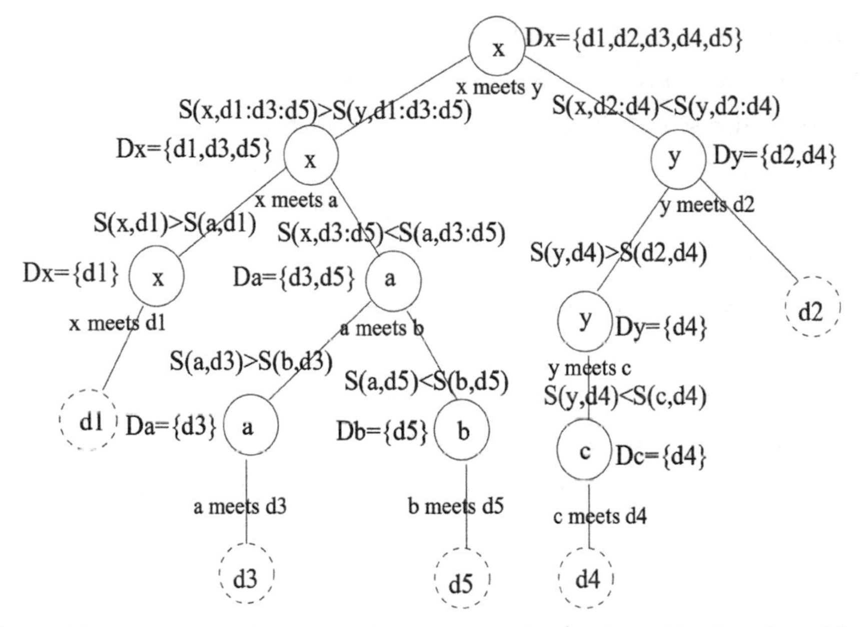

Figure 12.2 A tree showing the multicast process. The notation "$S(x, d_i : d_j : d_k) > S(y, d_i : d_j : d_k)$" means "$S(x, d_i) > S(y, d_i)$ and $S(x, d_j) > S(y, d_j)$ and $S(x, d_k) > S(y, d_k)$."

12.5 Analysis

In this section, we analyze the properties of our algorithms.

12.5.1 Property of Dynamic Social Feature Definition (12.2)

Theorem 1. *Suppose node x has met M_{xi} nodes with a certain social feature out of M_{xtotal} nodes it has met so far and node y has met M_{yi} nodes with the same social feature out of M_{ytotal} nodes it has met so far. Assume they have the same meeting frequency $p_i = M_{xi}/M_{xtotal} = M_{yi}/M_{ytotal}$ with these nodes, and $M_{xtotal} \leq M_{ytotal}$. According to Definition (12.2) of the dynamic social features, $x_i = (\frac{M_{xi}+1}{M_{xtotal}+1})^{p_i} * (\frac{M_{xi}}{M_{xtotal}+1})^{1-p_i}$ and $y_i = (\frac{M_{yi}+1}{M_{ytotal}+1})^{p_i} * (\frac{M_{yi}}{M_{ytotal}+1})^{1-p_i}$. Then, $x_i \leq y_i$. That is, Definition (12.2) breaks the tie of the same frequency by favoring the more active node.*

Proof. To prove $x_i \leq y_i$, it is equivalent to proving that $x_i - y_i \leq 0$. Expanding x_i and y_i and replacing M_{xi} by $p_i M_{xtotal}$ and M_{yi} by $p_i M_{ytotal}$, it is to prove that

$$\frac{(p_i M_{xtotal}+1)^{p_i} M_{xtotal}^{1-p_i}}{M_{xtotal}+1} - \frac{(p_i * M_{ytotal}+1)^{p_i} M_{ytotal}^{1-p_i}}{M_{ytotal}+1} \leq 0.$$

Multiplying the two sides by $(M_{xtotal}+1)(M_{ytotal}+1)M_{xtotal}^{p_i}M_{ytotal}^{p_i}$, we get $(p_i M_{xtotal} + 1)^{p_i} M_{xtotal}(M_{ytotal} + 1)M_{ytotal}^{p_i} - (p_i M_{ytotal} + 1)^{p_i}$ $M_{ytotal}(M_{xtotal} + 1)M_{xtotal}^{p_i} \leq 0$. Rearranging the inequality, it is to prove that

$$\left(\frac{p_i M_{xtotal} M_{ytotal} + M_{ytotal}}{p_i M_{xtotal} M_{ytotal} + M_{xtotal}}\right)^{p_i} \leq \frac{M_{xtotal} M_{ytotal} + M_{ytotal}}{M_{xtotal} M_{ytotal} + M_{xtotal}}.$$

Since $M_{ytotal} \geq M_{xtotal}$, $\frac{p_i M_{xtotal} M_{ytotal} + M_{ytotal}}{p_i M_{xtotal} M_{ytotal} + M_{xtotal}} \geq 1$. So the left side is a non-decreasing function with the increase of p_i. The maximum p_i is 1, so the maximum value of the left side is $\frac{M_{xtotal} M_{ytotal} + M_{ytotal}}{M_{xtotal} M_{ytotal} + M_{xtotal}}$, which is the right side. Therefore, the left side is less or equal to the right side. This proves the theorem.

12.5.2 The Number of Forwardings

Theorem 2. *In our routing framework, if there is only one destination d in the destination set D, the expected number of forwardings to reach the destination from source s is* $\ln g + 1$, *where g is the social similarity gap from s to d.*

Proof. The source node s has a social similarity gap g to the destination d. To reach d, the message will be delivered to a node with a smaller gap to d in each forwarding. For the convenience of later deduction, we set the gap from source s to d to be 1 and define the gap within which to reach d in one hop to be β as shown in Figure 12.3(a). In other words, if the message holder is within gap β to d, that node can deliver the message to d in one hop. Since the gap length is β and the gap from s to d is 1, the probability of a node falling in such a gap is β. So β is also the probability to reach d in one hop. Relative to the original gap g between s and d, gap β is equal to $\frac{1}{g}$.

Now, let us calculate the probability to reach d in h hops from s. If $h = 1$, that means d can be reached from s in one hop. That probability is β according to the above explanation. If $h = 2$, that means d can be reached from s in two hops. Then, there should be a relay lying between s and d as shown in Figure 12.3(b). Assume the gap from s to the relay is x and the gap from the relay to d is $1-x$. Now, the probability to reach d from s in two hops becomes $\frac{\beta}{1-x}$. Since x is in the range of $[0, 1-\beta]$, the overall probability to reach d

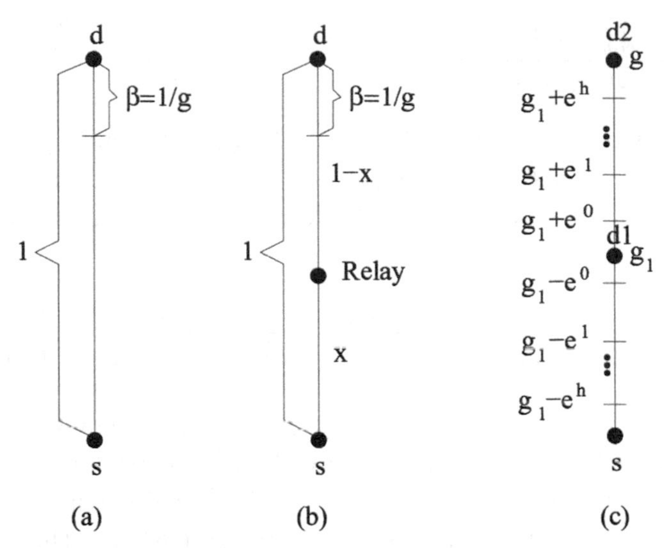

Figure 12.3 (a) One destination d, whose gap to source s is 1. The range to reach d in one hop is $\beta = 1/g$. (b) Reaching d in 2 hops via the relay node. The gap from s to the relay is x, and the gap from the relay to d is $1 - x$. (c) Two destinations d_1 and d_2, whose gaps to s are g_1 and g, respectively. We construct the range $[g_1 - e^h, g_1 + e^h]$ around g_1 to calculate the expected number of extra forwardings to reach d_1 after splitting.

in two hops should be $\int_0^{1-\beta} \frac{\beta}{1-x}dx$. The same reasoning can be extended to calculate the probability to reach d in h hops.

Therefore, the probability to reach d in
1 hop from s is: β,
2 hops from s is: $\int_0^{1-\beta} \frac{\beta}{1-x}dx = \beta \ln \frac{1}{\beta}$,
3 hops is: $\int_0^{1-\beta} \int_{x_1}^{1-\beta} \frac{\beta}{(1-x_1)(1-x_2)} dx_2 dx_1 = \frac{\beta}{2!}(\ln \frac{1}{\beta})^2$,
$\ldots$,
h hops is: $\int_0^{1-\beta} \int_{x_1}^{1-\beta} \cdots \int_{x_{h-1}}^{1-\beta} \frac{\beta}{(1-x_1)(1-x_2)\cdots(1-x_{h-1})} dx_{h-1} \cdots d_{x_1}$
$= \frac{\beta}{h!}(\ln \frac{1}{\beta})^h$, and so on.

These probabilities form a distribution as their summation $\sum_{h=0}^{\infty} \frac{\beta}{h!}(\ln \frac{1}{\beta})^h$ is 1 using the Taylor series for the exponential function e^x. Therefore, the expected number of forwardings is as follows: $\beta \cdot 1 + \beta \ln \frac{1}{\beta} \cdot 2 + \frac{\beta}{2!}(\ln \frac{1}{\beta})^2 \cdot 3 + \cdots = 1 + (\ln \frac{1}{\beta}) \sum_{h=1}^{\infty} \frac{\beta}{(h-1)!}(\ln \frac{1}{\beta})^{h-1}$. Using the Taylor series for e^x again, it is equal to $1 + \ln \frac{1}{\beta} \cdot \beta \cdot e^{\ln \frac{1}{\beta}} = 1 + \ln \frac{1}{\beta} = \ln g + 1$.

Theorem 3. *The expected number of forwardings in our routing framework with $k(k > 1)$ destinations is $\sum_{i=1}^{k-1} \ln(min(g-g_i, g_i)) + \ln g + O(k)$, where $g_i(1 \leq i \leq k-1)$ is the social similarity gap from source s to destination d_i and $g_k = g$ is the social similarity gap from the source to the farthest destination d_k.*

Proof. In our routing framework, the rule of compare-split is that when a message holder with k destinations meets another node, a destination d_i should be carried by the node that has a smaller social similarity gap to that destination. Let us first look at the 2-destination case as shown in Figure 12.3(c). Assume the social similarity gaps from source s to the farther destination d_2 and to the closer destination d_1 are $g_2 = g$ and g_1, respectively. We know from Theorem 12.5.2 that the expected number of forwardings to reach d_2 is $\ln g + 1$. Now, let us calculate the extra number of forwardings needed to reach d_1 after the two destinations split. From Theorem 12.5.2, the expected number of forwardings h to reach a destination with gap g from the source is $\ln g + 1$. So $g = e^{h-1}$. That means, if the message holder meets a node within the range of $[g_1 - e^0, g_1 + e^0]$, the expected number of hops to reach d_1 is $1(h = 1)$. If the message holder meets a node within the range of $[g_1 - e^1, g_1 + e^1]$ but not within the range of $[g_1 - e^0, g_1 + e^0]$, the expected number of hops to reach d_1 is $2(h = 2)$. In general, if the message holder meets a node within the range of $[g_1 - e^h, g_1 + e^h]$ but not within the range of $[g_1 - e^{h-1}, g_1 + e^{h-1}]$, the expected number of hops to reach d_1 is $h + 1$ and the probability to meet such a node is $\frac{2e^h}{g-g_1+e^h}$ from the gap range. Now, we discuss two cases: (1). $g_1 \leq \frac{g}{2}$ and (2). $g_1 > \frac{g}{2}$.

In case (1), if the two destinations split at the $h + 1$ ($h \geq 0$) hop, the expected number of extra forwardings to reach d_1 is

$$1 \cdot \frac{2e^0}{g-g_1+e^0} + 2 \cdot (\frac{2e^1}{g-g_1+e^1} - \frac{2e^0}{g-g_1+e^0}) + 3 \cdot (\frac{2e^2}{g-g_1+e^2} - \frac{2e^1}{g-g_1+e^1})$$
$$+ \cdots + \lceil \ln g_1 \rceil (1 - \frac{2e^{\lfloor \ln g_1 \rfloor - 1}}{g-g_1+e^{\lfloor \ln g_1 \rfloor - 1}}) = \lceil \ln g_1 \rceil - \sum_{h=0}^{\lfloor \ln g_1 \rfloor - 1} \frac{2e^h}{g-g_1+e^h}.$$

From $g_1 \leq \frac{g}{2}$ and $e^{\lfloor \ln g_1 \rfloor} \leq g_1$, we have

$$\sum_{h=0}^{\lfloor \ln g_1 \rfloor - 1} \frac{2e^h}{2(g-g_1)} \leq \sum_{h=0}^{\lfloor \ln g_1 \rfloor - 1} \frac{2e^h}{g-g_1+e^h} \leq \sum_{h=0}^{\lfloor \ln g_1 \rfloor - 1} \frac{2e^h}{g-g_1}.$$

That is, $\frac{1}{2} \frac{2(g_1-1)}{(g-g_1)(e-1)} \leq \sum_{h=0}^{\lfloor \ln g_1 \rfloor - 1} \frac{2e^h}{g-g_1+e^h} \leq \frac{2(g_1-1)}{(g-g_1)(e-1)}$.

Again from $g_1 \leq \frac{g}{2}$,

$\frac{1}{2} \cdot \frac{2}{e-1} \leq \sum_{h=0}^{\lfloor \ln g_1 \rfloor - 1} \frac{2e^h}{g-g_1+e^h} \leq \frac{2}{e-1}$.

This means that $\sum_{h=0}^{\lfloor \ln g_1 \rfloor - 1} \frac{2e^h}{g-g_1+e^h}$ is a constant. So the expected number of extra forwardings to reach d_1 is $\ln g_1 + O(1)$.

In case (2), if the two destinations split at the $h+1$ ($h \geq 0$) hop, the expected number of extra forwardings to reach d_1 is $1 \cdot \frac{2e^0}{g-g_1+e^0} + 2 \cdot (\frac{2e^1}{g-g_1+e^1} - \frac{2e^0}{g-g_1+e^0}) + 3 \cdot (\frac{2e^2}{g-g_1+e^2} - \frac{2e^1}{g-g_1+e^1}) + \cdots + \lceil \ln(g-g_1) \rceil (1 - \frac{2e^{\lfloor \ln(g-g_1) \rfloor - 1}}{g-g_1+e^{\lfloor \ln(g-g_1) \rfloor - 1}}) = \ln(g-g_1) + O(1)$.

Combining the cases (1) and (2), the expected number of extra forwardings to reach d_1 is $\ln(min(g-g_1, g_1)) + O(1)$. Adding the expected number of forwardings to reach d_2, the total expected number of forwardings to reach the two destinations is $\ln(min(g-g_1, g_1)) + \ln g + O(1)$.

We extend the same analysis idea to the k-destination case. The expected number of forwardings to reach the farthest destination d_k is $\ln g + 1$, and the expected number of extra forwardings to reach each other destination $d_i (i \neq k)$ is $\ln(min(g-g_i, g_i)) + \ln g + O(1)$. Then, the total expected number of forwardings to reach all of the k destinations is $\sum_{i=1}^{k-1} \ln(min(g - g_i, g_i)) + \ln g + O(k)$.

12.5.3 The Number of Copies

Theorem 4. *The number of copies produced by our routing framework is k, where k is the number of destinations in the multicast set.*

Proof. It is trivial to see that each split of the destinations will produce one extra copy. There are k destinations, so it takes $k-1$ splits to separate the k destinations into individual ones. Adding the original one copy, the number of copies produced by our routing framework is k.

12.6 Simulations

In this section, we evaluate the performance of our multicast algorithms by comparing them with their variations and the existing ones using a custom simulator written in Java. The simulations were conducted using a real conference trace [26] representing an OMSN created at INFOCOM 2006. The trace dataset consists of two parts: *contacts* between the iMote devices that were carried by conference participants and the self-reported *social features* of the participants, which were collected using a questionnaire form. The six social features considered were *Affiliation*, *City*, *Nationality*, *Language*, *Country*, and *Position*.

12.6.1 Algorithms Compared

We compared the following related multicast protocols.

1. *The Flooding Algorithm* (Flooding) [16]: The message is spread epidemically throughout the network until it reaches all of the destinations.
2. *The Social Profile-based Multicast Routing Algorithm* (SPM) [14]: The multicast algorithm based on static social features in user profiles.
3. *The Multi-SoSim Algorithm* (Multi-SoSim): Our multicast algorithm based on dynamic social feature Definition (12.1) and using the Euclidean social similarity metric.
4. *The E-Multi-SoSim Algorithm* (E-Multi-SoSim): Our multicast algorithm based on dynamic social feature Definition (12.2) and using the Euclidean social similarity metric.
5. *Variation* 1 *of the Multi-SoSim Algorithm* (Multi-FwdNew): This algorithm is similar to Multi-SoSim, but a message holder only forwards the message to a newly met node whose destination set is empty.
6. *Variation* 2 *of the Multi-SoSim Algorithm* (Multi-Unicast): The message to multiple destinations is delivered by multiple independent unicasts (from the source to each of these destinations), where each unicast is conducted using dynamic social features.

12.6.2 Evaluation Metrics

We used three important metrics to evaluate the performance of the multicast algorithms. Since a multicast involves multiple destinations, we define a *successful multicast* as the one that successfully delivers the message to all of the destinations. The three metrics are as follows: (1) *delivery ratio*: the ratio of the number of successful multicasts to the number of total multicasts generated; (2) *delivery latency*: the time between when the source starts to deliver the message and when all of the destinations have received the message; and (3) *number of forwardings*: the number of forwardings needed to deliver the message to all of the destinations.

12.6.3 Simulation Setup

In our simulations, we divided the whole trace time into 10 intervals. Thus, 1 time interval is 0.1 of the total time length and 10 time intervals make up the length of the whole trace. For each of the algorithms compared, we tried the sizes of the destination sets to be 5 and 10. In each experiment, we randomly generated a source and its destination set. We ran each algorithm 300 times and averaged the results of the evaluation metrics.

12.6.4 Simulation Results

The simulation results with 5 and 10 destinations are shown in Figures 12.4 and 12.5, respectively. For the flooding algorithm, as expected, it achieves the highest delivery ratio and the lowest delivery latency (almost close to 0 compared with others in the figures) at the cost of sending a copy to any newly met node. Thus, it has the highest number of forwardings. The Multi-SoSim algorithm outperforms SPM in having a higher delivery ratio and lower latency with a little increase in the number of forwardings. This is because the dynamic social features in Multi-SoSim can more accurately capture node encounter behavior than the static social features in SPM so that multicast efficiency can be improved. The little increase in the forwardings indicates that Multi-SoSim is more actively delivering the message to the destinations.

Figures 12.6 and 12.7 show the zoom-in simulation results of Multi-SoSim, Multi-Unicast, and Multi-FwdNew algorithms with 5 and 10 destinations. There is not much difference in delivery ratio and latency between Multi-SoSim and Multi-Unicast in this simulation as their curves are overlapped in

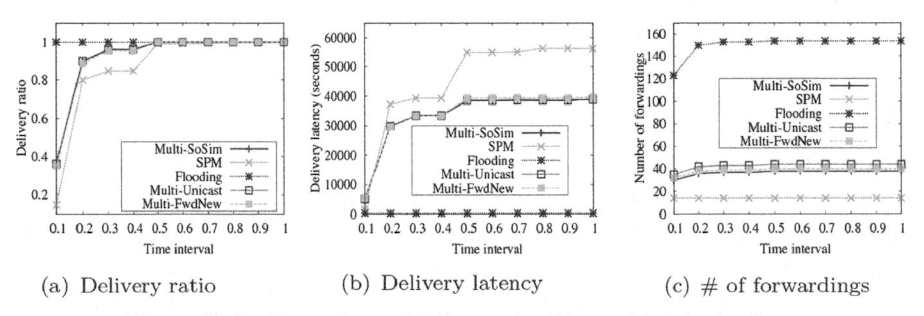

(a) Delivery ratio (b) Delivery latency (c) # of forwardings

Figure 12.4 Comparison of different algorithms with 5 destinations.

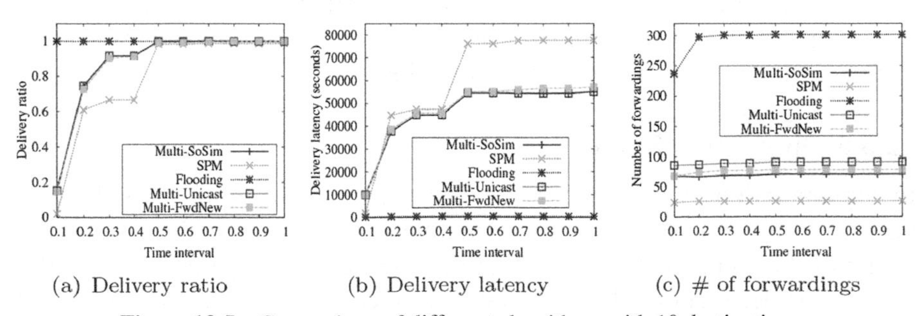

(a) Delivery ratio (b) Delivery latency (c) # of forwardings

Figure 12.5 Comparison of different algorithms with 10 destinations.

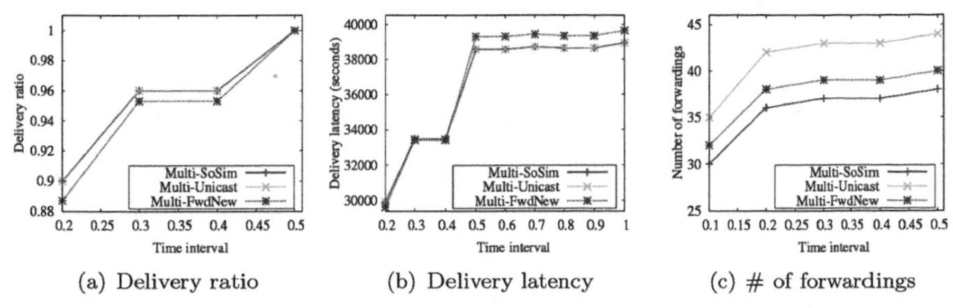

(a) Delivery ratio (b) Delivery latency (c) # of forwardings

Figure 12.6 Comparison of Multi-SoSim, Multi-Unicast, and Multi-FwdNew with 5 destinations.

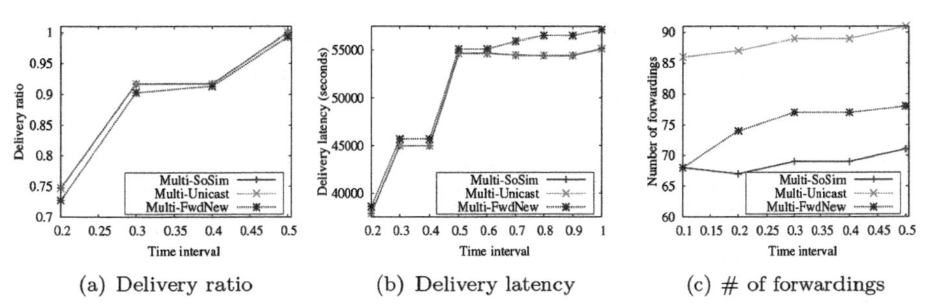

(a) Delivery ratio (b) Delivery latency (c) # of forwardings

Figure 12.7 Comparison of Multi-SoSim, Multi-Unicast, and Multi-FwdNew with 10 destinations.

the figures. But Multi-SoSim decreases the number of forwardings in Multi-Unicast by 16.7% and 29.9% with 5 and 10 destinations, respectively. This is because letting the destinations share the path in Multi-SoSim can reduce the number of forwarding nodes, especially when the number of destinations is increased. Multi-SoSim outperforms Multi-FwdNew in delivery ratio, latency, and the number of forwardings. With 5 destinations, the Multi-SoSim algorithm increases the delivery ratio by 1.5%, decreases latency by 2.0%, and decreases the number of forwardings by 6.7% compared with Multi-FwdNew. With 10 destinations, the Multi-SoSim algorithm increases the delivery ratio by 2.8%, decreases latency by 3.9%, and decreases the number of forwardings by 11.6%. This is because Multi-SoSim selects a better forwarder for each of the destinations whenever a message holder meets another node while Multi-FwdNew does that only when a newly met node is encountered.

Figures 12.8 and 12.9 present the comparison of Multi-SoSim and E-Multi-SoSim algorithms with 5 and 10 destinations. With 5 destinations, the

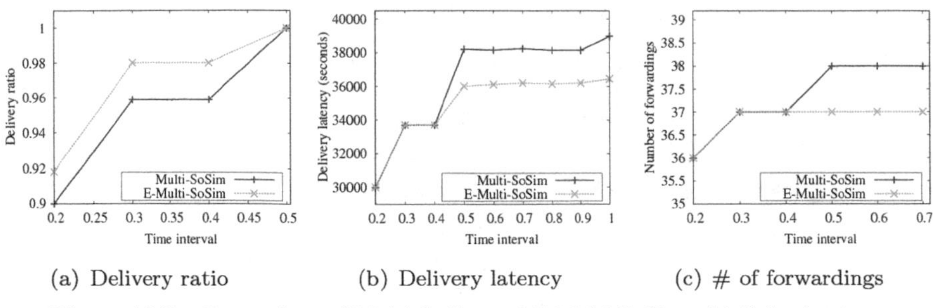

(a) Delivery ratio (b) Delivery latency (c) # of forwardings

Figure 12.8 Comparison of Multi-SoSim and E-Multi-SoSim with 5 destinations.

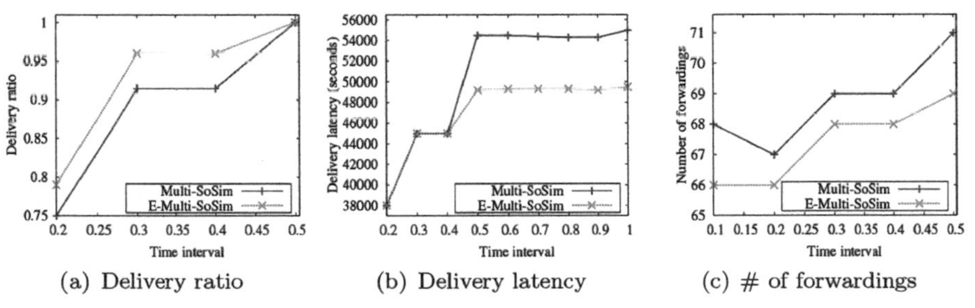

(a) Delivery ratio (b) Delivery latency (c) # of forwardings

Figure 12.9 Comparison of Multi-SoSim and E-Multi-SoSim with 10 destinations.

E-Multi-SoSim algorithm increases the delivery ratio by 2.1%, decreases latency by 6.4%, and decreases the number of forwardings by 2.7% compared with Multi-SoSim. With 10 destinations, the E-Multi-SoSim algorithm increases the delivery ratio by 4.3%, decreases latency by 2.9%, and decreases the number of forwardings by 10.6%. This is because the enhanced dynamic social features in E-Multi-SoSim can more accurately capture nodes' dynamic contact behavior to improve multicast efficiency.

12.7 Conclusion

In this chapter, we have proposed a novel social similarity-based multicast framework for OMSNs where node connections are established opportunistically. We have instantiated this framework with two algorithms Multi-SoSim and E-Multi-SoSim based on a compare-split scheme to select the best relay node for each of the destinations in each hop to improve multicast efficiency and dynamic and enhanced dynamic social features to capture nodes' contact behavior. We have conducted a theoretical analysis of our proposed algorithms and evaluated their performance by comparing them with other related algorithms through simulations using a real trace representing an OMSN. The simulation results have verified the advantages of the dynamic social features over the static ones and the appropriateness of the compare-split scheme adopted in our multicast algorithms. In our future work, we plan to test our algorithms using more traces in OMSNs as they become available.

Acknowledgements

This research was supported in part by NSF CNS grant 1305302 and NSF ACI grant 1440637.

References

[1] Fan, J., Chen, J., Du, Y., Gao, W., Wu, J., and Sun, Y. (2013). Geo-community-based broadcasting for data dissemination in mobile social networks, *IEEE Trans. Parallel Distributed Syst.*, 24 (4), 734–743.

[2] Wu, J., and Wang, Y. (2014). *Opportunistic mobile social networks*, Taylor & Francis.

[3] Xiao, M., Wu, J., and Huang, L. (2014). Community-Aware Opportunistic Routing in Mobile Social Networks, IEEE Trans. on Computers 63 (7), 1682–1695.

[4] Zhou, H., Chen, J., Fan, J., Du, Y., and Das, S. K. (2013). ConSub: Incentive-Based Content Subscribing in Selfish Opportunistic Mobile Networks. *IEEE J. Selected Areas Commun.*, 31 (9), 669–679.

[5] Zhou, H., Chen, J., Zhao, H. Y., Gao, W., Cheng, P. (2013). On exploiting contact patterns for data forwarding in duty-cycle opportunistic mobile networks, *IEEE Trans. Veh. Tech.*, 62 (9), 4629–4642.

[6] DTN Research Group, http://w5, www.dtnrg.org/

[7] Guo, B., Zhang, D., Yu, Z., Zhou, X., and Zhou, Z. (2012). Enhancing spontaneous interaction in opportunistic mobile social networks. *Commun. Mobile Comput.*, 1, 1–6.

[8] Jedari, B., and Xia, F. A Survey on Routing and Data Dissemination in Opportunistic Mobile Social Networks, http://arxiv.org/abs/1311.0347.

[9] Lee, U., Oh, S. Y., Kang-Won, L., and Gerla, M. (2008). Relaycast: scalable multicast routing in delay tolerant networks. In *Proceedings of IEEE ICNP*, pp. 218–227.

[10] Mongiovi, M., Singh, A. K., Yan, X., Zong, B., and Psounis, K. (2012). Efficient multicasting for delay tolerant networks using graph indexing. In *Proceedings of IEEE INFOCOM*.

[11] Wang, Y., and Wu, J. (2012). A dynamic multicast tree based routing scheme without replication in delay tolerant networks. *J. Parall. and Distribut. Comput.* 72 (3), 424–436.

[12] Xi, Y., and Chuah, M. (2009). An encounter-based multicast scheme for disruption tolerant networks. *Comput. Commun.* 32 (16), 1742–1756.

[13] Zhao, W., Ammar, M., and Zegura, E. (2005). Muticasting in delay tolerant networks: semantic models and routing algorithms. In *Proceedings of ACM WDTN*, pp. 268–275.

[14] Deng, X., Chang, L., Tao, J., Pan, J., and Wang, J. (2013). Social profile-based multicast routing scheme for delay-tolerant networks. In *Proceedings of IEEE ICC*, pp. 1857–1861.

[15] Mei, A., Morabito, G., Santi, P., and Stefa, J. (2011). Social-aware stateless forwarding in pocket switched networks. In *Proceedings of IEEE INFOCOM*, pp. 251–255.

[16] Vahdat, A., and Becker, D. (2000). Epidemic routing for partially connected ad hoc networks. Technical Report, CS-200006, Duke University.

[17] Chuah, M., and Yang, P. (2009). Context-aware multicast routing scheme for disruption tolerant networks. *J. Ad Hoc Ubiquitous Comput.,* 4 (5), 269–281.

[18] Gao, W., Li, Q., Zhao, B., and Cao, G. (2009). Multicasting in delay tolerant networks: a social network perspective. In *Proceedings of ACM MobiHoc*.

[19] Hu, J., Yang, L. L., and Hanzo, L. (2014). Distributed cooperative social multicast aided content dissemination in random mobile networks. *IEEE Trans. Vehicular Technol.* 64 (7) 3075–2229.

[20] Hu, J., Yang, L. L., Poor, H. V., and Hanzo, L. (2015). Bridging the social and wireless networking divide: Information dissemination in integrated cellular and opportunistic networks. *IEEE Access* 3, 1809–1848.

[21] Tanimoto Coefficient, https://docs.tibco.com/pub/spotfire/7.0.1/doc/html //hc/hc_tanimoto_coefficient.htm

[22] Cosine Correlation, https://docs.tibco.com/pub/spotfire/7.0.1/doc/html //hc/hc_cosine_correlation.htm

[23] Euclidean Distance, https://docs.tibco.com/pub/spotfire/7.0.1/doc/html //hc/hc_euclidean_distance.htm

[24] Rothfus, D., Dunning, C., and Chen, X. (2013). Social-similarity-based routing algorithm in delay tolerant networks. In *Proceedings of IEEE ICC*, pp. 1862–1866.

[25] Han, J. W., Kamber, M., and Pei, J. (2012). *Data Mining: Concepts and Techniques*, Morgan Kaufmann, MA, USA.

[26] Scott, J., Gass, R., Crowcroft, J., Hui, P., Diot, C., and Chaintreau, A. Crawdad trace cambridge/haggle/imote/infocom2006(v.2009-05-29), http://crawdad.cs.dartmouth.edu/cambridge/haggle/imote/infocom2006 (May 2009).

13

Ensuring QoS for IEEE 802.11 Real-Time Communications Using an AIFSN Prediction Scheme

Estefanía Coronado, José Villalón and Antonio Garrido

High-Performance Networks and Architectures (RAAP),
Albacete Research Institute of Informatics (I3A),
University of Castilla-La Mancha, Albacete, Spain

Abstract

The incessant development of High Quality (HQ) multimedia contents and the trend towards the use of wireless technologies have as a consequence the need for providing the users with an adequate level of Quality of Service (QoS) in IEEE 802.11 networks. The IEEE 802.11e amendment aims to overcome this situation by introducing the Enhanced Distributed Channel Access (EDCA) access method. This new method is characterized through a group of Medium Access Control (MAC) parameters, which are able to classify and prioritize the different types of traffic. In this regard, the most determining parameter is the Arbitration Inter-Frame Space Number (AIFSN). On this basis, and with the aim of improving the performance of the voice and video applications offered by EDCA, we propose a new AIFSN adaptation scheme. This scheme makes use of a *J48* tree classifier and a *M5* regression model. Our proposal is able to determine dynamically the optimum AIFSN values with regard to the network conditions, maintaining the backward compatibility with the stations that use the original IEEE 802.11 standard. The prediction algorithm is only queried by the Access Point (AP), without introducing additional control traffic into the network, making it possible to use it in real time. With respect to the standard EDCA values, the results show an enhancement in the quality of the voice and video communications and a significant reduction in the number of the retransmission attempts.

Keywords: QoS, 802.11e, EDCA, Artificial Intelligence.

13.1 Introduction

Over the past few years, wireless technologies have become imperative in the context of networking and communications. Their simplicity of deployment, multimedia content support and lower cost are displacing the traditional wired networks. In order to define this networking model, the Institute of Electrical and Electronics Engineers (IEEE) developed the 802.11 standard [1], which introduces the set of media access and physical layers specifications for implementing Wireless Local Area Networks (WLANs). However, the use of the Internet and the consumption patterns are changing rapidly, especially those related to multimedia contents. The nature of such contents involves temporal restrictions that require Quality of Service (QoS) mechanisms to ensure an adequate level of satisfaction in the user perception. For this reason, the IEEE 802.11e amendment [2] was released with the aim of classifying and prioritizing the different types of traffic in wireless networks.

One of the most relevant features introduced by the IEEE 802.11e amendment is the possibility of differentiating traffic flows and services. As a consequence, the QoS level of the wireless network is notably improved. For this purpose, this amendment defines a new contention-based channel access method called Enhanced Distributed Channel Access (EDCA), which allows for the prioritization of the different types of traffic, making use of a set of user priorities. To this end, several values for the medium access parameters are defined. These parameters include the value of the Arbitration Inter-Frame Spacing Number (AIFSN), the size of the Contention Window (CW), and the Transmission Opportunity interval (TXOP). Nevertheless, in some research it has been demonstrated that EDCA does not provide the required QoS level for real-time applications. This situation worsens in the cases in which the network is partially or fully composed of stations that only support the IEEE 802.11 standard and the traffic load level increases.

The application of Artificial Intelligence (AI) and Data Mining (DM) techniques allows traffic patterns to be found in a network, making it possible to prioritize each traffic flow accordingly. As a result, a higher level of QoS and a general improvement in the performance are achieved. In this chapter we propose a new prediction scheme for the AIFSN priority values based on network conditions, focusing on the temporal restrictions of the voice and video transmissions and enhancing their normalized throughput. This operation can be performed dynamically in the Access Point (AP) of the network without altering the remaining devices, therefore maintaining full compatibility. Furthermore, the adjustment of these parameters does not introduce additional control traffic into the network. In short, the main contribution of the proposed model is to address the existing limitations of the IEEE 802.11e amendment, providing QoS mechanisms for multimedia transmissions and maintaining the compatibility with existing devices.

The remainder of this chapter is organized as follows. Section 13.2 reviews the IEEE 802.11e amendment and gives some background information on the points that aim to improve the QoS level offered by EDCA. In Section 13.3, the process of supervised learning of the chosen predictive models is described. In Section 13.4, we present the proposed prediction scheme and the process followed throughout its design. The results of the performance evaluation and a comparison with the AIFSN configuration defined in the IEEE 802.11e amendment are described in Section 13.5. Finally, Section 13.6 provides some concluding remarks on our proposal.

13.2 QoS in IEEE 802.11 Networks

Initially, the original IEEE 802.11 standard introduced two medium access functions: the Distributed Coordination Function (DCF) and the Point Coordination Function (PCF). However, these medium access functions are not able to differentiate traffic flows and provide the required QoS level. Therefore, the IEEE formed a working group with the task of developing the IEEE 802.11e amendment that considered these aspects.

13.2.1 IEEE 802.11e

The IEEE 802.11e amendment was developed with the aim of providing QoS support and meeting the voice and video streams requirements over IEEE 802.11 WLANs [2]. As backward compatibility must be kept, a distinction is drawn between the stations that support QoS (QSTAs) and the stations that do not offer such support (nQSTAs), only using DCF. For this purpose, the 802.11e amendment implements the Hybrid Coordination Function (HCF) and, thus, its two contention-based channel access methods: HCF Controlled Channel Access (HCCA) and EDCA. According to this amendment, the implementation of the HCF coordination function is mandatory for all the QSTAs. Nevertheless, only EDCA is supported by the commercial network cards on current devices as a method for accessing the wireless medium.

The EDCA channel access method distinguishes between eight different User Priorities (UPs). Moreover, four Access Categories (ACs) are defined, which are derived from the UPs and are able to classify and prioritize the traffic streams. In this way, in order from highest to lowest priority, Voice (VO), Video (VI), Best Effort (BE) and Background (BK) access categories are considered, as shown in Figure 13.1. Each one of these ACs works on its own transmission queue and is characterized by an EDCA parameter set. This EDCA parameter set specifies a priority level by using an AIFSN value, a specific duration for the CW and a TXOP interval. Thus, the AP sends this EDCA parameter set via beacon frames to the stations of a Basic Service Set (BSS). The IEEE 802.11e amendment allows the APs to modify the aforementioned values. However, no mechanism is considered in this amendment for carrying out this task and most commercial devices do not implement such a service.

The AIFSN determines the Arbitration Inter-Frame Spacing (AIFS), which is the period of time that a station has to wait until it is allowed to initiate a new transmission. The AIFS for each AC is shown in Equation (13.1), where the *SlotTime* denotes the duration of a slot according to the physical layer, and the Short Inter-frame Space (SIFS) refers to the amount of time used by high priority actions that require an immediate response.

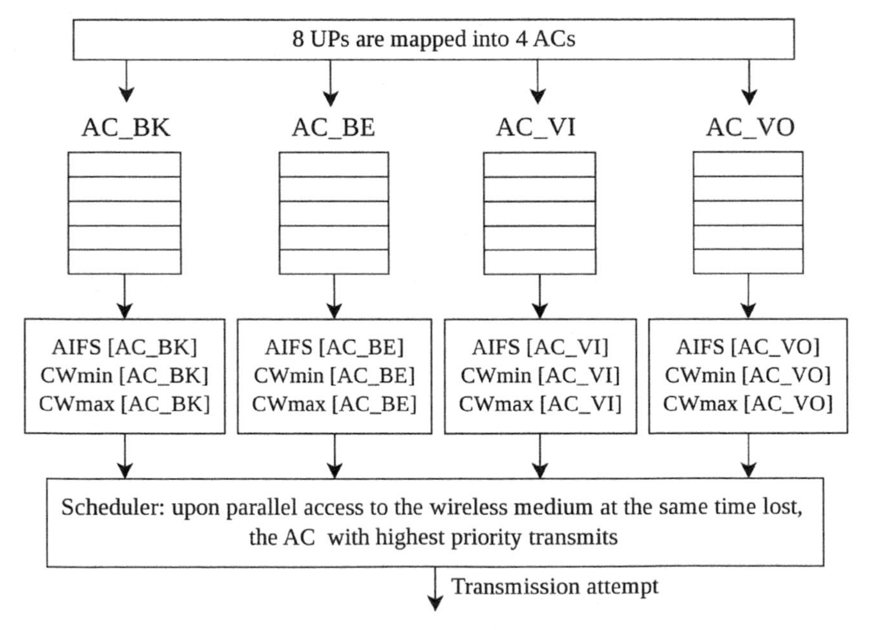

Figure 13.1 EDCA access categories mapping.

$$AIFS[AC] = AIFSN[AC] \cdot SlotTime + SIFS \tag{13.1}$$

Moreover, the stations are assigned an AIFSN value according to their priority, which must be higher than or equal to 2. In order to provide a fair transmission for the legacy stations, the IEEE 802.11e amendment defines a standard combination of AIFSN parameters, as shown in Table 13.1. Meanwhile, the CW size determines the length of time that a station must wait until it is able to conclude the Backoff algorithm. In this way, the CW values are assigned in the inverse order to that of the priority of the corresponding AC. TXOPs allows it to transmit multiple streams without gaining medium access every time that a frame is transmitted and are usually used for real-time applications.

With regard to these parameters, the AIFSN plays the most important role in order to ensure optimum traffic differentiation. In [3], Villalón et al. show several scenarios in which a set of configurations for the AIFSN and CW are taken into account. In this case, they prove that the AIFSN has a greater relevance when identifying priorities than the CW. This conclusion was also reached by Hui et al. in [4], who proved that both the collisions and access media delay decrease, allowing for an improvement in the global throughput of the network.

13.2.2 Dynamic Adaptation in IEEE 802.11e

Wireless network conditions, such as the traffic load, can change over time. Consequently, several dynamic proposals that consider the aforementioned circumstances have emerged. Their main aim is to adapt the EDCA parameter set, i.e., to identify the optimal values for the AIFSN, CW_{max}, CW_{min} and TXOP parameters.

An approach with this same goal is presented in [5], where He et al. take into account three possible load levels, showing the behavior of the proposed scheme under different network conditions. This proposal achieves a reduction in the number of retransmission attempts and an enhancement in the network performance. In spite of this, there is a drop in the amount of voice and video information transmitted, which impairs its temporal restrictions.

Table 13.1 Default EDCA parameter set for IEEE 802.11g PHY layer

AC	CW_{min}	CW_{max}	AIFSN	TXOP
AC_BK	aCW_{min}	aCW_{max}	7	–
AC_BE	aCW_{min}	aCW_{max}	3	–
AC_VI	$(aCW_{min}+1)/2-1$	aCW_{min}	2	3.008 ms
AC_VO	$(aCW_{min}+1)/4-1$	$(aCW_{min}+1)/2-1$	2	1.504 ms

In [6] Nilsson et al. introduce an adaptation scheme by using the CW size, achieving better results than EDCA. However, compatibility with legacy DCF stations is not considered. Banchs et al. introduce in [7] a new way of offering backward compatibility with the DCF stations. This algorithm is able to prioritize the voice and video frames over the others. As the priority of the DCF stations cannot be modified by updating the EDCA parameter set, the CW size is increased by retransmitting packets that are properly received by the DCF stations. In this way, the priority of the stations that use this medium access function decreases. Nevertheless, unnecessary traffic is introduced into the network.

The design of an analytical model to improve the network performance has also been taken into account. Nevertheless, most of these models make assumptions that may not be fulfilled in real transmissions. In [8] Gallardo et al. define a model by using Markov chains. However, they consider the same bit rate for all the stations. In a similar way, the mathematical model presented in [9] by Banchs et al. is only tested under network saturation conditions. In spite of being less frequent than Markovian models, p-persistent models have been also taken into account to overcome this situation. In [10], Mackenzie et al. make an accurate time-domain analysis to verify the performance of the CSMA/CA algorithm. Moreover, the operating mode of EDCA is modeled from a different point of view. After this model is designed, it is properly validated via simulation. Nevertheless, it is assumed that the stations always have some information to be transmitted and that no errors occur in the wireless channel.

As can be seen, most of these approaches are not able to keep backward DCF compatibility and simultaneously provide a dynamic adaptation of the EDCA parameter set without introducing additional traffic in the network. Furthermore, in the existing literature on the topic, there is no evidence of other research work that makes use of AI techniques to improve the QoS level by tuning such a parameter set.

13.3 Supervised Learning

In supervised learning [11, 12], the information relative to objects or instances is represented by a set of *n* input features, $X = (X_1, \ldots, X_n)$, and an output variable, *Y*. The process consists of learning a model, $h_\Theta(x)$, from a training dataset, (X, Y), which contains the information relative to several objects whose current outputs are already known (that is why it is called *supervised*).

The model is then used to predict the output value y for new cases when only the values of their input features (x) are known. In case of regression problems, $Y \in \mathbb{R}$, and therefore $h_\Theta(x) \in \mathbb{R}$. In classification, however, the goal is to determine the class of a certain instance. In such a case, $Y \in \{c_1, \ldots, c_K\}$.

In this work, both classification and regression algorithms are used. Accordingly, X represents the configuration of the different parameters used for managing the multimedia traffic in a network, whereas the output Y represents the throughput achieved by the setting. Thus, $h_\Theta(x) \in \mathbb{R}$ returns the predicted throughput of the network, y, given the parameter configuration x.

There is a large number of supervised learning models for regression, such as Linear Regression [13], Neural Networks [14], Support Vector Machines [15], or Regression Trees [16]. The choice of a certain model depends on several factors. Thus, some are more powerful than others, i.e., achieve more precision and can detect more relevant patterns in data. However, the ease in which they can be interpreted can be an issue in some scenarios. Models such as Neural Networks are considered *Black Box* models, as the information related to underlying patterns in data cannot be drawn from them. In contrast, regression trees are very easy to interpret and provide useful information on the relation between input and output features. Another important issue concerns computational complexity. For instance, obtaining y from x with a Neural Network implies some matrix multiplications and can be too slow in some settings. However, processing a regression tree might only require a few comparisons.

However, the requirements of the issue in question determine directly the type of learning model that must be selected. In particular, the application domain of this work requires the models to be used in real-time. For this reason, they must not be computationally complex. Furthermore, to fit the problem, models must also be analyzable, interpretable and modifiable after they have been learned. On this basis, the prediction scheme is finally composed of a *J48* decision tree classifier and a *M5* regression model. The most relevant features of these models and the possibilities that they may offer are detailed below.

13.3.1 *J48* Decision Tree Classifier

The *J48* decision tree classifier is based on the *C4.5* algorithm, which is the successor to the ID3 algorithm [17]. This tree can be found in the *weka* package for machine learning [18]. This model aims to design a decision tree that is as short as possible.

The algorithm follows a recursive procedure by means of a heuristic greedy search to obtain the final model. In this way, it selects every attribute according to its gain ratio (see Equation (13.2)). This guideline expressly refers to the information gain obtained as a result of the classification made and the entropy of the predictive variable, X_i.

$$Gain\ ratio = \frac{I(C, X_i)}{H(X_i)} = \frac{H(C) - H(C/X_i)}{H(X_i)}. \qquad (13.2)$$

The information gain is given by the expression $I(C, X_i)$, which calculates the amount of mutual information between X_i and C. In other words, the algorithm evaluates the potential uncertainty when classifying an attribute X_i on a set C. This, in turn, is calculated as the difference between the entropy of the different outputs of the set C, $H(C)$, and the entropy obtained after using a certain attribute X_i, $H(C, X_i)$. In order to prevent the variables with a wider range of possible values from being the biggest beneficiaries when carrying out the classification, the information gain is weighted with the entropy of the predictive variable, $H(X_i)$.

The algorithm divides the training data set into several subsets that are as pure as possible until a leaf node is reached. In this respect, the following internal node to be selected is the attribute which maximizes the aforementioned gain ratio. Once the final tree has been modeled, this algorithm also incorporates a pruning technique to reduce its size and complexity.

In the context of this work, an example of a subtree of the *J48* classifier designed can be seen in Figure 13.2. In this way, it is possible to calculate the label for a certain parameter combination which is situated at the leaf nodes. This label is obtained by using the rest of the parameters as an entry point for the tree.

13.3.2 *M5Rules*

The *M5* algorithm [16] represents $h_\Theta(x) \in \mathbb{R}$ as a regression tree, and it is very similar to its counterpart, *c4.5* [17], which is used for classification problems. In fact, given the requirements of this work, the similarities of both algorithms and their different point of view to address a problem have also contributed to select them as prediction models for the adaptation scheme.

A regression tree represents a partition of the input space. Each node contains a condition defined over some input attribute X_i. For instance, if a node of the tree is defined by the condition `[VI_channel_occupancy <= 0.341]`, it represents the one branch which would be used to process all objects so

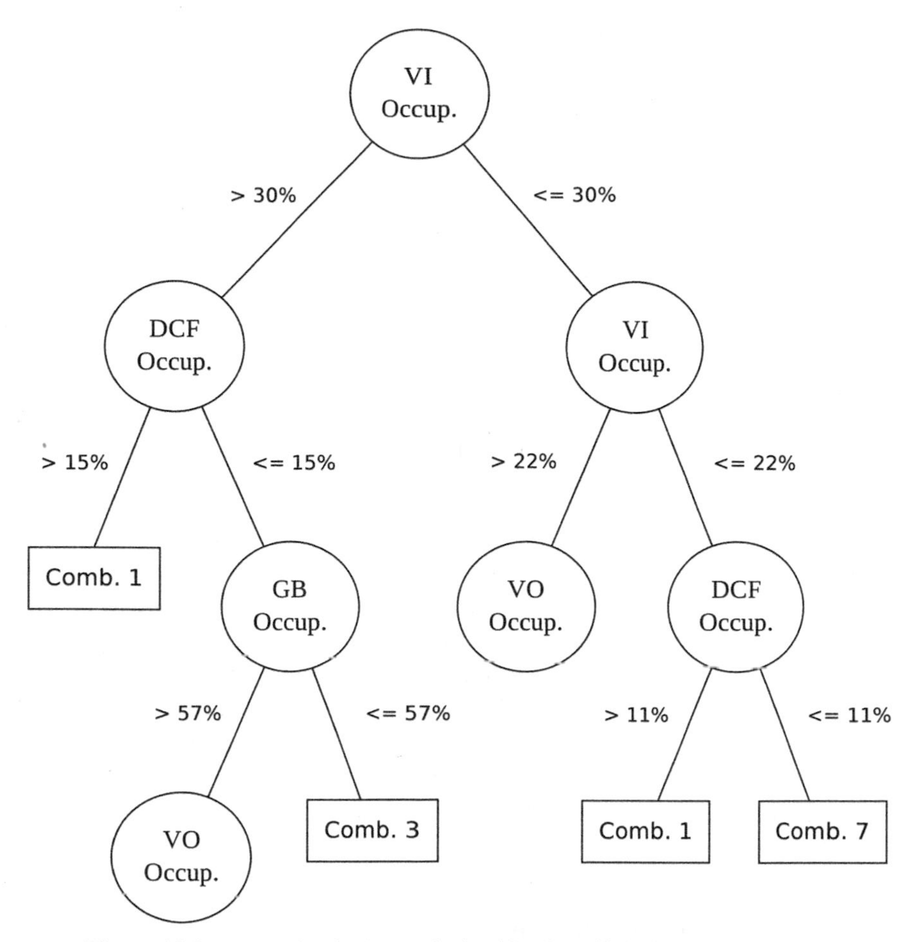

Figure 13.2 Example of subtree obtained by the *J48* decision tree classifier.

that their value for variable `VI_channel_occupancy` is smaller than 0.341, whereas the other branch would be used to process the rest of the cases. Each leaf represents an input subspace and corresponds to all the cases which fit the conditions represented by the path from the root of the tree to the leaf.

In *M5*, there are two possibilities to obtain the output values for the cases falling into a leaf of the tree. The first one, namely *regression tree*, uses the mean output value of the training data falling into that leaf as default prediction. The second one, namely *model tree*, learns a multivariate linear regression equation from the training data corresponding to the leaf, and uses it to predict the output values.

```
Rule: 1
IF
VI_channel_occupancy <= 0.341
THEN

max_th[0] =
-0.0449 * global_channel_occupancy
+ 0.0701 * DCF_channel_occupancy
+ 0.1152 * BE_channel_occupancy
- 0.0392 * VI_channel_occupancy
- 0.279 * VO_channel_occupancy
+ 2.0059 [151/2.844\%]
```

Figure 13.3 Example of rule induced by *M5Rules*.

The algorithm used in this work, namely *M5Rules*, is included in the *weka* package for machine learning [19]. It first learns a regression (or model) tree from training data by means of the implementation of *M5* included in this package, namely *M5P*, and then extracts a set of rules. Figure 13.3 shows an example of the obtained rules. So that the value of all objects for variable `[VI_channel_occupancy <= 0.341]`, and the output value is obtained as a linear expression from the rest (5) of the variables. The information `[151/2.844\%]` indicates that 151 objects of the training dataset fall into that leaf, and that the relative error obtained with the linear expression for those objects is 2.844%.

13.4 AIFSN Tuning Scheme

13.4.1 Proposal Description

Recently, the consumption of multimedia contents through wireless networks has shown a considerable increase. In this regard, the research on QoS has become especially relevant since the IEEE 802.11e amendment was published. However, there has been a recurring problem in such research due to the existence of stations that only support the original IEEE 802.11 standard. Most of these efforts are focused on improving the features of the EDCA channel access method. However, even though the different EDCA parameters can be adapted to network conditions, no changes can be made in the case of DCF stations.

In IEEE 802.11e, EDCA allows it to adapt the access channel parameters over time in a dynamic way. Nevertheless, this feature is not used in commercial APs due to the complexity involved in determining the network conditions. Furthermore, the process in charge of carrying out this task should be as simple as possible due to the fact that the updating of the network information must be carried out in real-time. For this reason, the standard values of the EDCA parameter set specified in the aforementioned amendment are usually considered by the stations that use EDCA, regardless of the network saturation level.

The main goal of our proposal is to enhance the offered QoS level and maximize the performance of the voice and video traffic. To achieve this goal, our scheme aims to identify the optimal AIFSN values and adapt them to the network conditions in order to enhance the performance offered by the standard EDCA parameter set. At the same time, it seeks to ensure backward compatibility between the stations that use EDCA and DCF in the BSS. Our main aim is to enhance the audio and video performance by decreasing the collisions between these types of applications. Accordingly, a reduction in the global retransmission attempts and an increase in the overall performance of the network are achieved.

When a transmission takes place, there is a large number of variable parameters that may determine the wireless channel conditions. As a consequence, deploying an adaptive scheme for the priority setting through the AIFSN combination is not a simple task. The main conditioning factors are described below.

- *Number of active applications of each type of traffic*. This is a parameter that can be identified in a simple way by the AP. However, this value at a particular moment in time is insufficient. That is because it cannot provide further information about the current conditions of the network, i.e., the scheme will not be allowed to obtain real information about the current occupancy of the wireless channel.
- *Applications bit rate*. Linked to the previous one, this factor provides more detailed information about the state of the wireless medium. Unfortunately, it is difficult to calculate in real-time. To identify these values it is necessary to introduce periodical control traffic in the network. Nevertheless, this feature is not typically used in IEEE 802.11e.
- *Transmission rate*. Every single station may carry out its transmissions by using a different transmission rate. Therefore, the specific period of time that each of them keeps the channel busy is different. This parameter would be a good way of estimating the network conditions. Nonetheless, this value needs to be used jointly with the above factors.

- *Presence or absence of DCF legacy stations.* The existence of stations without QoS support restricts the use of priority parameters in EDCA due to the fact that these values cannot be duly adjusted for these stations.
- *Occupancy level of the wireless medium.* A good approximation of the network status can be obtained by estimating the period of time that every type of application makes use of the wireless channel. This value provides in a simple way even more detailed levels of information regarding the combined use of all the factors described above.

Due to the inherent variability of the aspects that are part of a wireless network transmission, we must consider a scheme with low computational complexity and capacity to adapt itself to changes over time. On the basis of these requirements, artificial intelligence techniques are used in order to identify and interpret traffic patterns. Furthermore, such techniques are capable of making decisions based both on their previous decision and on the behaviour of the network.

In order to address such a problem, we have considered the design of two different predictive models. This will allow us to carry out a comparative evaluation of the capabilities that each of them could offer in the final scheme. These models are a *J48* decision tree classifier and a *M5* regression model. Before deciding on the use of these classifiers, many others, such as the Naive Bayes classifier, have been taken into account. However, their main features are their low computational complexity, their self-explanatory capacity and their high degree of adaptability to the problem, as is described in Section 13.3. In [3, 4], it is concluded that the AIFSN is the most important factor in the EDCA parameter set. As a consequence, the main function of the designed models is to identify the AIFSN combination that achieves the highest voice+video normalized throughput in every single moment regardless of network saturation.

In this context, 9 sets of AIFSN values are selected as candidates to be considered by the model and can be seen in Table 13.2. These values have been chosen by gradually increasing the difference in slots of time between the different ACs with the aim of reducing the external collisions among the traffic transmissions. For that reason, the AIFSN value related to each AC is suitably separated from each other. However, in those cases in which the AIFSN for video traffic is higher than 2, its priority to access the wireless channel is reduced with regard to the legacy stations. Given this situation, the usage of the default AIFSN values is more advisable rather than the tuning

Table 13.2 Set of AIFSN values analyzed

	S0	S1	S2	S3	S4	S5	S6	S7	S8	S9
BK	7	8	9	8	9	12	10	12	14	14
BE	3	4	5	4	5	6	6	8	10	12
VI	2	2	2	3	3	3	4	5	6	7
VO	2	2	2	2	2	2	2	2	2	2

of this parameter. Moreover, to carry out a fair performance evaluation of our model, this combination must be also evaluated independently. Thus, this configuration is also included in the aforementioned table under the name *S0*. The proposed combinations aim to outperform the results offered by EDCA, enhancing the voice+video normalized throughput and the overall network performance, mainly by reducing the external collisions among the traffic of different ACs.

In order to design an accurate classifier, a large amount of information must be provided during its construction. The information must contain a wide range of different network conditions in order to acquire enough knowledge. For this reason, previous to the learning process, a huge set of tests is carried out by considering several factors that may compromise the network performance and that are described in Section 13.4.2. As part of these tests, the aforementioned 9 sets of AIFSN values are considered with the aim of finding an alternative value to the standard one in order to enhance the performance of the network.

The described parameters are used as an entry point for the predictive models and must be calculated periodically. In our case, they will be calculated once per second. After this period of time, the classifier checks whether the previously selected AIFSN values are already the most favorable combination or whether they need to be modified. Once the optimal AIFSN set has been calculated by the AP, it is responsible for distributing these values embedded in an EDCA parameter set. Distribution is handled through the beacon frames and, therefore, no additional control traffic is introduced into the network. This behaviour is shown in Figure 13.4.

Thus, the proposed schemes have low complexity due to the fact that the AP only has to perform simple operation checks by using the described model, and there is no need to transmit additional control traffic. Furthermore, this scheme only requires making a few minor adjustments to the APs and no changes are made to the commercial network cards. Therefore, total compatibility with existing devices is maintained at the same time an enhancement in network performance is made possible, especially for voice and video traffic.

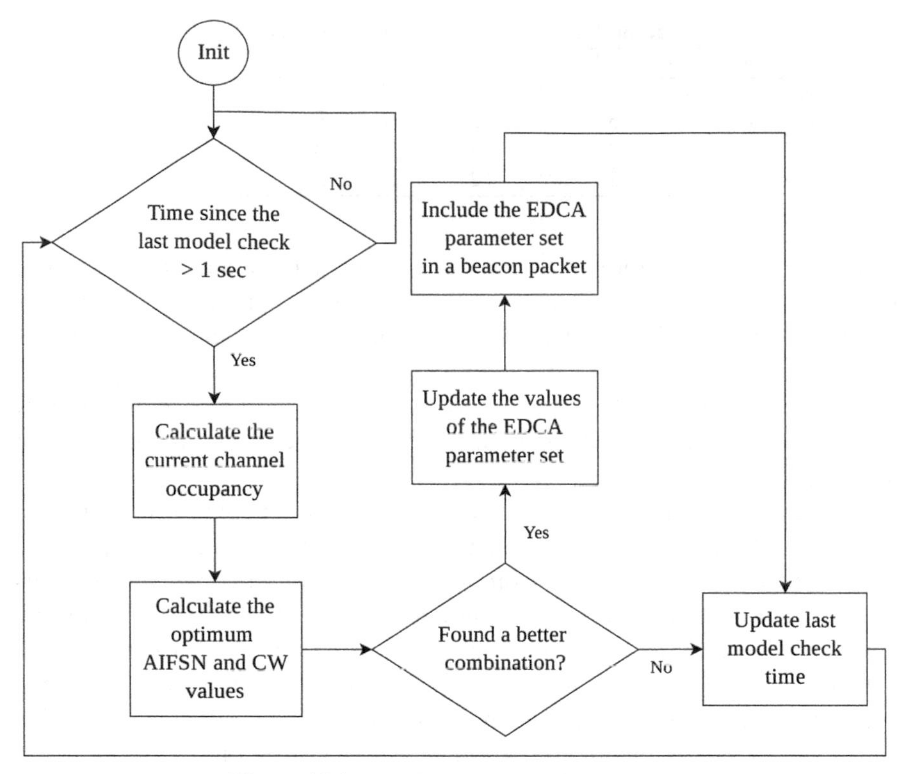

Figure 13.4 MAC parameter update process.

13.4.2 Design of the Predictive Models

The design and construction of the *J48* and the *M5* predictive models need to have a considerable amount of proper information that must be acquired during the training period. In our case, this information is obtained from the results of a set of tests that is previously performed, aiming to cover a wide range of possible scenarios with different network conditions. This process makes the construction of the most accurate model possible. In this regard, a group of 18 scenarios has been designed and tested via simulation by using Riverbed Modeler 18.0.0 [20]. These scenarios take into account applications that utilize EDCA and those that only make use of DCF as a channel access method. A detailed description of the traffic distributions used in this step can be observed in Table 13.3. In this way, our proposal can guarantee full compatibility with those stations that only support the original IEEE 802.11 standard.

Table 13.3 Traffic proportions used for the training step

# Scenario	VO	VI	BE	BK	DCF
1	20%	20%	20%	20%	20%
2	60%	10%	10%	10%	10%
3	30%	30%	10%	10%	20%
4	10%	60%	10%	10%	10%
5	30%	20%	10%	10%	30%
6	10%	10%	10%	10%	60%
7	10%	10%	20%	20%	40%
8	10%	10%	40%	30%	10%
9	50%	50%	–	–	–
10	10%	10%	30%	30%	20%
11	60%	40%	–	–	–
12	40%	60%	–	–	–
13	–	–	40%	30%	30%
14	30%	–	20%	20%	30%
15	–	30%	20%	20%	30%
16	–	–	50%	50%	–
17	20%	–	40%	40%	–
18	–	20%	40%	40%	–

The aforementioned scenarios are made up of a variable number of stations, considering different percentages of uplink transmissions of every type of traffic (BK, BE, VI and VO). On this basis, the traffic load level is increased in every scenario in steps of 10 stations, causing the number of stations to range from 10 to 80. As a result, eight combinations of the same scenario with different traffic load levels are tested. In order to ensure that the proposed scheme is able to adapt itself to the network conditions independently of the transmission rate of the stations in the BSS, two different values for this factor have been considered to carry out the tests. In this way, all the applications share a transmission rate of 12 Mbps and 36 Mbps, regardless of the type of traffic that they transmit. In order to provide a further evaluation, all the tests have been carried out by using 60 different random seeds.

Each station of the BSS transmits a different type of traffic whose characteristics are shown in Table 13.4. Furthermore, the bit rate of the different types of traffic is modeled by using a group of probability distributions. Table 13.4 shows that the stations that only support DCF and those that use EDCA to transmit BK and BE traffic use the same transmission source. This source is modeled by making use of a Pareto distribution with a location of 1.1 and a shape of 1.25. By contrast, voice traffic represents a

Table 13.4 Traffic parameters used for predictive models design

	Packet Size	Data Rate
DCF	552 bytes	512 Kbps
BK	552 bytes	512 Kbps
BE	552 bytes	512 Kbps
VI	1064 bytes	800 Kbps
VO	104 bytes	20 Kbps

Constant Bit Rate (CBR) service using the G728 codec [21], whereas video applications transmit H.264 [22] streams. In addition, multimedia applications have temporal restrictions that are important to be modeled in the scheme. For this purpose, deadline periods of 10 ms and 100 ms are considered for voice and video transmissions, respectively. In this way, when any packet remains in the transmission queue for longer than the indicated thresholds, it is discarded.

The different tests designed to evaluate the models are performed by making use of the proposed AIFSN configurations that can be seen in Table 13.2. Among the values that this table contains, the AIFSN combination proposed in the IEEE 802.11e amendment has also been considered due to the fact that it achieves the highest performance in certain cases. Therefore, it is a suitable combination to be chosen by the classifier. Furthermore, the results of this combination are compared with the ones achieved by our proposal. During the execution of each test, the characteristics of every scenario, i.e., the parameter combination and the traffic distribution that determine such a scenario, remain static. These values are modified according to a given order until all possible combinations of such parameters have been considered. In this way, the models are allowed to acquire real knowledge. If variable information were provided to the classifier during its development, the learning process would be unfeasible.

Once the results of the aforementioned tests have been obtained, they must undergo a significant pre-processing in order to extract the desired information. Initial tests included several outcomes, such as the number of applications of every type of traffic or the percentage of occupancy of the wireless medium. Nevertheless, this fact is unacceptable since part of our main aim is to develop a prediction scheme as simple as possible. Due to the wide variety of resulting parameters, it was necessary to perform a supervised variable selection to discard those that were unrelated. After carrying out this variable selection, only the global and the particular level of occupancy of each type of traffic are considered by the final models.

Despite the pre-processing phase, there are certain differences in the way of providing the information to the *J48* classifier and the *M5* model. To model the *J48* tree classifier, only the data relating to the AIFSN configurations that achieve the highest performance of the multimedia applications have been used. By contrast, the *M5* model takes the information of all the AIFSN combinations, regardless of the throughput achieved by each of them. The purpose of this last model is to improve the voice+video normalized throughput by using a group of regression functions. For this reason, the result of the voice+video normalized throughput is added as a parameter to the model due to the fact that this is the factor that the regression model must maximize. Furthermore, the final model is made up of ten groups of sub-models, i.e., it contains one group of rules per AIFSN tested combination, which attempt to achieve the highest voice+video performance. The few factors considered by both the models are able to provide a good approximation of the network conditions while allowing the construction of a simple and accurate adaptation scheme.

Moreover, both the data pre-processing and the design of the predictive models are carried out using Weka 3.7.0 [18]. During the aforementioned design, a 10-fold cross-validation process is performed in order to guarantee that both the training and the testing data sets are totally independent. The *J48* classifier achieves a hit rate of 94.90% in this process, meanwhile in the case of the *M5* regression model, an average correlation coefficient of 0.8916 and a mean absolute error of 0.0554 are obtained. The above values show the accuracy of the proposed models and the high relation between the parameters involved.

The training phase of classification trees and regression models usually involve a computational complexity that could be as high as $O(pN^2)$ for an input with N instances and p attributes [23, 24]. However, due to the wide variety of available data, the predictive models are built offline and only once. For that reason, in our case, this complexity is only focused on the computational requirements during the execution of the algorithms in the AP, which are practically negligible. In this device, as the computational complexity does not depend on any parameter, and the number of performed operations is usually the same, the execution time is $O(1)$ for both models. Nevertheless, the computational time of these algorithms is not the same. In contrast to the *J48* tree, which only makes a few comparisons, the *M5* model requires to evaluate the voice and video traffic performance achieved by every AIFSN configuration. For that reason, the computational time used by the first of them is lower.

13.5 Performance Evaluation

In this section, we carry out a performance analysis in order to verify the proposed schemes via simulation, making use of Riverbed Modeler 18.0.0. This suite of protocols provides the user with a wide range of models that shows the behaviour of different kinds of networks by means of event simulation. Moreover, Modeler allows the users to develop, evaluate and test new wireless protocols and technologies, thus contributing to make progresses in networks research. In this regard, a set of 20 scenarios have been designed, covering a wide range of different network conditions. In this way, the main features of the evaluation conducted and the results obtained during this process are shown below.

During the performance evaluation, both stations that use DCF and EDCA have been included. The first twelve scenarios take into account both types of stations while in the remaining eight only EDCA stations can be found. All these scenarios are made up of 100 stations, involving an equal proportion of applications of each type of traffic, i.e., 20 stations per type of traffic are included in the BSS. Despite this fact and with the aim of considering a wireless network as real as possible, all the stations are not active at the same time. Instead, a specific transmission probability has been assigned to every station according to its AC, as shown in Table 13.5. These probabilities and the variable number of active stations allow for the evaluation of our proposal under different network saturation conditions.

The scenarios have a duration of 600 seconds and are divided into two periods. During the first one, the stations that are not transmitting any information try to start a new transmission every 30 seconds with a probability associated with their AC (see Table 13.5). During the second one, the transmitting applications attempt to stop the transmission every 30 seconds with the same probability as that previously used. With this approach, many scenarios with a multitude of traffic loads are considered. Due to all scenarios being simulated by using 60 different random seeds and each of them being divided into 20 time intervals, in the end, 24,000 different intervals have been tested.

The bit rate of all the applications used during the whole performance evaluation process is assigned according to their AC. These values are the same as the ones shown in Table 13.4. Moreover, the stations are randomly distributed over the network coverage of the BSS. In addition, different transmissions rates have been taken into account for all the applications regardless of the type of traffic they transmit. In this way, and with the aim of modeling signal propagation through the wireless medium, the Ricean [25]

Table 13.5 Description of the set of test scenarios

# Scenario	VO	VI	BE	BK	DCF
1	10%	1.5%	2%	2%	2%
2	10%	5%	2%	2%	2%
3	10%	7%	2%	2%	2%
4	8%	6%	3%	3%	7%
5	4%	2%	3%	3%	10%
6	3%	3%	4%	4%	8%
7	5%	3%	7%	7%	4%
8	6%	6%	10%	5%	5%
9	6%	9%	6%	6%	6%
10	8%	–	8%	8%	8%
11	–	6%	6%	6%	9%
12	6%	6%	6%	6%	6%
13	10%	8%	–	–	–
14	8%	4%	–	–	–
15	6%	10%	–	–	–
16	7%	7%	7%	7%	–
17	10%	–	8%	8%	–
18	–	8%	7%	7%	–
19	9%	8%	6%	6%	–
20	9%	7%	8%	–	–

model has been considered. This model is characterized by a factor, k, which determines the ratio between the power in the line-of-sight component and the power in the scattered paths. In our case, a k factor of 32 has been used. Furthermore, in all the analysis, IEEE 802.11g [26] defines the physical layer of the network. Therefore, regardless of the type of application, stations use of all the transmission rates defined in this amendment.

In order to evaluate the performed simulations, a large amount of statistical information has been obtained. However, only some of the factors have been selected in order to consider only those that are able to summarize the main results. The metrics that have finally been considered include the voice+video normalized throughput, the number of retransmission attempts, the overall throughput of the network and the normalized throughput achieved by the stations that use DCF. As the main aim of our proposal is to enhance the performance of the voice and video applications, the first of these statistics refers to the sum of the normalized throughput of such applications.

In Table 13.6, the voice+video normalized throughput results for the 24,000 simulated intervals are shown. This table presents the percentage of 30-second transmission intervals during which our proposal has experienced losses or gains of voice+video normalized throughput with regard to the

Table 13.6 Voice+Video normalized throughput improvements in 30 s intervals

	With DCF Traffic		Without DCF Traffic	
	J48	*M5*	*J48*	*M5*
Unaltered	49.42%	52.11%	35.52%	31.30%
Losses	4.49%	5.48%	2.55%	1.44%
Gain [1%–5%]	27.20%	23.37%	23.64%	17.35%
Gain [5%–10%]	8.67%	12.78%	6.68%	5.93%
Gain [10%–15%]	6.06%	2.86%	7.41%	6.89%
Gain [15%–20%]	2.01%	1.52%	4.48%	4.55%
Gain [from 20%]	2.16%	1.90%	19.73%	32.55%

existence of DCF traffic. These values have been calculated in comparison with the results obtained from the standard AIFSN combination. Moreover, this table includes the cases in which the results are unaltered. We have considered as unaltered the results in which the gains or the losses are lower than 1%. Furthermore, those cases in which our proposal experiences a decrease in the level of performance higher than 1% have been designated as losses.

It can be observed in Table 13.6 that in a large number of cases, the results of our schemes remain unaltered. This situation is a consequence of taking into account low traffic load levels in a significant amount of the tested scenarios where all the AIFSN combinations achieve the highest performance. In particular, in the presence of DCF traffic and depending on the network saturation, the percentage of unaltered scenarios is higher for the scenarios with DCF traffic than for those that only take into account EDCA stations.

There is also a small percentage of cases where our proposal experiences small losses in performance. These losses represent 4.49% and 5.48% of the cases (for the *J48* and the *M5* models, respectively) in the presence of stations that make use of DCF, while this value is lower when only stations that use EDCA are considered (2.55% and 1.44% of the intervals, respectively). This situation is due to a group of wrong decisions made during the test by the predictive models. Furthermore, the usage of a single parameter allows it to have a good approximation of the network conditions, but it does not allow it to identify the network conditions completely. Nevertheless, the number of scenarios in which this situation occurs is much lower than those in which our proposal improves upon the performance offered by EDCA. In fact, the results show that the performance improvement is from 20% in many cases. Finally, it can be clearly seen that the gains achieved by the *J48* tree are from 20% in 19.73% and in 32.55% of the scenarios by the *M5* model in the absence of DCF traffic.

Figures 13.5, 13.6 and 13.7 shows a representative subset of the tested scenarios where the traffic proportion becomes more problematic for EDCA usage, i.e., they present the scenarios in which EDCA begins to suffer performance losses. The first eight scenarios of this subset take into account both stations without QoS support and those that make use of EDCA. However, in the remaining four scenarios of these figures only stations which use EDCA are considered. Meanwhile, in Figure 13.8 only a subgroup of scenarios with both types of stations is considered due to the fact that the DCF normalized throughput is evaluated.

During the simulations, twenty intervals with many different traffic load levels are taken into account. The first and the last five intervals have the lowest traffic load due to the fact that all the stations are starting or ending their transmissions. When the traffic load of the network is low, all the tested AIFSN combinations offer the highest throughput. In this way, the results of both the standard AIFSN combination and those of the proposal are identical. For this reason, only the ten remaining intervals are shown in Figure 13.5, in which the standard values start to suffer traffic losses. In this figure, the voice+video normalized throughput is shown. It can be observed that in all cases the throughput achieved by our proposal is higher than when using the standard AIFSN values. Furthermore, it is shown that the difference is even greater in scenarios without DCF traffic. In these cases, an improvement of from 35% can be obtained.

The improvement achieved by our proposal is a direct consequence of decreasing the amount of collisions in the network. Furthermore, the proposed models offer a reduction in the number of retransmission attempts as can be seen in Figure 13.6. These decreases have a direct impact on the improvement of the global throughput of the network, which is illustrated in Figure 13.7.

The gradual separation of the AIFSN values assigned to the different ACs, especially in the cases in which the AIFSN value for video traffic is higher than 2 slots, allows the stations that only support the original IEEE 802.11 standard to be given a higher priority to access the wireless channel. Therefore, our models are not only able to maintain the compatibility with the aforementioned stations, but also to enhance their offered performance. In this way, the eight most representative scenarios in which DCF transmissions take place have been selected in order to show this improvement. The results of these scenarios can be observed in Figure 13.8. In spite of providing a higher priority to the legacy stations and improving their performance, our schemes do not only not penalize the voice and video applications, but also enhances the throughput offered, as can be seen in Figure 13.5.

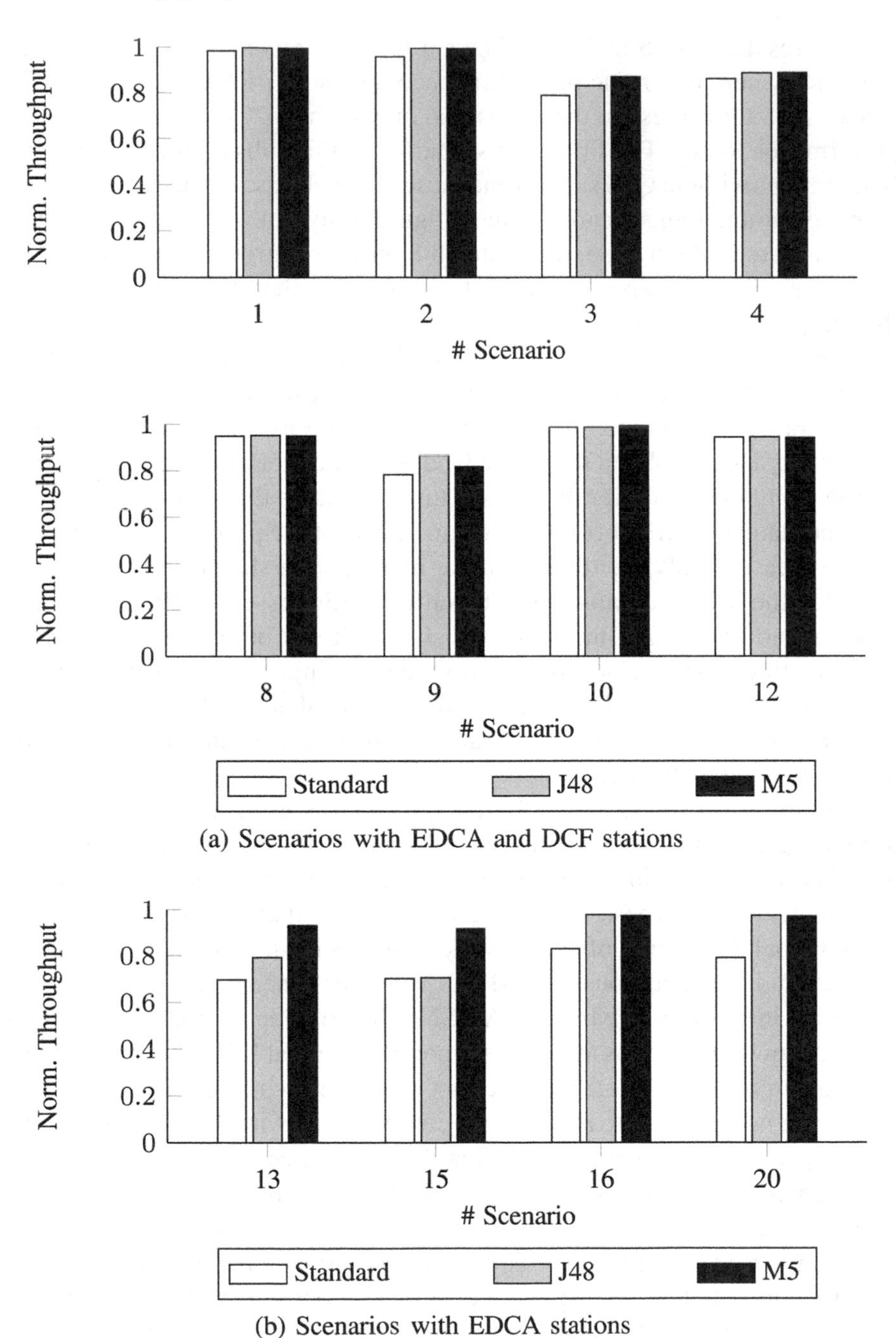

Figure 13.5 Voice+Video normalized throughput.

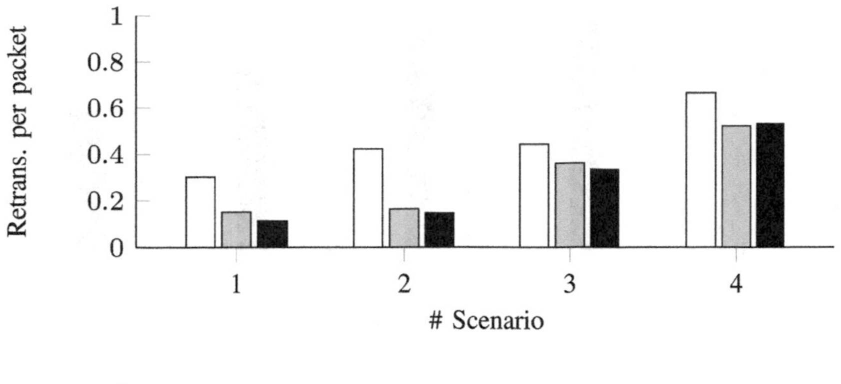

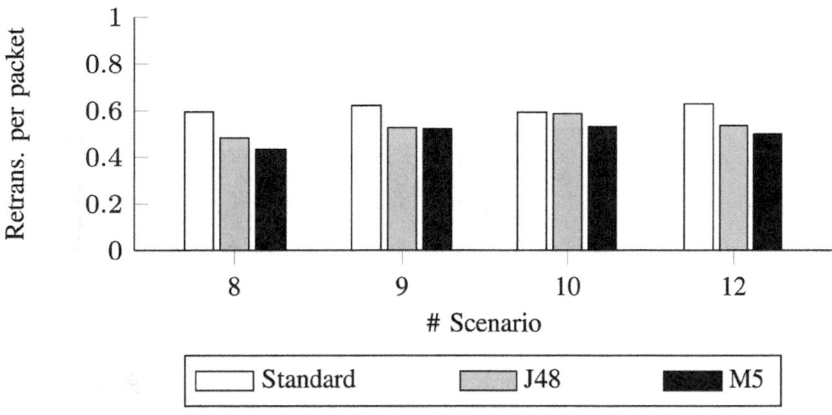

(a) Scenarios with EDCA and DCF stations

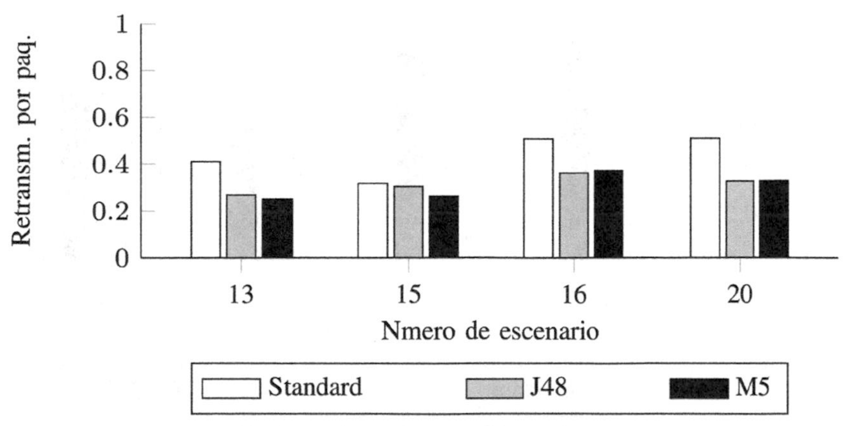

(b) Scenarios with EDCA stations

Figure 13.6 Overall retransmission attempts.

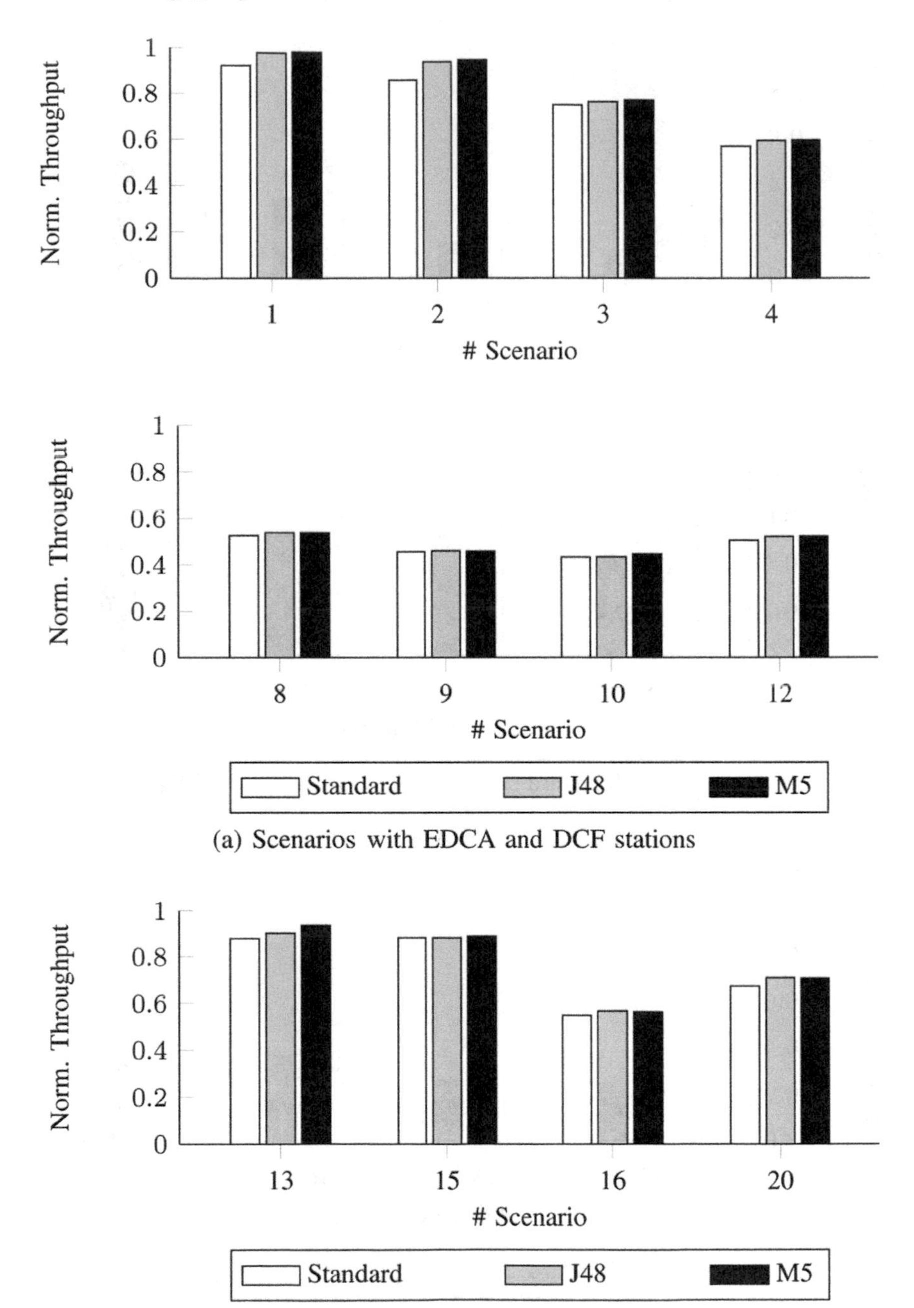

(a) Scenarios with EDCA and DCF stations

(b) Scenarios with EDCA stations

Figure 13.7 Global norm. throughput.

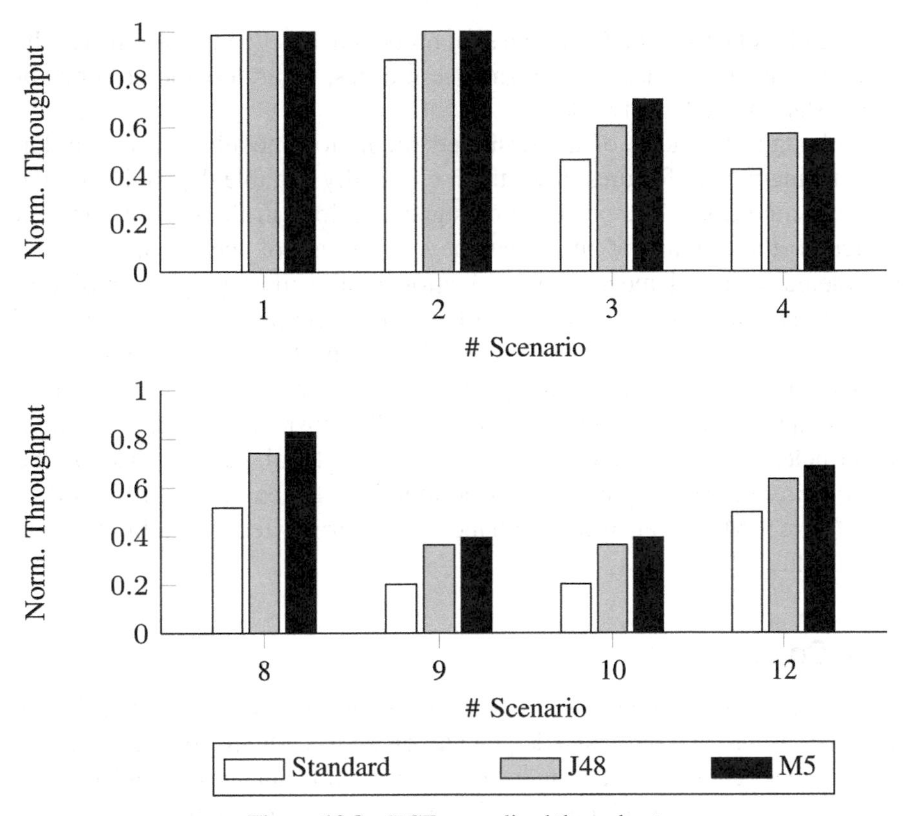

Figure 13.8 DCF normalized throughput.

Suitable selection of the AIFSN values contributes not only to enhancing the performance offered by the voice and video applications, but also to improving the throughput of the remaining types of traffic and the overall quality of the network. This can be verified in Table 13.7, which summarizes the average enhancements that the designed models achieve in all the scenarios evaluated. According to the statistic information shown in this table, both the

Table 13.7 Overview of the performance evaluation results

	With DCF Traffic			Without DCF Traffic		
	Standard	*J48*	*M5*	Standard	*J48*	*M5*
VO+VI Norm. Th.	0.9259	0.9440	0.9438	0.7759	0.8790	0.9338
Retrans. per Pkt	0.5447	0.4459	0.4271	0.4604	0.3428	0.3329
Global Norm. Th.	0.6014	0.6199	0.6225	0.6776	0.6965	0.7010
DCF Norm. Th.	0.5498	0.6790	0.7118	–	–	–

J48 tree classifier and the *M5* regression model are able to improve the results obtained for all the selected performance metrics, regardless the presence of legacy stations in the network.

Although the nature of the designed prediction models is very similar, their distinguishing features make them especially suitable depending on the network conditions. The presence of legacy stations in the network and, in particular, the amount of time that they make use of the channel can be considered a key element in the selection of the most precise prediction scheme. The *M5* regression model offers more accurate results in terms of voice+video throughput in cases where the traffic load of the stations that use DCF is low. By contrast, the prediction given by the *J48* tree classifier is more appropriate when this amount of DCF transmissions becomes high. Nevertheless, the *M5* model outperforms the global performance of the network in almost the previously aforementioned scenarios with regard to the *J48* tree. Moreover, it also enhances the normalized throughput of the DCF traffic.

13.6 Conclusions

The demand for multimedia services is growing fast, especially in real-time applications which require an adequate level of QoS. On this basis, the use of AI techniques contributes to find traffic patterns and enhance the performance of the network. In this chapter, we have proposed a prediction scheme for improving the voice and video communications over WLANs, making use of a *J48* tree classifier and a *M5* regression model previously designed. With this aim, it is able to dynamically adapt the standard AIFSN combination defined in IEEE 802.11e, while allowing for the compatibility with the stations that only support the original IEEE 802.11 standard. This scheme is only queried by the AP of the BSS, which transmits the calculated values to all the remaining stations without introducing additional control traffic into the network.

The experimental results show that our proposal outperforms the voice+video normalized throughput of the standard AIFSN combination defined in IEEE 802.11e, achieving in that way an improvement of from 20% in some scenarios. It is also shown that a suitable separation of the AIFSN values from each other for every AC leads to a reduction in the amount of external collisions between the traffic of different ACs. As a consequence, the global throughput of the network is also enhanced. Furthermore, and mainly due to its simplicity, the proposed scheme is able to be executed in real-time.

The IEEE 802.11e amendment seeks to ensure QoS support to delay-sensitive multimedia applications in any type of scenario. As this amendment states that compatibility with the legacy stations must be kept, the AIFSN combination that it defines must suit a wide range of traffic conditions, including those in which stations without QoS support are considered. For this reason, and especially under these traffic conditions, there is considerable scope for improvement of the QoS level. Accordingly, the improvements achieved by our proposal were greater in the absence of applications that use DCF.

The designed prediction models achieve different results depending on the status of the network traffic. Therefore, it may be beneficial to attempt the design of a new dynamic scheme that is able to combine the strengths of each approach in order to select in every moment the model that more suits the network conditions.

Acknowledgments

This work has been jointly supported by the Spanish Ministry of Economy and Competitiveness and the European Commission (FEDER funds) under the projects TIN2012-38341-C04-04 and TIN2015-66972-C5-2-R, and the grant BES-2013-065457.

References

[1] *Wireless LAN Medium Access Control (MAC) and Physical Layer (PHY) Specifications*, ANSI/IEEE Std 802.11, LAN/MAN Standards Committee of the IEEE Computer Society Std., 1999.

[2] *Wireless LAN Medium Access Control (MAC) and Physical Layer (PHY) Specifications. Amendment 7: Medium Access Control (MAC) Quality of Service (QoS)*, ANSI/IEEE Std 802.11e, LAN/MAN Standards Committee of the IEEE Computer Society Std., 2005.

[3] Villalón, J., Cuenca, P., and Orozco-Barbosa, L. (2007). On the capabilities of IEEE 802.11e for multimedia communications over heterogeneous 802.11/802.11e WLANs. *Telecommun. Syst.*, pp. 27–38.

[4] Hui, J., and Devetsikiotis, M. (2005). A unified model for the performance analysis of IEEE 802.11e EDCA. *IEEE Trans. Commun.*, pp. 1498–1510.

[5] He, R., and Fang, X. (2009). *A fair MAC algorithm with dynamic priority for 802.11e WLANs*, pp. 255–259.

[6] Nilsson, T., and Farooq, J. (2008). A novel MAC scheme for solving the QoS parameter adjustment problem in IEEE 802.11e EDCA. *World of Wireless, Mobile and Multimedia Networks*, pp. 1–9.

[7] Banchs, A., Serrano, P., and Vollero, L. (2010). Providing service guarantees in 802.11e EDCA WLANs with legacy stations. *IEEE Trans. Mobile Comput.*, pp. 1057–1071.

[8] Gallardo, J. R., Cruz, S. C., Makrakist, D., and Shami, A. (2008). Analysis of the EDCA access mechanism for an IEEE 802.11e-compatible wireless LAN. *IEEE Symposium on Computers and Communications, 2008. ISCC 2008.*, pp. 891–898.

[9] Banchs, A., and Vollero, L. (2006). Throughput analysis and optimal configuration of 802.11e EDCA. *Comput. Netw.*, pp. 1749–1768.

[10] Mackenzie, R., and O'Farrell, T. (2012). Achieving service differentiation in IEEE 802.11e enhanced distributed channel access systems. *Communications, IET*, pp. 740–750.

[11] Mitchel, T., *Machine Learning*. McGraw-Hill, 1997.

[12] Duda, R., Hart, P., and Stork, D. (2001). *Pattern classification,* 2nd ed. Wiley.

[13] Weisberg, S. (2005). *Applied linear regression*, 3rd ed. Hobo-ken, NJ: Wiley.

[14] Haykin, S. (2009). *Neural networks and learning machines*, 3rd ed. Prentice Hall.

[15] Boser, B. E., Guyon, I. M., and Vapnik, V. N. (1992). A training algorithm for optimal margin classifiers. In *Proceedings of the Fifth Annual Workshop on Computational Learning Theory*, pp. 144–152.

[16] Quinlan, R. J. (1992). Learning with continuous classes. in *5th Australian Joint Conference on Artificial Intelligence*, pp. 343–348.

[17] Quinlan, J. R. (1993). *C4.5: Programs for machine learning*. Morgan Kaufmann Publishers Inc.

[18] Machine Learning Group at the University of Waikato, "Weka 3.7.0," 2014.

[19] Hall, M., Frank, E., Holmes, G., Pfahringer, B., Reute-mann, P., and Witten, I. H. (2009). The weka data mining software: An update. *SIGKDD Explor. Newsl.*, 11 (1), 10–18.

[20] Technology, R. (2014). Riverbed Modeler 18.0.0.

[21] ITU-T. (2013). Coding of speech at 16 Kbit/s Uing Low-delay Code Excited Linear Prediction.

[22] ISO/IEC and ITU-T. (2003) Advanced Video Coding for Generic Audiovisual Services. ITU-T Recommendation H.264 and ISO/IEC 14496-10 (MPEG-4 AVC).

[23] Miu, T., and Missier, P. (2012). Predicting the execution time of workflow activities based on their input features. In *Proceedings of the 2012 SC Companion: High Performance Computing, Networking Storage and Analysis*, ser. SCC '12. IEEE Computer Society, pp. 64–72.

[24] Hastie, T., Tibshirani, R., and Friedman, J. (2009). *The elements of statistical learning: data mining, inference and prediction*, 2nd ed. Springer.

[25] Punnoose, R. J., Nikitin, P. V., and Stancil, D. D. (2000). Efficient simulation of ricean fading within a packet simulator. In *Vehicular Technology Conference, 2000. IEEE-VTS Fall VTC 2000. 52nd*, pp. 764–767, vol. 2.

[26] *Wireless LAN Medium Access Control (MAC) and Physical Layer (PHY) Specifications. Amendment 4: Further Higher Data Rate Extension in the 2.4 GHz Band*, ANSI/IEEE Std 802.11g, LAN/MAN Standards Committee of the IEEE Computer Society Std., 2003.

Index

About the Editors

Dr. Kewei Sha is an Associate Director of Cyber Security Institute at University of Houston, Clear Lake (UHCL). He is also the Assistant Professor of Computer Science at UHCL. Before he moved to UHCL, he was the Department Chair and Associate Professor in the Department of Software Engineering at Oklahoma City University. He received Ph.D. in Computer Science from Wayne State University in 2008. His research interests include Internet of Things, Cyber-Physical Systems, Mobile Computing, Data Analytics, and Network Security and Privacy. Dr. Sha has served as the secretary of Technical Committee on the Internet of the IEEE Computer Society (IEEE-CS TCI). He also served as Editor or Guest Editor of many international journals, and served as Chairs of many conferences and workshops, including Technical Program Chair of ICCCN 2015 and MedSPT 2015. His research has been supported by NSF, CNSF, Oklahoma City University and UHCL.

Prof. Aaron Striegel is an Associate Professor and serves as Associate Chair in the Department of Computer Science & Engineering at the University of Notre Dame. He received his Ph.D. in December 2002 in Computer Engineering at Iowa State University under the direction of Dr. G. Manimaran. Prof. Striegel's research interests focus on instrumenting the wireless networked ecosystem to gain insight with respect to user behavior and global network performance. Further research interests of Prof. Striegel include computer security and the adaptation of low-cost gaming peripherals for rehabilitation. Prof. Striegel has received several best paper awards including USENIX LISA, IEEE Healthcom, and HotPlanet. Prof. Striegel has received various research and equipment funding from NSF, DARPA, Sprint, Intel, Google, and Alcatel-Lucent. He has also been the recipient of a NSF CAREER award in 2004 and has been a recent participant in NAE symposia on Engineering Education and the Informed Brain in the Digital World.

Prof. Min Song served as Program Director with the NSF from 2010 to 2014 and is currently the Founding Director of Institute of Computing and Cybersystems, Dave House Professor and Department Chair of Computer Science, and Professor of Electrical and Computing Engineering at Michigan Tech.

Through his outstanding contributions in promoting NSF's international leadership, Min received the prestigious NSF Director's award in 2012. Min's research interests include design, analysis, and evaluation of wireless communication networks, cognitive radio networks, network security, cyber physical systems, and mobile computing. His research has been supported by NSF, DOE, NASA, and private Foundations. Min's professional career comprises 26 years in industry, academia, and government, and has held various leadership positions and gained substantial experience in performing a wide range of duties and responsibilities. Min launched and served as Editor-in-Chief of two international journals. He also served as Editor or Guest Editor of 14 international journals, and served as Chairs of many conferences, including General Chair of IEEE INFOCOM 2016 and Technical Program Vice-Chair of IEEE GLOBECOM 2015. Min is currently serving as the IEEE Communications Society Director of Conference Operations. Min was the recipient of NSF CAREER award in 2007.